Dark Matter of the Mind

Dark Matter of the Mind

The Culturally Articulated Unconscious

DANIEL L. EVERETT

The University of Chicago Press
Chicago and London

The University of Chicago Press, Chicago 60637
The University of Chicago Press, Ltd., London

Paperback edition 2017
Printed in the United States of America

23 22 21 20 19 18 17   3 4 5 6

ISBN-13: 978-0-226-07076-6 (cloth)
ISBN-13: 978-0-226-52678-2 (paper)
ISBN-13: 978-0-226-40143-0 (e-book)
DOI: 10.7208/chicago/9780226401430.001.0001

Library of Congress Cataloging-in-Publication Data

Names: Everett, Daniel Leonard, author.
Title: Dark matter of the mind : the culturally articulated unconscious / Daniel L. Everett.
Description: Chicago : London ; The University of Chicago Press, 2016. | Includes bibliographical references and index.
Identifiers: LCCN 2016013247 | ISBN 9780226070766 (cloth : alk. paper) | ISBN 9780226401430 (e-book)
Subjects: LCSH : Subconsciousness. | Knowledge, Theory of. | Context effects (Psychology) | Cognition and culture. | Language and culture. | Philosophical anthropology.
Classification: LCC BF315.E84 2016 | DDC 154.2—dc23 LC record available at http://lccn.loc.gov/2016013247

♾ This paper meets the requirements of ANSI/NISO Z39.48-1992 (Permanence of Paper).

For Shannon, Kristene, and Caleb
To whom I read by kerosene lamp in the Amazon

And for Linda
My companion in the woods

Real bands are made primarily from the neighborhood. From a real time and real place that exists for a little while, then changes and is gone forever. They're made from the same circumstances, the same needs, the same hungers, culture.

—BRUCE SPRINGSTEEN, *E Street Band induction speech, Rock and Roll Hall of Fame*

Contents

Preface

> Between stimulus and response, there is a space. In that space is our power to choose our response. In our response lies our growth and our freedom.
>
> VIKTOR FRANKL, *Man's Search for Meaning*

In 1959, Edward T. Hall first published *The Silent Language*. As he said, "It wasn't just that people 'talk' to each other without the use of words, but that there is an entire universe of behavior that is unexplored, unexamined, and very much taken for granted. It functions outside conscious awareness. . . . What is most difficult to accept is the fact that our own cultural patterns are literally unique and therefore *not* universal" (Hall [1959] 1973, vii). Hall is not merely offering here the banal observation that there is an unconscious. He is talking about cultural tacit knowledge—a more articulated notion. It is this cultural articulation of unspoken or ineffable values, knowledge, and roles that this book addresses.

Much science has transpired since Hall wrote those words. But his prescient exploration of what we know but don't know we know—"unknown knowns"—was one of the first to put "dark matter of the mind" on the intellectual agenda clearly enough for us to see how one might build a theory of it. The words that follow owe a tremendous debt to Hall.

My thoughts on these matters extend back nearly forty years to my initial field research in Mexico and Brazil, when I first confronted alternatives to my way of thinking I had never imagined possible. As my thinking developed, I decided to write this book in order to articulate a perspective on the unconscious, the ineffable, the unspoken, our cultures, and the structured knowledges, values, and roles of the individual. I certainly realize that all psychologists (and just about everyone else) are aware of and acknowledge the importance of the unconscious, just as I am aware that anthropologists know about culture, and all linguists language. It isn't via the terms *culture*, *language*, or *unconscious* alone that this book attempts to contribute, however. Rather, this contribution is a new theory of how these things work together.

In particular, it is about how our unconscious is structured and infused with meaning by our individual experiences and social living. With more than two thousand years of writing in both the East and the West on the mind, society, and the individual, the individual ideas are less likely to be novel than the way that they are fitted together.

Anyone who has lived in another culture, learning to maneuver through a different language or alternate set of cues and clues of values, knowledge, food, social interactions, smells, sights, and so on, has been at once exhilarated, exhausted, stymied, and challenged by the newness and strangeness of their novel environment. In effect, they are faced with the task of learning to live all over again. And how do we do that? How do we come to understand the cues and clues of the world around us?

For example, if you give a lecture, how might you know from people's faces whether they are understanding you? When you use a concept, why do you believe that you understand it? Why do you like the music that you like? How do you know that the cry you heard is from your own child? How can people tell without looking whether someone is running upstairs or downstairs? How do you know what your mother looks like? What does tofu taste like? Why do you say "red, white, and blue" instead of "white, blue, and red"? We come to know these things, though often not how to express them.

There are many things, in fact, that we know but are unable to communicate effectively, if at all. Such unconscious knowledge is referred to in philosophy, psychology, computer science, linguistics and other fields as "tacit knowledge." I refer to it here, however as "dark matter" because I believe that the phenomenon is more structured and nuanced that the more familiar term perhaps suggests. But whatever we call it, comprehending this covert knowledge is crucial for our understanding of ourselves and others. This comprehension is my goal in what follows.

Although I grew up nearly bicultural less than ten miles from the California-Mexico border, I began my first serious trek through the worlds of novel cultures as a Christian, evangelical missionary with Wycliffe Bible Translators in the Brazilian Amazon. In that peculiar role, an invisible chasm separated me from the people to whom my missionary organization had assigned me, the Pirahãs. From my middle-class, US, industrialized, evangelical society to a hunter-gatherer group of the densest rain forest on earth, I traveled with a "calling" to translate the Bible, to produce the appropriate *perlocutionary* effects in the receivers—that is, to do a translation so effective that it would produce the same response in the Pirahãs that it did for those hearing the message for the first time in ancient Palestine. I wanted to transmit to this group of jungle dwellers legends from the first-century, Middle Eastern culture of the Bible. I naively supposed that this material could be effectively translated into the Pirahãs' language and culture. I was insufficiently daunted by the fact that this peaceful, semi-nomadic Amazonian community could hardly have been less like the violent, desert pastoralists of the first-century Middle East. The geography, climate, topography, languages, societies, and cultures of the first-century culture of Israel and twentieth-century Amazon were so different that in attempting to communicate the Bible to them, I might as well as have been delivering a letter from Mars via a beta version of Google Translate.

Theories eventually emerge in individual anthropologists, linguists, cognitive scientists, or philosophers from the experiences of their lives about what it is that they should be doing, how they should understand the world around them as this impinges upon their professional interests. Some construct grand theories. Others create little sets of theories. Others may just have hunches. There are some who gravitate toward broad generalizations, while others are content to link understandings of the particulars of their experience into narratives that are less sweeping.

In what follows, I propose a model of how we become who we are as individuals and societies, based on the acquisition and organization of particulars. But these particulars do not include the building blocks of some grander theories—I am not concerned directly with such familiar anthropological themes as totemism, animism, ethics, religion, folk theories of health and reproduction, and so on. That is because I believe that none of these are basic, but derivative, based upon more primitive building blocks that emerge naturally from living.[1] Rather, these particulars require no psychic unity of man, no nativism, and, especially, they require no innate content or concepts. This is a rather bold proposal, so we should get right to it.

Acknowledgments

I spent the better part of thirty-five years conducting field research in Mexico (four months among the Tzeltales) and Brazil (mainly among the Pirahãs, but also with time spent among or with specialists working with more than a dozen other groups (including the Sateré, Banawá, Wari', Jarawara, Jamamadi, Suruí, Deni, Karitiana, and Kīsedje, among others). The days were humid, hot, and bug filled. I had typhoid fever, amebic dysentery, intestinal parasites that were never diagnosed, malaria, and more malaria; survived near misses with poisonous snakes, attacks by anaconda in the river, tarantulas and tarantula-eating vespa wasps, giant centipedes, jaguars, pumas, and ocelots; and spent nights alone in a small lean-to in the jungle, hearing the heavy steps of something walking around my small, solitary camping spot. I have seen my children nearly die from malaria. And my life has been threatened by drunken Brazilians and drunken aborigines. I often wondered why I did this. One of my highest-ranked values is comfort. Yet an even more highly ranked value is understanding. The only thing that really kept me in the jungle all those years was a desire to understand others.

I never achieved understanding of anything alone. Over the years, I have benefited from conversations with many people, from their comments on my ideas, both verbal and written, from their examples, and from the advice, help, and excellent instruction I received from members of every culture and language I have studied. In addition to the indigenous communities of Brazil where I have worked, I also want to thank a number of individuals by name.

So it is my pleasure to thank Geoffrey Pullum, Yaron Senderowicz, Sascha Griffiths, Brian MacWhinney, Caleb Everett, Robert Van Valin, Ted Gibson,

Richard Futrell, Steve Piantadosi, Keren Madora, David McNeill, Nick Enfield, Daniel Ezra Johnson, Phil Lieberman, an anonymous reviewer for the University of Chicago Press, and especially T. David Brent—my editor at the University of Chicago—for his guidance, insightful comments, and encouragement throughout this project.

Introduction

> When making a decision of minor importance, I have always found it advantageous to consider all the pros and cons. In vital matters, however, such as the choice of a mate or a profession, the decision should come from the unconscious, from somewhere within ourselves. In the important decisions of personal life, we should be governed, I think, by the deep inner needs of our nature.
>
> SIGMUND FREUD

Why This Book?

This introduction lays out the ground rules and provides a map for the entire discussion that follows. The tripartite thesis of this book is (i) that the unconscious of all humans falls into two categories, the unspoken and the ineffable; (ii) that all human unconscious is shaped by individual apperceptions in conjunction with a ranked-value, linguistic-based model of culture; and (iii) that the role of the unconscious in the shaping of cognition and our sense of self is not the result of instincts or human nature, but is articulated by our learning as cultural beings. I refer to this more nuanced conceptualization of the unconscious as "dark matter," which I define as follows:[1]

> Dark matter of the mind is any knowledge-how or any knowledge-that that is unspoken in normal circumstances, usually unarticulated even to ourselves. It may be, but is not necessarily, *ineffable*.[2] It emerges from acting, "languaging," and "culturing" as we learn conventions and knowledge organization, and adopt value properties and orderings. It is shared and it is personal. It comes via emicization, apperceptions, and memory, and thereby produces our sense of "self."

It sometimes happens in nature that things that are not seen may be more important than things that are seen. Atoms come to mind. And space. Astronomers claim that the matter of our universe that can be seen accounts for only some 5 percent of the material of the entire universe, whereas "dark energy" accounts for as much as 68 percent and "dark matter" some 27 percent. "We are much more certain what dark matter is not than we are what it is. First, it is dark, meaning that it is not in the form of stars and planets that we see" (Ericksen 2015). In physics there is thus a place for explanations that involve things that appear to be unseeable in principle.

As a young missionary among the Pirahãs, with unspoken and ineffable values and beliefs absorbed from my own social activities, enveloping culture, and psychology, my faith in God and my mission, I thought only that "nothing is too hard for a dedicated missionary." Looking back, I can identify many of the hidden problems it took me years to recognize, problems based in contrasting sets of tacit assumptions held by the Pirahãs and me. A representative sample of those includes the following:

1. Pirahãs have no concept of God—certainly no "Supreme Being."
2. Pirahãs do not like for any individual to tell another individual how to live.
3. Pirahãs do not feel spiritually lost.
4. Pirahãs do not have a concept of spirituality that matches any other I am familiar with.
5. Pirahãs do not fear (or seek) death or "the afterlife."
6. Pirahãs do not see the messages of a foreign culture as relevant.
7. Pirahãs do not talk about or believe in things that they have not seen or for which there is no firsthand witness.
8. It is difficult for Pirahãs' jungle culture to make space for tropes and values based on images of deserts, dryness, sand, camels, and other geographical aspects of biblical culture.
9. Pirahãs do not normally list names of dead people or find lasting lessons from the past deeds of the dead, other than in their role of transmitting culture from their generation to the new generation.
10. Pirahãs believe that their way of life is the best one for them and that non-Pirahãs' beliefs and way of life are best for non-Pirahãs.
11. Pirahãs do not practice torture or capital punishment; for example, death by crucifixion is alien to them. Moreover, the concept of a society sanctioning punishments (especially death) for its members is incomprehensible.
12. Although most American missionaries believe that God has "prepared" every culture to understand the "gospel" (the good news, i.e., to understand that God's son, Jesus, died on a cross for their sins), the Pirahãs find the concepts of savior, sin, and salvation incomprehensible.
13. In spite of American missionaries' belief that people like the Pirahãs are afraid of a dark, threatening evil spiritual world and that many of them will be overjoyed at the missionary's arrival with the news that Jesus has freed them from this fear, the Pirahãs fear nothing and were uninterested in the missionary message.
14. American missionaries believe that all languages will be able to understand all the concepts necessary to express the full New Testament message. It is the job of the translator to find the appropriate words and phrases in the target language and then to match them with the appro-

priate Greek, Aramaic, or Hebrew concepts. This is false. (See chap. 8 for extended discussion of some practical problems of translation.)

15. American missionaries believe that their targeted people will (and should) respect and perhaps even love them for giving up their own homes and families to travel to tell them about Jesus. Yet the Pirahãs never put much distance or time between themselves and their families and have a hard time understanding why one would.
16. American missionaries believe that all people will or should believe the miracles recorded in the Bible, yet the Pirahãs don't believe in the supernatural, because they have no experience of it.

More important than even these differences between the Pirahãs and me, however, is the fact that they were unspoken. All of us were guided through our encounters by the invisible hands of conflicting cultural values. Only years later did I understand that the Pirahãs' unspoken beliefs and knowledge not only did not correspond to the beliefs of the average American missionary, but were in direct conflict with many of the values, beliefs, and knowledge that were so important for my evangelization objectives. Such considerations perhaps explain why I was a failure as a missionary in the sense that I produced no converts. This eventually led me to abandon Christianity altogether. Conflicting beliefs and values could also explain why, in my experience, so few missionaries actually produce long-lasting change (such that it survives their departure or death, for example) in the groups to which they "minister" (Hefner 1993). Pirahã values also played a role in my own conversion from Christian to atheist.

Another set of phenomena that made it difficult for me to communicate with the Pirahãs was the development of our bodies as well as different fears based on those bodies. Our bodies are built from different diets, different activities, different genes, and so on. I smelled different. The Pirahãs found it difficult to relate to a man who spent most of his day sitting and writing. They couldn't understand why I wouldn't give away all my canned food to them and supply my family's needs by fishing and hunting—like any man should. They were much better able to withstand the bugs that were constantly biting all of us. They also knew their way around the jungle. They were not afraid of jungle cats, snakes, and so on, though they had a healthy respect for them. They saw me, therefore, as having little of relevance to say because of the apparent lack of relevance of what I did, what I feared, what I ate, and how I thought.

Though these differences between me and the Pirahãs only scratch the surface of the deep, wide gap I had to bridge if I were to communicate with them about values, the Bible, my own life, and so on, notice that the differences do not fall into a single category. Nor can we say that such differences are primarily due to contrasting sets of "memes" or ideas. Memes and ideas alone are unable to account for these differences for the simple reason that not all of these or other differences are able to be made explicit; not all of the differences I encountered are reducible to propositions; and many of the differences are negative, rather than positive—learning what not to do, rather than what to do. And many of our different conceptions came from different bodies and physical experiences. So whatever cultural contrast is, it ranges beyond a set of explicit ideas.

Perhaps more important than anything else I learned was that while I am able to summarize some Pirahã beliefs and values, as I partially do above, there is a huge amount of individual variation. And how does the individual engage with his society? Does an individual behave in a particular way because of his or her culture or because of his or her personal psychology? What is the role of culture in accounting for the behavior of the individual? Or for the individual's psychology? Does each Pirahã participate in a collective intention to "live according to Pirahã values"? How is culture even possible? Or is it possible at all? Is a society living by a certain culture following rules like a football team? Or are they playing notes in harmony like an orchestra? How do cultures hang together at all, if they do?

Because I had no answer to these questions, I was ill equipped to do what I had set out to do. And largely (and fortunately, in retrospect), I failed. But my early years among the Pirahãs were by no means wasted. From these experiences came my desire and first empirical work to understand the intersection of personal psychology and cultural knowledge and values.

Eventually, I want(ed) to reach a theoretical understanding of the unspoken nature of culture and psychology—what Sapir called the "unconscious patterning" of society or Hall ([1959] 1973) referred to as "the silent language." This book is the result of a long quest. My ideas come from field research on more than a dozen Amazonian and other societies, as well as from reading Kenneth Pike, Edward Sapir, Aristotle, Clifford Geertz, Robert Brandom, Michael Polanyi, and myriad other anthropologists, linguists, psychologists, philosophers, and biologists.

Though I knew what I wanted to write, I was nonetheless surprised as I reached the conclusion of this book to learn that the theory I have developed here of the self is mildly reminiscent of the Buddhist notion of *anatman*, the idea that humans have no nature and no self apart from the experiences they have united in their memories. As Flanagan (2013) and Albahari (2006) have shown, Buddhism is built on and develops a serious alternative to Western ideas on the philosophy of human selfhood and human nature. My conclusion is far from spiritual, however. It is that the "non-self" is fragile, as it follows experience, being a posteriori rather than a priori and thus can take many unforeseen twists and turns.

Another crucial component of the thesis of this study is that minds do not experience and do not know, and minds are not the repositories of tacit information. Individuals are. By this I mean that our brains are just organs of our bodies. (I will use *mind* and *brain* interchangeably, but with the caveat that the mind is just a way of talking about the brain.) The brain cannot escape its diet, its sleep patterns, unsafe conditions, its hormones, or its body.[3] Our bodies learn (for example, with muscle memory; e.g., Bruusgaard et al. 2010) from our brains to our fingertips. Failure to take this fact into theorization has been one of the greatest shortcomings of the cognitive sciences. To repeat: Minds do not learn. Brains do not learn. Societies do not learn. Cultures do not learn. Only individuals learn. And what individuals learn is largely in the form of a culturally articulated dark matter. Brains are part of our bodies, so they play a role in the entire body's ability to learn. It is the body that learns in this sense.

Consider the kinds of dark matter that direct human activities. One example, from Gellatly (1986), involves poultry farming. It was noticed early on that if it were possible to determine the sex of chicks as they hatch, this would be of economic benefit in sorting chickens into laying hens vs. edible roosters, and so on. According to Gellatly (1986, 4), "[In the] 1920s, Japanese scientists discovered a method by which this could be done based on subtle perceptual cues with a suitably held chick. It was, nevertheless, a method that required a great deal of skill, developed through practice. After four to six weeks of practice, a newly qualified chick-sexer might be able to determine the sex of 200 chicks in 25 minutes with an accuracy of 95 per cent, rising with years of practice to 1,000–1,400 chicks per hour with an accuracy of 98 per cent."

Or consider remarks in a similar vein by Polanyi:

> Following the example set by Lazarus and McCleary in 1949, psychologists call the exercise of this faculty [apprehending "the relation between two events, both of which we know, but only one of which we can tell"] a process of "subception." These authors presented a person with a large number of nonsense syllables, and after showing certain of the syllables, they administered an electric shock. Presently, the person showed symptoms of anticipating the shock at the site of "shock syllables"; yet, on questioning, he could not identify them . . . He had acquired a knowledge similar to that which we have when we know a person by signs which we cannot tell. ([1966] 2009, 8ff)

Based on this type of case, some researchers on the mind likewise have argued that many of our thoughts and actions are heavily influenced by things we do not know we know, do not know how to say, or are simply unable to talk about. For example, Freud claimed that much of what guides the workings of the mind is unconscious, while Chomsky refers to the inborn, tacit knowledge of universal grammar that all Homo sapiens are born with.

In my own work, I have long referred to the invisible forces that act on our mind as the "dark matter of the mind." Recently I was pleased to read that some psychologists, such as Joel Gold (Gold 2012; see also Gold and Gold 2014), also use this phrase:

> The conscious mind—much like the visible aspect of the universe—is only a small fraction of the mental world. The dark matter of the mind, the unconscious, has the greatest psychic gravity. Disregard the dark matter of the universe and anomalies appear. Ignore the dark matter of the mind and our irrationality is inexplicable.

My view of dark matter is quite different, however, from Freud's, Polanyi's, Chomsky's, or Gold's, though there is some overlap. For example, Gold takes the dark matter to be essentially psychological. For me, however, it is found in the individual, in the individual's nurturing culture, and in the connections between the two. Psychology alone is insufficient.

Dark matter is recognized in one way or the other by all who study humans. In fact, in one sense, all of psychology is a sustained attempt to understand the dark matter of the mind. For example, consider the recent research of Susan Carey (2009) on concepts, of Elizabeth Spelke (2013) on a range of inborn knowledge, and Alison Gopnik (2010) on children's ability to learn things that may seem innate but are not, just to name a few. One of the pioneers of the study of dark matter, a source we will return to many times in the course of this book, was Edward Sapir (1884–1939), who spent his life researching the connection of individual psychology to cultural patterning. Unfortunately, psychology as it is currently practiced largely ignores that aspect of dark matter which most exercised Sapir and interests me—namely, how culturally directed unspoken knowledge, along with the structuring of apperceptions (the ways by which we process, make sense of, and assimilate our experiences), explicit learning, body memory, and so on, combine to form the individual. Culturally directed psychology was addressed perhaps most insightfully by Sapir, and the formation of the individual by Buddhism.

Philosophers have also written a great deal on tacit knowledge (similar to but not identical to my concept of dark matter), going back at least to the seminal work of Michael Polanyi. The works that deal with tacit knowledge directly range from Searle's (1978) discussion of the "background," to work by John McDowell (2013) and Robert Brandom (1998) on the implicit vs. explicit in concept formation, Bourdieu's (1977) proposals on "habitus," and others that we discuss in the course of this book.

Let me underscore again the caveat that dark matter is not to be confused with tacit knowledge. This will become more important as we proceed. Of course, some work in philosophy could be interpreted as making a case that tacit knowledge cannot exist. For example, according to Koster's (1992) interpretation of Wittgenstein, knowledge is not a representation in the mind that can either be talked about or be ineffable, but it is action—what we "know" is a matter of what we do. I think that this position that knowledge is action has a good deal to commend it, but I also believe that falls short of a full theory for largely failing to recognize the continuity between apperception and memory (on which more below). And it seems to fail to tell us why and how actors act.[4]

Within my own area of specialization, linguistics, Noam Chomsky—one of the founders of the cognitive sciences and generative linguistics—was among the first intellectuals to develop a theory of the structure, meaning, and importance of dark matter, as he introduced the theories of deep structure and universal grammar.[5] Another influential linguist was Kenneth L. Pike. Although Chomsky has been more influential across a broader swath of intellectual areas, linguist Kenneth Pike's (1967) ideas on unspoken knowledge come closer to the dark matter that this essay addresses. Crucial to this study is Pike's notion of *emicization*.

Emicization emerges from Pike's work on the *emic* vs. *etic*. He coined these words based upon the widely used linguistic terms phon*etic* vs. phon*emic*. Phonetics (articulatory, acoustic, or auditory) is the study of speech sounds from the perspective of a non-native speaker, say, a physicist or linguist. Phonemics is the study of the set of phonetic sounds that native speakers perceive as single sounds—that is, the sounds that are important from the perspective of a native speaker, an insider.[6] For example, English speakers all hear one sound, /p/, in the words *park*, *spark*, and *carp*, when in fact there are at least three sounds, all written as *p* in these words, namely, [p^{h}], [p], and [p̚], respectively.[7] Native speakers thus know less *explicitly* about the sounds of their language than they tacitly know about them, since speakers in general never perceive the separate etic sounds but only the single emic sound that an etic sound is associated with. Yet they never confuse etic sounds in use. Thus even though native speakers lack overt knowledge of the distribution of the etic sounds of their language—for example, the three separate manifestations (technically, *allophones*) of /p/ in the examples just given—their own emic knowledge produces behavior that can be described as: "Use [p] in syllable-medial positions, [p^h] in (some) syllable-initial positions, and [p̚] in phrase-final positions."

Extending this etic/emic contrast to culture, Pike (1967) makes a case—one that we will draw upon repeatedly throughout this book—that the insider (emic) vs. outsider (etic) perspective on cultural events, perception, and myriad other aspects of human behavior are possible only because of the crucial use of tacit knowledge, as this term will be developed here.[8]

What is needed, and what is attempted below, is a sustained argument in support of the hypothesis that our actions, beliefs, desires, values, and other behavioral or mental markers of the self emerge from the implicit knowledge and apperception that we acquire as members of particular social groups, from our families and tribes to our societies and nations. Unlike Hall's silent language or Searle's "background," dark matter is multilayered, differentially manifested, and variously derived from the experiences of living.

Our discussion here is a microcosmic culture of its own—an arrangement of knowledge and apperceptions that are not quite like those of any other discussion. From this microculture, I hope that answers emerge for several questions of interest to cognitive scientists. Perhaps the broadest question it attempts to answer is how cultures and individuals shape one another. This is an old question but one still worth trying to come to grips with, in my opinion. Clearly, if you or I had been born in very different countries by the same parents or to different parents in the same country, we would not be "you" or "I," but quite different people. In the former case I might still be redheaded and pale-skinned, but perhaps I would be taller or shorter, fatter or thinner, stronger or weaker, smarter or dumber, more tolerant or less tolerant, and so on, than I currently am. I would likely prefer different foods and react differently to pain, and different things would disgust me or please me. Most would acknowledge that none of us would be the same as we now are, physically, mentally, or morally, if raised in another culture. The crucial question is, how different would we be?

Our investigation of dark matter of the mind sets out to answer the question of what it means to be human from our perspective of apex social primates. It asks whether humans are so constrained by instincts or physics that our freedom is an illusion. It interrogates the very notions of culture, society, and the mind. For some (e.g., Tooby, as discussed below), culture is little more than collection of oddities—a lexicon of cues, values, and bits and pieces of knowledge. To these people, if we strip away this lexicon like some agglomeration of mental barnacles, we will find the same cognitive and emotional structures and functions in all humans. I disagree. I want to push back against the idea that humans are more alike around the world than they are different. In addressing the issues, many subsidiary questions and problems arise, such as how—if at all—cultural variation interacts with emotions, physical development, morality, death rates, cognition, and so on. Concerns with the cognitive-cultural contribution to a theory of Homo sapiens have been discussed for a long while. Edward Sapir was a pioneer of the study of mind-culture interaction. His premature death in 1939 lead to a decades-long hiatus in such studies.

To review a bit of the post-Sapirian work on cognition that emerged in the 1950s, there was a partial return to the study of the mind, but now as a computer rather than as part of a larger culture. The date most associated with this new "mental turn" is September 11, 1956. On that day, a gathering of researchers at the Massachusetts Institute of Technology focused on the nature of the human mind, an event that Gardner (1987) and others refer to as the birth of the "cognitive revolution" (Boden [2006] provides a superb and comprehensive history of the cognitive sciences). I believe that this assessment is incorrect for various reasons, however. First, it was *not* a revolution in any sense, however popular that narrative has become. As I just stated, Sapir explicitly studied cognition and culture decades before this conference, no less insightfully than studies introduced in 1956 and subsequent years. Moreover, the "revolution" that emerged from this question asked fundamentally the wrong question, focusing on the mind as a disembodied knower (in the unfortunate Cartesian tradition). Nevertheless 1956 was unarguably a watershed year, a rebirth of studies of the mind, at least on the US side of the Atlantic. The personalities and works associated with the MIT conference were deeply influential in the revival of interest in the mind. The presenters at the conference included George A. Miller, Noam Chomsky, Nobel Prize winner Herbert Simon, and Allen Newell. Many other philosophers, anthropologists, psychologists, computer scientists, and linguists subsequently flocked to identify with the emerging cognitive sciences.

Following the 1956 conference, funding began to emerge for studies of the mental or, in the new buzz phrase, the "cognitive sciences." In the early rounds of funding in the mid-1970s by the Alfred P. Sloan Foundation, grants were awarded to the University of California, San Diego; the University of Texas at Austin; MIT; Yale University; Brown University; and Stanford University. Later grants to further develop efforts were also awarded to Carnegie Mellon University; the University of Pennsylvania; the University of Chicago; the University of California, Berkeley; the University of Rochester; and the Cognitive Neuroscience Institute. In these early awards and efforts by the different institutions (these grants coincided with my entry into linguistics as a student), there were conflicts over the very definition of cognitive science. What was cognitive science? Was it singular or plural (several cognitive sciences vs. a single cognitive science)? Who was *really* doing cognitive science as it should be done?

My own view is that no one was. In all of this funded research, this new energy, these brilliant ideas, there was nothing in the cognitive sciences that suggested that people were asking serious questions about the context in which the mind is formed—particular individuals situated in particular cultures. Culture was examined as a manifestation of the mind in most cases, even in so-called cognitive anthropology. Unfortunately, these studies were thus largely unidirectional, *mind→culture*, and they were dramatically myopic in their obsessive focus on computation and the metaphor of the mind as a computer. In retrospect, it seems that because of the newness of the computer at the time—the metaphor that drove the '50s cognitive studies—continuing on to the present, these otherwise superb studies all failed to consider human emotions and the role of the individual as a whole (body and culture) in cognition, instead focusing narrowly on what they saw as the "computational" aspects of thinking. Yet emotions, muscles, hormones, even bacteria and the body—that is, *the individual* (if one believes as I do that there is nothing to an individual but one's body)—are the portals to reasoning and cognition.[9] No theory of cognition can hope to succeed without a focus on the entire individual—not merely their "minds" and their place in society. Cognitive scientists never examined in any detail the foundational relationship of culture to the mind, the mind as an outgrowth of culture. The reason seems to follow from the misleading idea that the mind is a digital computer, an evolved software running presently (but not necessarily) on neurological hardware. This metaphor is fragile, though. For example, unlike the brain and body, computer software doesn't grow biologically from its hardware (see below and Dreyfuss 1965, 1994; Haugeland 1998; among others). Nor do computers possess emotions—one of the primary drivers of human cognition. And we cannot overlook the fact that the mind is shaped by its environment even when it is not attending to its environment per se, an ability beyond any current computer. These are just a few of the serious shortcomings—or at least, I will try to show below that they are—of the digital computer theory of the mind. Thus, from the perspective here, the entire cognitive sciences "revolution" took the wrong road, the wrong "turn," as philosophers occasionally use that term.

I expect that this book may bother some because it makes the case that standard cognitive psychology comes up short in understanding the human mind, for the reasons just given. Although the study of dark matter is vital to understanding how the mind works, all knowledge is itself a product of living culturally—structuring one's life around ranked—but, crucially, violable—values, experiences, apperceptions, and the like, that are learned, and only occasionally taught, as a member of a society.

The two most visible names associated with tacit knowledge over the past sixty years have been Michael Polanyi and Noam Chomsky. Their work offers different conceptions of the nature and sources of this knowledge. The tradition in which Polanyi's work is situated focuses on tacit knowledge that is learned, internalized, and forgotten until called upon, such as how to play a song on the guitar or ride a bike. Contra to this tradition is the nativist idea, associated most frequently with Chomsky, but in fact running throughout Western thought from Plato through Bastian's "psychic unity of mankind." Nativism is the idea that humans share some knowledge because it is programmed into all of us innately: instincts, moral principles, rules of grammar, and a number of congenital concepts. Other prominent exemplars of purported innate tacit knowledge include Freud's notion of the unconscious; Campbell's idea of the universal mythic structure, "monomyth"; Jung's theory of archetypes; and the work of Cosmides, Tooby, Fodor, and Pinker—what some call the "massive modularity" of evolutionary psychology.[10] The theses of learned tacit knowledge and nativism need not be opposed, of course. It is possible that both learned and innate forms of tacit knowledge are crucially implicated in human cognition and behavior. What we are genuinely interested in is not a false dichotomy of extremes but in a continuum of possibilities—where do the most important or even the most overlooked contributions to knowledge come from?

I am here particularly concerned with difference, however, rather than sameness among the members of our species—with variation rather than homeostasis. This is because the variability in dark matter from one society to another is fundamental to human survival, arising from and sustaining our species' ecological diversity. The range of possibilities produces a variety of "human natures" (cf. Ehrlich 2001). Crucial to the perspective here is the concept-apperception continuum. Concepts can always be made explicit; apperceptions less so. The latter result from a culturally guided experiential memory (whether conscious or unconscious or bodily). Such memories can be not only difficult to talk about but often ineffable (see Majid and Levinson 2011; Levinson and Majid 2014). Yet both apperception and conceptual knowledge are uniquely determined by culture, personal history, and physiology, contributing vitally to the formation of the individual psyche and body.

Dark matter emerges from individuals living in cultures and thereby underscores the flexibility of the human brain. Instincts are incompatible with flexibility. Thus special care must be given to evaluating arguments in support of them (see Blumberg 2006 for cogent criticisms of many purported examples of instincts, as well as the abuse of the term in the literature). If we have an instinct to do something one way, this would impede learning to do it another way. For this reason it would surprise me if creatures higher on the mental and cerebral evolutionary scale—you and I, for example—did not have fewer rather than more instincts. Humans, unlike cockroaches and rats—two other highly successful members of the animal kingdom—adapt holistically to the world in which they live, in the sense that they can learn to solve problems across environmental niches, then *teach* their solutions and *reflect* on these solutions. Cultures turn out to be vital to this human adaptational flexibility—so much so that the most important cognitive question becomes not "What is in the brain?" but "What is the brain in?" (That is, in what individual, residing in what culture does this particular brain reside?)

The brain, by this view, was designed to be as close to a blank slate as was possible for survival. In other words, the views of Aristotle, Sapir, Locke, Hume, and others better fit what we know about the nature of the brain and human evolution than the views of Plato, Bastian, Freud, Chomsky, Tooby, Pinker, and others. Aristotle's tabula rasa seems closer to being right than is currently fashionable to suppose, especially when we answer the pointed question, what is left in the mind/brain when culture is removed?

Most of the lessons of this book derive from the idea that our brains (including our emotions) and our cultures are related symbiotically through the individual, and that neither supervenes on the other. In this framework, nativist ideas often are superfluous. Of course, I maintain (D. Everett 2012a) that in order to see this, we must understand the *platforms* (universal) of human cognition, the *nature of the tasks* humans have to perform, and the ways in which humans *live culturally* and come to acquire the cerebral dark matter that ultimately shapes who they are and how they think about and relate to the world around them. These arguments extend the case for culturally derived tacit knowledge begun in D. Everett (2012a) for language, to a fuller spectrum of cultural and cognitive determinants of individual identity.

The flexibility and cognitive resources of humans are most concentrated in the dark matter. This matter itself emerges from many sources. Culture is but one of those. Emotionally driven goals are another. Material environment is another. The nature of the tasks to be performed is another. But the recognition of culture that plays a role, even in domains where it was once considered irrelevant, is vital to the understanding of ourselves and our species. So here we want to consider the case that what we do and what we come to know are shaped primarily by the greatest distinctive feature of our species: culture.

To understand what is at stake, let's consider again the phrase "human nature." There are many definitions and conceptions of human nature, and we explore them in more detail in the final chapter. We find them in biology (Wilson 1978; Ehrlich 2001), in philosophy (Plato), in psychology (Freud [1916] 2009); Pinker 1995), in most major religions, in neuroscience (Paul Churchland 2013; Patricia Churchland 2013), in ecology (Cashdan 2013), in theology (Calvin [1536] 2013), and in literature (Twain [1916] 1995), among other places. For some theists, human nature is the propensity to rebel against God and do wrong, because we were damaged by original sin. In Hinduism all humans are defined by the *atman*, the true soul or essence of the person, which is in need of self-knowledge (that is, "liberation"). In Buddhism there is *anatman*, "no-self," meaning that there is no essence of an individual human, just the set of experiences they pass through, their apperceptive histories (or in Buddhist terms, the *skandhas*: form, sensation, perception, mental formations, and consciousness).

Human communities produce unseen forces that shape the way we live, including the ways we think, communicate, make moral judgments, conduct science, and find happiness, through the activity some anthropologists refer to as "culturing"—acting in a community, constrained by others' values and concepts (Latour 1986, 2007). The forces that shape us are variously known as values, implicit information, culture, background, and so on. But these forces are more occult and powerful than the average description of them might lead us to believe. It is these forces that I intend by "dark matter." Although my concept of dark matter is related to Michael Polanyi's tacit knowledge or "personal knowledge," Polanyi's focus was unlike mine in that it was not so much on culture as on subroutines and components of large intentional acts (e.g., to ride a bike, we need to first learn to keep our balance on the bike, learn to pedal, learn to brake, and so on—subroutines that are ultimately forgotten [but still present] in the single mature intention of "ride a bike"). My concept of dark matter, on the other hand—to slightly paraphrase George Harrison's quasi-eponymous song—is "within us and without us," at once embodied in individual humans at the same time that it serves as the unseen connective force between members of a given society. It includes our tacit collective intentions to maintain cultural values and knowledge that binds cultures together. If correct, this view presents a challenge to the past sixty years of study in the cognitive sciences, because these sciences have failed to account for the nature, origins, and effects of this dark cultural matter on the formation of human identity.

For example, some evolutionary psychologists appeal to specific ranges of predispositions to define innate human nature. A common example of such dispositions emerges from research that suggests that many humans react to risky behavior as "sexy." But reactions to risky behavior are not uniform across all types of risk. According to the relevant research (Wilke et al. 2006), the kind of risky behavior that is most likely to attract, say, women to men involves risk that is part of the evolutionary history of the human brain. So a female human could be "turned on" by a man washing windows on a skyscraper or a man swimming in deep water because deep water and high places are part of the primeval fears that evolved in our species—perhaps the entire Homo genus. If this were correct, then we would all share a biologically determined attraction to people who show evolutionary superiority over primordial risks. And for this very reason, a woman would not be aroused by the sight of Homer Simpson working in a nuclear factory, because radioactivity was not part of her species' evolutionary development, even though radioactivity is much more dangerous than heights (radioactivity can kill many more). In this view, there is a human nature that is formed by the sum of our evolutionary predispositions, which may go by various names—one of the most common being innate "mental modules," or simply "the innate mind" (Carruthers, Laurence, and Stich 2005, 2007, 2008).

Our discussion here examines and rejects this view. As a glimpse of why, consider an example of dark matter that is often overlooked by psychologists and anthropologists, perhaps because it is thought to be too obvious: cultural knowledge. So imagine that you are walking in the Amazon rain forest, accompanied by someone raised there; someone who has survived by means of their understanding of the local flora and fauna. As you make your way through the humid green growth of the climax forest, out of the corner of your eye you notice a branch move. You walking partner notices it at the same time. What you have both seen can in principle be measured—the speed of the branch in motion, the distance from the resting state that the branch moved, how high off the ground the branch is, the branch's color, whether fruits or nuts are growing on the branch and so on. What is measurable is external to both of you. It is both epistemically and ontologically objective (Searle 1997).

On the other hand, your interpretation of this movement is internal and cannot be measured. It is ontologically subjective. But—at least from the perspective of the local—the experience is epistemically objective. Thus it can be studied, even though it is an ontologically subjective experience. That is, it is by its nature something that you alone can know (i.e., your own interpretation). But the experience is one someone can have objectively—others can see what you saw. Although you may have no interpretation of what you saw at all, other than "the branch moved," perhaps you wonder why the branch moved, or maybe you think it was moved by an animal, by the wind, or by a falling object. You lack a specific hypothesis or knowledge about the cause behind the movement of the branch. This is understandable. Your perception and your interpretation both suffer from a lack of background knowledge. Did you notice whether other tree branches on the same or different trees were also moving? Did you notice what direction from you (north, south, east, or west; downriver, upriver; uphill, downhill; etc.) the tree sat in relation to your vantage point? Did you notice the species of the tree whose branch was moving? Perhaps you did. Likely not, though. On the other hand, your companion very probably did and does have an interpretation of what you both just saw, based upon an automatic and tacit environmental hermeneutics—his emicized culture. And this interpretation along with its cultural foundations can both be studied to some degree, meeting the condition of epistemic objectivity.

Your companion can tell you immediately whether the wind moved the branch (were other branches moving at the same time on other trees?) or whether an animal or falling fruit caused the movement. He or she knows what animals live in that species of tree, whether they are likely to be found around the position of that branch (in the case of a very tall tree, such as a Brazil nut tree), whether they are eating, hiding, hunting, and so on. Your companion's tacit knowledge—like yours of your native environment—is hard-won and largely independent of explicit instruction.

As another example, imagine that you are teaching your daughter how to pilot a motorboat in a switchback river. You are both sitting back at the transom; she has her hand on the control arm. You tell her that she has to anticipate the turns by beginning to turn ever so slightly toward them, then correcting slightly—never wait until the last second, never turn abruptly. And yet as you approach turns, you find that the rear end of the boat begins to come around faster than it should, moving you sideways down the river, threatening to capsize. You forgot to tell her that movements must be even slower the lower the boat sits in the water or when the weight is distributed unevenly in the boat. Some of what you know can be easily spoken. But there is a "feel" to the actions you are explaining for which you cannot find words. Is this "feel" of the action something you *know*? Is it related to knowledge or something completely different? To some this is like asking whether one "knows" how to like lemons. Tasting consists of at least two components under this view. First, tasting is knowledge acquired by doing. Second, tasting is a preference resulting from the implicit ranking of tastes with higher ranking "tasting better" to you.

Many philosophers, going back to Socrates, refer to knowledge as "warranted true belief." In most cultures, therefore, taste is not considered knowledge—because it isn't "true" in any objective sense and because it has no external warrant—you like what you like. Thus my enjoyment of Mexican food is perhaps not best characterized as warranted true belief. But if taste is not knowledge in our culture, we need to ask what it is. Perhaps it is a form of intuition. But then we need (as we attempt here) to try to come to grips by what we mean by intuition. If we say that knowledge is using concepts and accept Robert Brandom's (1998) concept of concepts, then we agree that nothing is known conceptually unless we can use it in an inference—that is, make it explicit. Some things that shape us cognitively, however, such as taste, cannot be made explicit. Therefore we describe so many exotic meats with the phrase "tastes like chicken," even when we know they don't in fact taste like chicken. We lack sufficient comparators in our apperception set, nor word in our vocabulary.

A recent article on the difficulties of expressing tacit knowledge about food is found in the popular press, where one author (Fleming 2014) asserts that "the English language doesn't offer a specific vocabulary for describing food aromas. Despite the fact that smell is the dominant force in flavour perception, English speakers refer to aromas by the names of the foods they are most commonly associated with. Aniseed, citrus or nutty, for instance." This is because perception follows at times from either knowledge or apperception, and we do not expect to find words for all cases from the latter source, because some apperceptions are themselves ineffable. Therefore we will encounter common human experiences for which no culture has words.

Our experience and appreciate of taste is an aggregate of sorted apperceptions, as well as the biology of taste sensors. How we come to have taste experiences and how these ultimately give us our sense of what we like to eat, contributing to our very sense of "self" and personality illustrates what we are trying to get at here. In other words, self is largely a memory of *skandhas* (apperceptions) that forms the self or "nonself" (the *anatman*) as these are ordered and selectively recalled/stored by our episodic and short-term memories. This is not unrelated to Hume's quote that all experience is a sequence of concepts. Here the claim is that all self is a sequence of experiences. So if this were correct, it would be a recursive definition of the self.

Interestingly, however, among the Pirahãs of the Brazilian Amazon, a taste for something *does* seem to be classified as a kind of knowledge. Thus when they offer a stranger food, they ask, "Do you know how to eat this?" If it looks unappealing, one can simply reply, "I do not know how to eat this." No one loses face, and one's inability is chalked up to an experientially based ignorance.

One way to build the case for dark matter would be to construct our understanding of it via an exploration of our knowledge of language. By this tack we can build on what is known about our most significant and largest set of intuitions and expert knowledge to offer an account of intuitions and expert and tacit knowledges across domains. This is not a new approach. Claude Lévi-Strauss used linguistic principles learned from Ferdinand de Saussure and Roman Jakobson, to invent "structuralist anthropology," while Marvin Harris, Clifford Geertz, and others appealed to Pike's concepts of etic vs. emic to construct very different approaches to the study of culture. Sapir, Whorf, Boas, and the early American anthropologists of course saw linguistics as one branch of anthropology, so it is not surprising that there would be some mutual influence of thinking about culture on thinking about language and vice versa, though history for some reason shows a fairly unidirectional flow during a period of time, from language to culture. In *Language: The Cultural Tool* (D. Everett 2012a), I attempted to show new ways in which cultural reflection might influence linguistic theorizing and language understanding, arguing that cultural values affect and effect some linguistic structures. And, of course, in modern times perhaps no one has been more insightful in analyzing the connections between culture and language than Michael Silverstein (see the various references to him in the bibliography and throughout the text), though there are many other superlative researchers.

Kenneth L. Pike provided anthropologists and linguistics with the basic conceptual opposition that has affected both the study of culture and the study of language profoundly. Pike's etic vs. emic dichotomy is crucial throughout the present study. As Pike put it:

> The etic approach treats all cultures or languages—or a selected group of them—at one time. It might well be called 'comparative' in the anthropological sense (cf. M. Mead, 1952: 344) were it not for the fact that the phrase 'comparative linguistics' has a quite different usage already current in linguistic circles . . . The emic approach is, on the contrary, culturally specific, applied to one language or culture at a time. (1967, 37)

But the etic-emic relationship is a difficult empirical issue, he points out: "Regardless of how much training one has . . . emic units of a language must be determined during the analysis of that language; *they must be discovered, not predicted* [emphasis mine]—even though the range of kinds of components of language has restrictions placed upon it by the physiology of the human organism" (37ff).

In a sense, then, we are engaged here in an exploration of how the "emic" in each of us shapes our language, our culture, and our construction of personal identity. It argues for "emicization"—the construction of an insider point of view, particularly the "dark matter" that makes us and our societies who and what they are. Dark matter is then emicization via aggregated apperceptions (personal interpretations of experiences), acquired concepts, and their cultural-internal interpretations. It is partially a function of the body (brain and fingers, at least—there is a sense in which my fingers know how to play the guitar) and culture. Dark matter is unique to every individual as every culture is unique for the group possessing it. This is inherent in the notion of emicization.

In my sense, "mind" is an oblique reference, largely inaccurate, to the individual's capability to know and arrange what they know. I take the peculiar view that knowledge is not merely a state of mind, but stored and created at least in part by actions, experiences, memory, relationships, and orientations.

By "culture" I roughly mean (see chap. 2) what we know, our values, and the systematicization of our values, knowledge, and apperceptions *as members of a given society*. In other words, culture is societal systematicization, the hypernymization of terms; connection of facts to individual roles, and cultural objectives. Cultural systematicization often brings recursive, hierarchical structure to knowledge across all the mental dimensions just mentioned. It is not merely all our knowledge. Rather, it is the *arrangement of our knowledge and values and roles*.

This all gets us to the primary thesis of this entire work: there is no useful notion of "human nature." Rather than "human nature," people are controlled by dark matter, acquired and shaped via culture, biology, and individual psychology. This leads to a replacement of the idea of human nature with the alternative concept of "personal nexus" or constructed self.

Let me point out what I intend as the novel contribution of this book: from it emerges a new, linguistically inspired synthesis of philosophy, psychology, anthropology, and linguistics to understand the self without nativist incantations.[11] It is the proposal that tacit knowledge and the construction of the cultural-psychological-physical nexus just are what we know as "self." There is no human nature if by this we mean a kind of a priori knowledge common to all and only humans.

The chapters each build a distinct layer of the story. In the first four chapters, part 1, we lay out the three foundational components of my model of dark matter: culture-as-values, values-as-emicization, and culture-as-grammar. These points are preceded in chapter 1 with a definition and discussion of the nature of dark matter, and its pedigree in two principal lines, the Platonic and the Aristotelian. In chapter 2, we discuss a novel, ranked-value theory of culture. To do this we examine a number of historic and well-known definitions of culture offered over the past hundred or so years, concluding that all are inadequate and offering a simpler, yet more comprehensive and satisfying definition of culture that immediately incorporates the idea that it is the source and object of dark matter. Chapter 3 explains how culture and dark matter are acquired, focusing especially on the development of concentric circles of attachment in the Amazonian society of the Pirahãs. Chapter 4 examines the fundamental contribution of dark matter to our interpretation of the world around us and our ability to navigate through it.

In part 2, beginning with chapter 5, we look at dark matter in grammar, texts, gestures, translation, individual values, and the interpretation of lived situations. In this chapter's discussion of the presupposed dark matter of texts, the focus is on discourse and how it is constructed by and imbued with dark matter, tacit information reflecting cultural values, knowledge, and roles. Chapter 6 examines the emergence of grammar from culture and individual apperceptions via emicization. Chapter 7 reviews the role of gestures in bringing culture and grammar together, while simultaneously providing a separate level of organization that—in conjunction with prosody—enables hearers and speakers to better understand or communicate themes, discourse flow, and content. This chapter also argues that in spite of claims to the contrary, gestures support the Aristotelian, empiricist view of language ontogeny and phylogeny over the nativist, Platonic conception of dark matter. It does this in part via a critique of Goldin-Meadow's (forthcoming) work on "homesigns," gestures invented by what she believes are children without linguistic input—the nonhearing children of hearing parents who invent signs in order to communicate. In chapter 8 we continue with this theme in the realm of translation, looking closely at how dark matter enables us to reinterpret texts, individuals, and the world around us, and to communicate our own interpretations to others, intra- and interculturally.

Moving on to part 3, we discuss the implications of the theory developed to this point for two issues of fundamental and urgent importance for understanding human psychology and human evolution. In chapter 9 it is argued that there is nothing like instincts or modules in our higher-level cognitive abilities (e.g., language), the interpretative principles of the world around us, Bastian's elementary ideas, Freud's tripartite mind, Campbell's monomyth, and so on. This chapter considers and rejects several proposed instincts, specifically considering, as a concrete, detailed case study, why recent work on a "phonology instinct" seems to be mistaken. The final chapter, 10, argues that human nature is reducible to an understanding of our bodies, our memories, our intelligence, and our apperceptions. There is beyond these matters no "psychic unity of mankind"—no single human nature.

PART ONE

Dark Matter and Culture

1

The Nature and Pedigree of Dark Matter

> Adherence to an epistemology is not something which merely 'happens to a person but instead it reflects a component of his moral development. In some sense he is . . . morally responsible for adopting an epistemology even though it can neither be proved nor disproved to the satisfaction of those who oppose it.
>
> KENNETH PIKE, *With Heart and Mind: A Personal Synthesis of Scholarship and Devotion*

This chapter provides a definition and pedigree of the notion of dark matter, tracing two broad clades of descent—the Platonic and the Aristotelian. I align my own work with the Aristotelian, empiricist line, arguing in favor of this tradition throughout the remainder of the book.

The first section of the chapter looks at kinds of knowledge that linguists, philosophers, psychologists, and anthropologists have proposed that are relevant for our present concerns. I give a definition of dark matter, its major divisions into unspoken and ineffable, and how it is distinct from other types of knowledge, including tacit knowledge. We look at knowledge-how vs. knowledge-that, for example, arguing, along with others, that no clear dividing line exists between such knowledges.

In this regard, I also discuss knowledge as "particle, wave, and field," borrowing these three perspectives from the work of Kenneth Pike. I look at linguistic knowledge of sound units—phonemes and allophones (the local variants of phonemes)—in order to exemplify these epistemological perspectives.

Next we move to a discussion of the Platonic tradition of innate knowledge, continuing from Plato to Chomsky and beyond, via many others. This section also looks closely at the influence of the thesis of the "psychic unity of mankind" first proposed by Adolf Bastian. Following this, we survey different works in the Aristotelian tradition of dark matter. We then reach the conclusion of this chapter, also providing a survey of the major lessons learned here.

Kinds of Knowledge

As we have seen, over the last few decades, the two biggest proponents of tacit knowledge have arguably been Michael Polanyi and Noam Chomsky, though with radically different conceptions of the nature and source of this knowledge. Chomsky's conception of tacit knowledge descends from Plato; Polanyi falls closer to Aristotle's ideas. In contemporary research, forms of the debate over tacit knowledge are to be found in the contrasts between evolutionary psychology vs. embodied cognition, questions about whether computers can replicate human cognition in any significant sense, on the nature of concepts (what does it mean to hold a concept), in translation theory, expert systems, the assumptions that drive business decisions, the behavior of populations vs. individuals, and so on. The belief that humans share basic concepts and other knowledge innately has been a popular one in the history of Western thinking. Some of the more prominent examples beyond the two just mentioned include Freud's notion of the unconscious forces that drive human psychology; Jung's theory of "collective unconsciousness"; Joseph Campbell's "monomyth"; the work of Cosmides, Tooby, and Pinker in evolutionary psychology; and Jerry Fodor's work on mental modules.

I want to argue, to the contrary, that humans are designed to be flexible, that "human nature"—when characterized as inborn cortical, hardwired information, and instincts—is incompatible with flexibility. Humans adapt as well as upright primates might be expected to have done in the world in which they live. We are able to vary behaviors in similar environments and to respond effectively to environmental change, in the sense that their psychologies and cultures can solve problems of various types encountered by the species worldwide. Cultures are evolution's ultimate solution to the problem of providing adaptive flexibility—so much so that the most important cognitive question, as others have put it, becomes not "What is in the brain?" but "What (culture/society/environment) is the brain in?"

Our minds and our cultures are constructed symbiotically. The relationship between culture and psychology is not supervenience, though each influences, shapes, and enhances the other. Because of this I want to argue here that nativist ideas are for the most part superfluous in understanding human cognition, once we recognize the universal platforms of human cognition, the nature of the tasks humans have to perform, and the ways in which humans "live culturally" and come to acquire the dark matter of their minds, which ultimately shapes who they are as well as how they think about and relate to the world around them.

The flexibility and cognitive resources of humans are found particularly in the dark matter of our minds—the tacit knowledge we acquire from lived experience. This knowledge comes from many sources. Culture, or "culturing," is but one of those. Material environment is another. The nature of the task another. But the role of culture in domains where it was once considered irrelevant is vital to the understanding of ourselves and our species.

The knowledge I am concerned most with is any knowledge-how or any knowledge-that that is unsaid in normal circumstances, unarticulated even to ourselves or, at the extreme, ineffable (Levinson and Majid 2014). It comes from personal observation and social expectations (standards, values, and so on). It is shared and it is simultaneously personal. It is acquired via emicization and apperceptions. The sense of self is just these plus memory.

Variation is where the action is at in understanding ourselves, our conspecifics, and our species as a whole. We should not be too quick to look for the invariant, the abstract generalizations. We should first attempt to fully describe (Geertz 1973) and reflect on the particular (James [1906] 1996). And variation operates at multiple levels simultaneously. The two broad intersection planes of variation that concern us here are individual psychology and group culture.

Another crucial idea is that human cognition and behavior are not law-like. When we attempt to understand human cognition and human actions, we are not doing physics. We do not expect exceptionless laws. We expect rather a confluence of dynamic, shifting properties. Answers will not be simple many times and they will focus as much on variation, if not more so, than they will on constancy. I am less interested in universals and more in understanding, and I do not believe that the former is always the basis of the latter.

In fact it seems that understanding is often list-like and descriptive, rather than abstract and general (see James [1906] 1996 as well as Fodor's (1998, 40ff) lexical semantics). The best way to judge the success of our proposals is their consonance with what we know more generally about the world, and the inclusivity and utility for those trying to deepen their understanding of the cognitive life of Homo sapiens.

I am certainly not the first to develop the thesis that nurture is at least as important as nature in understanding the thought and behavior of humans and other animals. Among the books that take on this subject from different angles are Paul Ehrlich's (2001) *Human Natures* (biology); Alison Gopnik's (2010) *The Philosophical Baby* (child development); Jesse Prinz's (2014) *Beyond Human Nature* (philosophy), and Philip Lieberman's (2013) *The Unpredictable Species* (neuroscience and linguistics). But the approach to the issue of human nature and knowledge is different here. First, it is based on anthropological and linguistic field research on some highly interesting and distinctly un-Western cultures. Second, a major—and unique—concern of this study is a new understanding of culture, and the interaction of culture with

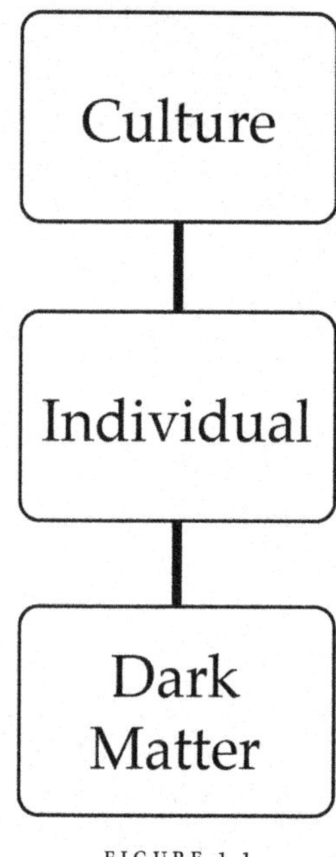

FIGURE 1.1

individual psychology. This interaction, as mentioned, is referred to in this work as "emicization," borrowed from Pike (1967).

For convenience, let's repeat the earlier definition of dark matter:

> Dark matter of the mind is any knowledge-how or any knowledge-that that is unspoken in normal circumstances, usually unarticulated even to ourselves. It may be, but is not necessarily, *ineffable*. It emerges from acting, "languaging," and "culturing" as we learn conventions and knowledge organization, and adopt value properties and orderings. It is shared and it is personal. It comes via emicization, apperceptions, and memory, and thereby produces our sense of "self."

Since the literature is already rich with discussions of the unconscious, culture, and tacit knowledge, I offer figures 1.1 and 1.2 as a way of understanding how dark matter is different. Dark matter is a combination of culture and individual psychology, produced by the culturing of the individual as well as the individual's apperceptions, interpretations of others, and personal psychology. The basic theses to be explored in regard to this idea include the following: First, we experience the world culturally, in a specific temporal-geographical-axiological context. Second, we *slot* our experiences into a "cultural field" or "table," a matrix of values and knowledges that establishes the

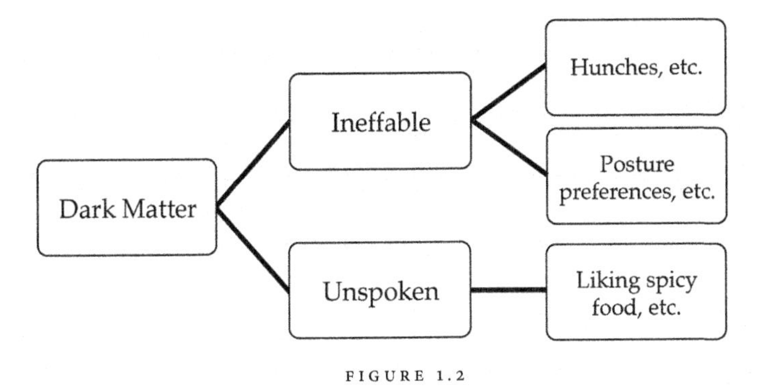

FIGURE 1.2

relationships of experiences to one another hierarchically and which is vital to our emerging sense of self and belonging. I call this (see chap. 2) *C(ultural)-grammar*.[1] Third, we interpret our experiences individually and jointly, based on the cultural concepts that we have internalized. Fourth, we employ episodic memory consolidation and reconsolidation to form our complete autobiographical memory and thus our identities.

As an attempt to illustrate the latter points, let's consider the components of bike riding, from learning to mastery. Ryle ([1949] 2002) distinguished between "knowing-how" and "knowing-that." As Polanyi ([1966] 2009) interprets Ryle, knowing-how consists in the ability to act. This ability, so it is claimed, cannot be made fully explicit. As Polanyi (ibid.) observes for bike riding, we acquire knowledge-how by *doing* something. Explicit instructions are less useful than examples.

Many philosophers, psychologists, physiologists, and others have written on knowing-how. Merleau-Ponty, Heidegger, and Nietzsche are but three of the more prominent philosophers who have discussed what it is to know how to do something. Gestalt psychologists have also written about the whole of a learned action being greater than the sum of its parts. For example, skiing is not merely moving your knee one way or the other, holding your body in a certain position, and so on. Rather, it is learing to "flow" with the act of skiing. In Pike's terms, an action has more than a static existence; it also has dynamic existence (how it is performed) as well as "field" existence (how it relates to other actions within a given cultural system).

For any knowing-how, there exists a process to be mastered, mainly by doing, usually learned in parts. Riding a bike entails learning to mount, pedal, steer, turn, dismount, brake, and so on, without falling. We cannot learn such things by reading about them or being lectured about them. We need (painful) practice, and we benefit from examples (partially due to mirror neurons that communicate with our muscles so as to prepare them to act as we perceive another's action).

Over time, our basal ganglia, synaptic connections, even our muscles—through specific patterns of use and excitation, even creating new fiber nuclei in some cases (Bruusgaard et al. 2010)—register these actions. We build up such memories by integrating individual apperceptions of parts of action into a whole. Then we have mastered bike riding. It has become an emic action rather than merely etic actions. The involvement of the muscles, nerves, and the body more generally, along with the seeming irrelevance of propositional knowledge, make bike riding a prototypical case of knowledge-how—the kind of knowledge one more effectively "demonstrates" than "tells."

But are these ingredients completely lacking from what is referred to as knowledge-that? Consider a simple proposition such has "John knows that an apple is a fruit." How can we determine whether John in fact knows the fruity nature of apples? We can only know it by what assertions he makes, how he responds to an apple in his environment, how he infers properties of the apple based on his knowledge and so on. And another question: Then how has John acquired this knowledge of apples? In just the same way, by doing and watching actions done. As John acts, he affects his synapses, his body (e.g., salivation, olfactory sense, hunger pangs), and his physiology and cognitive states more generally. In other words, to say that John knows *x*, we have to also know how John reacts to *x*, uses *x* in a sentence, draws inferences from *x*, and the like. We must know *how John knows how* to do these things. Yet if this is so, then the knowledge-how vs. knowledge-that distinction is at best weak, at worst nonexistent.

There is another sense in which the distinction between knowing-how and knowing-that is weak. Knowledge-how can be translated into knowledge-that. It is possible (and often done), for example, to develop a bike-riding algorithm that can enable a machine to try to ride a bike.[2]

When I began to learn to ride a bike, I remember focusing during each attempt on what I thought were the important steps—from my intentionally white-knuckled grip on the bar, to keeping up the minimum speed necessary to maintain forward motion without risking a lurch into the intersection, to avoiding objects by focusing ahead sufficiently to plan my steering. Conscious concentration on every subroutine of the larger act of "riding" tired me out. But as I internalized my knowledge by use of my body, before long, I "got it"—my brain's system of equilibrioception adjusted to this new form of locomotion, and I came to "know how" to ride a bike. Riding a bike is a paradigm example of knowing-how. But is this knowing-how utterly distinct from knowing-that? To answer this question, let's consider another problem that seems superficially like knowing-how: the knowledge of a linguistic sound system, a phonology. The phonological knowledge of native speakers is both knowing-how and knowing-that simultaneously. Before looking at phonology, however, we need to better understand "knowing-that."

Knowing-that is generally assumed to be quite different from knowing-how. The former is propositional knowledge, while the latter encompasses skills. Thus knowing-how is taken by some philosophers and psychologists not to be knowledge at all but only habits, skills, or capacities developed over time through practice, or repetition and imitation—muscle memory, largely—whereas knowing-that is taken to be part of cognition proper. Knowing how to ride a bike, again, in this view, is not properly speaking cognition, but merely muscle routines. And yet if knowing that a proposition means *x* requires that we know how to use it in *x*-related inferences, as Brandom (1998) urges, then the how/that distinction is again weakened, if not eliminated.

Because there is a clear sense in which I have "internalized" or "emicized" the bike riding process, yet I am nevertheless unable to give a fully adequate explicit description of what I am doing when so engaged, it may not be possible to reduce knowing-how to knowing-that. There is some part of all knowledge that is not reducible to propositions. But you might reply "It has already been demonstrated that we can devise a set of algorithms for bike riding that, when fed into a robot, enables it to actually ride a bike. Wouldn't that prove that bike riding is ultimately a knowing-that rather than a knowing-how, and thus that knowledge is not ineffable in principle?"

No. And for largely the same reason that Google Translate cannot be said to speak a language or Searle's (1980b) "Chinese room" cannot be said to speak Chinese.[3] Whatever algorithm the robot is following, I am not using these algorithms to ride my bike. How do I know? I know first because they are never conscious, and they are impossible to make conscious for the average bike rider. Second, no scan or other image of my brain is going to turn them up. They are literally not in my brain. But third and most important, the algorithms add nothing to our understanding of human bike riding. The way to understand that is to study the human body and learn what the muscles do, how the brain controls them in bike riding, and how the entire body adjusts to angles, wind, and so on, to maintain balance, speed, and directedness without overt commands. We do not linguistically represent all problems; we "resonate" with many of them (Gibson 1966, 1979). Our bodies adjust to the movement of the bike, the contour of the earth, the holes in the pavement, and so on. Perhaps the notion of resonance might even help us to understand why trying to make what we are doing explicit while riding the bike actually interferes with our riding and can make us fall. The algorithms describe the physics of bike riding, let us say, but not the contents of my brain or how I actually accomplish the physical task. Of course, I cannot escape the constraints of physics—speed, trajectory, energy, gravity, and so on. But if you ask me what I am doing, or if you were to look at an image (fMRI, CAT scan, etc.), no such algorithms turn up. They are not causally implicated in what I am doing. They may describe what I am doing abstractly, but when we externally describe the steps to accomplish a behavior, we are not describing the emicization of the behavior. There is of course no problem with that so long as we do not confuse the two. A robot riding a bike has not emicized bike riding. It has no conceptual choices and no sense of "this is the way *we* do it."

To see this more clearly, consider the difference between descriptions of riding a bike vs. riding a bike "very well." (Later I will contrast descriptions of grammar and language.) I can offer some account, however inadequate, of my bike riding, what I am doing, what I hope to do, what I think my legs, eyes, arms, hands, and so forth, are doing, but I cannot explain how to ride a bike perfectly, except to prohibit certain effects of poor bike riding: Don't fall down!" "Go faster!" "Turn sharper!"

And yet the question still lingers—namely, "Why do we pretend that the dividing line between knowing-how and knowing-that is so clear?" I think that the answer is straightforward: it's linguistic. For the past sixty years or so, cognitive scientists have adopted the idea that the brain represents knowledge in some way, most likely propositionally. Knowing-that knowledge is just that which can be attributed in propositional form to a cognizing being. Knowing-how, on the other hand, is behavior that cannot easily be summarized in this form and so is usually not considered to be knowledge. But what if this is just a distinction without a difference, simply a reflection of what we are able to describe rather than any meaningful distinction in the individual? Riding a bike and proving a theorem perhaps best represent the two poles of this purported knowledge distinction.

Setting aside this controversy, one way the two types of knowledge are the same is in cultural-activity slotting. Once we have learned something, whether knowing-how or knowing-that, the thing learned occupies a "slot" in our cultural matrix, a position in a grid of similar things known. This concept of matrix and slot are ideas from Pike's (1967) analysis of language (and, by extension, culture) as "particle, wave, and field."

Knowledge as Particle, Wave, and Field

A particle can be either or both a unit of language and behavior in Pike's theory. An easy example is a distinctive feature of sound, such as [+voiced], an indication that at some point during the sound in question, the native speaker notices that the vocal folds are vibrating.[4] This example will do double duty, reinforcing the idea that knowing-how and knowing-that are not easily distinguished, if at all, and corroborating the nature of a unit in Pike's sense. Part of linguists' understanding of the particle [+voiced] is its set of positive or active properties—"the vocal cords vibrate; the vocal cords are relaxed; the air pressure above the glottis is less than the air pressure below the glottis; the Bernoulli effect sets the vocal cords in rapid motion," and so on. The average speaker will have a less technical awareness, knowing merely that /b/ is not /p/, /t/ is not /d/, and so on. Like average speakers, a crucial part of the scientist's understanding of a particle, however, is what it is not. For example, if every sound in a language were [+voiced]—such as /b/, /g/, /a/, /i/, /m/.—this feature would not be distinctive, and it would not therefore be necessary to learn it as a particle of the language. To do so would be superfluous, adding nothing to our understanding of the language. On the other hand, if a language has both [+voiced] and [-voiced] sounds—the latter represented by sounds such as /p/, /t/, /k/, /s/, /h/—then we would need to learn, as part of learning this language, whether a sound is + or -voiced. Particles are thus understood positively and negatively, by what they *are* and what they are *not*. Properties of particles cannot be known in isolation from other particles of the language, but as part of a system.

Each particle in a language or a culture must also be studied as a "wave." This is the dynamic perspective, the particle-in-use perspective. Thus when we say something in which /b/ or /p/ is found in different contexts of use—for example, *ba*, *ab*, and *aba*; *pa*, *ap*, and *apa*—the voicing and articulation of the consonant and the vowel will look different spectrographically according to their relative positions in words and phrases. Thus, in action, details are revealed about the particles that cannot be seen in isolation. In particular, the vocal cords will begin to vibrate differently in different contexts. This differential onset of voicing is referred to as voice-onset timing (VOT), and it has played a fundamental role in our understanding of the phonetic differences between sounds within and across languages for decades.

For example, in English VOT begins later than in Spanish. In Spanish a *b* is fully voiced, by and large, from start to finish. But in English a *b* has a later VOT. Aspiration (a puff of air following a sound) is often found in languages like English because this makes it easier—especially in word or syllable-initial positions—to hear the difference between, say, *p* and *b*. Thus English, but not Spanish, has aspirated voiceless stops (spreading the glottis more widely during phonation, indicated as: [–voiced, + spread glottis]). Again this aspiration serves largely as an aid to hearing the difference between [–voiced] and [+voiced] segments. We can understand these features of individual sounds of English and Spanish only by seeing them used in context, a dynamic perspective. It is never enough to study them strictly in isolation.

The wave perspective of bike riding is how bike-riding varies in different occasions of riding, whether that is how different cultures or individuals ride, or how turning corners may vary, or the relative speed of the rider. It is the knowledge of bike riding in action.

And yet even once we understand a unit as a particle and as a wave, we still have not grasped the meaning of any behavioral particle until we see how it fits into the language or culture as a whole. How is it *slotted*? Slotting of units entails a field perspective. The easiest way to illustrate a field perspective would be to show a cultural or linguistic matrix. In the case of individual sounds we have what structuralist linguists referred to as a "phoneme chart." (table 1.1) Phoneme charts are iconic. Left to right, they represent the front of the mouth moving toward the back of the mouth. Reading vertically, we see differences in mode of articulation (voiced, voiceless, occlusive, fricative, etc.). Thus, in the chart, a *p* is contrasted in voicing with *b* and in place of articulation with *t*. All vowels are voiced, so that *i*, *a*, and *u* contrast in the rounding of the lips and their position in the mouth—front (*i*) to back (*u*), top to bottom (*a*).

Consider each position in each chart a "slot." Though the phonemic squares are not cultural slots per se, the entire chart is itself (at least par-

TABLE 1.1. Pirahã Phonemes

Consonants () = missing from women's speech

p	t	k	?
b		g	
	(s)		h

Vowels

i		u
	a	

tially) a cultural selection—the sounds a language has "decided" to use at some point in time. Consider each position in a given context a slot as well. Thus there are two kinds of slot that every behavioral particle must fit into—a matrix slot (i.e., within the phoneme chart) and a context of usage slot (where it goes in a word). These are also known as paradigmatic and syntagmatic slots, respectively. By slotting the behavior or the linguistic unit in this way, we understand it more effectively both within a given culture or language and across cultures and languages. For example, Pirahã syllable structure makes use of these phonemes in allowing only syllables that potentially allow a single initial consonant, followed by one or two vowels (with the added condition that a syllable must have at least one consonant and vowel or two vowels), given in the following moraic-phonotactic constraint:

> *Pirahã phonotactics:* IV(V)
>
> *Moraic constraint:* Pirahã syllables must be greater than one mora, where voiced consonants have half a mora; voiceless consonants are one mora in length, and vowels are one and one half moras (all relative durations, no absolute timing implied).[5]

Likewise, English words have a syntagmatic arrangement along the lines of [[Prefix-[[Root] Stem]-Suffix] Word], as in [[en [[light]en]]].

How might bicycle riding or any other nonlinguistic behavior relate to these concepts of particle, field, and wave? As a particle, bicycle riding can be described as "locomotion via a two-wheeled, manually powered vehicle that one sits on, requiring balance." But it is also understood as *not* being "motorcycle riding," "riding a unicycle," "riding a 'girl's bike' vs. a 'boy's bike,'" "riding a delivery bike," "tricycle riding," "walking," or "driving." It is a behavioral particle when seen as a single action according to the categorization of a particular culture.

Likewise, bicycle riding must be slotted in a chart of "means of locomotion" and another of "forms of recreation" at a minimum. Only by examining behavioral units in relation to other behavioral units of the same type can we understand ourselves or others. Of course, the meta-issue is that fields are themselves cultural constructs—cultures are not merely knowledge, values, actions, and so on, but the *arrangement* of these things.

Thus our tacit knowledge includes knowledge structures. If our dark matter is largely composed of what we learn as individuals via participation in a given culture, is there any other kind of dark matter? For example, are we born with innate dark matter as well as those ideas and lessons we glean from our lives? What are the conceptions of tacit knowledge that have dominated thought and discussion in history? Before returning to my empirical proposals, it is necessary to undertake a brief historical survey of the notions of tacit knowledge that have influenced current discussions and debate, in order to contextualize and thus better understand the discussion of the remainder of the book. I am going to argue later that such facts lead us to conclude that some forms of knowing, such as phonology, are at once knowing-how and knowing-that, an ability that embodies dynamic and static cognition (D. Everett 1994).

THE PLATONIC TRADITION OF INNATE KNOWLEDGE

In what follows, we survey different concepts of tacit knowledge. This survey reveals that concepts vary so extensively as to call for another term, my *dark matter of the mind*: knowledge and values relating to our environment, ranging across the unspoken and the ineffable, the bodily and the mental, engaging the full individual.

Moving toward our brief history of dark matter, we need to mention at the outset an underlying assumption upon which much theorizing about human knowledge is based—namely, the belief that all humans share a basic psyche. Adolph Bastian, of whom we will learn more presently, developed the influential idea of the "psychic unity of mankind" that profoundly affected the thinking of people such as Carl Jung, Joseph Campbell, Franz Boas, and others. This "unity" is one of the first modern proposals of nativism—in this case, that there are concepts universal to all humans. In modern linguistics, this idea has strongly influenced the development of the natural semantic metalanguage theory of Anna Wierzbicka (1996) and her colleagues (see chap. 9) and finds resonance in Chomsky's universal grammar, as well as compatible concepts of a variety of instincts proposed in recent literature (e.g., a moral instinct [Hauser 2006], religious instinct [Haidt 2013], art instinct [Dutton 2010], and language instinct [Pinker 1995]).

In the spirit of Bastian's proposal that all humans share innate tacit knowledge, beginning in the 1950s, Noam Chomsky pursued the then-moribund ideas of rationalism by proposing that humans possess an innate tacit knowledge of the rules of their native languages. This innate knowledge is in fact so rich that they do not "learn" their native languages but "acquire" them. From Chomsky's theory, this tacit knowledge—universal grammar (UG)—must arise by adjusting variable aspects of UG to local languages, with the range and ability to make such "adjustments" deriving from the human genome. According to the theory of UG, the dark matter of human minds is a reflex of the way that our minds are constructed phylogenetically, either via natural selection or some other set of principles such as Turing's self-organizational proposals (Turing 1952).

Also influential—and contrasting to some degree with Chomsky's strongly nativist work—is that of Michael Polanyi's, seen in such works as his *Tacit Dimension* ([1966] 2009) and *Personal Knowledge* (1974), which have exerted tremendous influence in philosophy, psychology, and even in business. Unlike Chomsky, Polanyi was concerned less with the *source* of this knowledge than with its effects on our day-to-day activity.

Taking a step back from more recent writings, the idea of "tacit" knowledge has been around in various forms for quite a while, much longer than either Chomsky or Polanyi. Since its history is implicated in the understanding the nature of dark matter, I want to trace its pedigree through its two major lines in what follows. Of course, the place to begin, as always, is with Plato. (I want to make it clear from the outset of this brief survey, however, that I am not summarizing the complete oevres of these philosophers, neither their views of "self," etc. I am, rather, looking exclusively at whether they postulated a priori concepts.)

In the Western world, a foundational discussion of tacit knowledge is Plato's *Meno* dialogue. The *Meno* is set at a dinner part where Socrates enters into a conversation with Plato's fifth-century BCE eponymous character. Meno, as with other poor souls who chose to debate Socrates, realizes before the end of his exchange that he has bitten off more than he can chew. The catalyst for their debate is a discussion about unlearned knowledge. At the dinner party, the host asks Socrates if virtue is acquired by teaching or practice, or whether it comes in some other way.

Socrates answers that virtue is not learned. This puzzles Meno, who then asks Socrates what he means by saying that people do not learn, and that what they call learning is only a process of recollection. Socrates answers by calling to himself one of Meno's servant boys and leading him through a series of questions about geometry, in order to establish that this uninstructed child apparently possesses surprising and accurate knowledge of the Pythagorean theorem. The philosophical question that Plato offers up for consideration is just this: How can people know so much about things they have neither been instructed in nor had any direct experience with? Knowledge that precedes experience is known as *a priori* knowledge. Socrates's questioning proves the existence of such knowledge, if, as Plato would have it, the slave knows geometry without having been instructed in it.

Plato's dialogue has provoked interest for more than two thousand years. Almost solely because of this dialogue have other thinkers in history debated the existence of a priori knowledge. In fact, it seems fair to say that all theories of the nature and origin of human language fall on one side or the other of the debate begun by Plato—that is, whether our knowledge of the world is at least a priori or a posteriori.

Another reason that Plato's ideas about ideas are so important to many is that if ideas are invariant in heaven, then they are timeless, pure, and free from contamination by the body and the world. The dualism of Descartes is thus prefigured by Plato. Yet—looked at with modern eyes—the *Meno* isn't very convincing. It should be obvious that Socrates is asking leading questions ("Do you know that a figure like that is a square?") designed to restrict possible responses, question-forms, by exploiting Greek interactional structures and cultural understanding. Knowing this, we can see that in fact no a priori knowledge is discovered. And yet Plato's concept of an innate endowment of universal ideas common to all humans still resonates. We see its guiding hand in the work of philosophers as disparate as Descartes and Kant, though disputed by others such as Locke and Hume. Chomsky goes so far as to call the "mystery" of the source of that knowledge which exceeds our experience (especially what he assumes to be evidence for a priori linguistic knowledge) "Plato's problem."

For Plato, all true knowledge is found in heaven. Like the Apostle Paul (whom he apparently influenced), Plato believed that we see only "shadows" of truth on earth or, as Paul put it, "as through a glass darkly" (I Corinthians 13:12). But for Plato, unlike Paul, we all have access to our inborn truth via reflection. Therefore, what we call "learning" is really only remembering.

If all humans are born with the same set of identical a priori ideas—even if the set of such ideas is much smaller than or quite different from that envisioned by Plato—then all humans partake of a universal, epistemological human psyche or human nature. Because of its intrinsic interest and the position of Plato in the history of the Western world, over the centuries a large number of distinct philosophical traditions have emerged from Plato's concept of innate, universal ideas. In what follows, I want to trace the ideas of Plato's main descendants (relative to the current discussion) in chronological order.

Perhaps the most influential of all who followed Plato in the belief in innate a priori knowledge was René Descartes (1596–1650). Descartes not only supports a version of the rationalism that was launched by Plato, but he adds a *pernicious* twist, one that has come to be known as "dualism"—the idea that mind and body are separate substances requiring separate understandings. This wasn't altogether new with Descartes, as we have just seen, but it took on a name and greater prominence in Western philosophy after his writing.

In many of the religions of the world, from Buddhism to Christianity to Islam, the body is opposed to the mind (and soul) and is a corrupting influence. Thus ascetics arose who purposely violated the basic biological values of the body, seeing a "pure" heart and "pure" thoughts by denying themselves satisfaction of their hunger, sexual needs, physical comfort, and so on. Such ideas also permeated philosophy, so that even philosophers as brilliant as Descartes and Plato, or mathematicians like Alan Turing, or computer scientists such as Herbert Simon, or linguists such as Chomsky, have reached the conclusion that it is possible to ignore the body in the study of the mind.

Descartes does this by focusing on human nature, the notion of self as the mental, and by disregarding the body as little more than a sheath for the mind. In spite of its popularity, however, I contend that dualism is one of the worst errors ever introduced into philosophy. This Cartesian misunderstanding of the mind has perhaps done as much to retard understanding of the individual and cognition as any proposal ever made—with the possible exception of Kant's transcendental arguments (see below). Dualism is pernicious because it has led to the deeply misguided view of the mind as software running on a physical hardware (hence the separation of emotions and the physical peculiarities of the human body from the study of human cognition, perhaps best illustrated in fictional stories about computers that think (e.g., *2001: A Space Odyssey*) or in the popular science fiction film *Transcendence*, where a mind is "uploaded" onto the Internet and then takes over the world).[6]

Descartes's dualism is also reflected in his utterly ignorant view of animals. To Descartes, animals were meat machines. They possessed no consciousness, no thought, no feelings, or the like. Moreover, his view of the mind of humans as special and disconnected from bodily experience led instinctively to a linguistic-based—that is, human-based—theory of cognition.

But as Paul Churchland aptly puts it:

> Both of these classical accounts [the syntactic and the semantic views] are inadequate, I shall argue, especially the older syntactic/sentential/propositional account [of explanatory understanding]. Among many other defects, it denies any theoretical understanding whatever to nonhuman animals, since they do not traffic in sentential or propositional attitudes. (2013, 22)

Any view of cognition that ignores nonhuman animals ignores evolution. Whether we are talking about the nature of ineffable knowledge or any other kind of cognitive or physical capacity, our account must be informed by and be applicable to comparative biology, if it is to have any explanatory adequacy. Animal cognition helps us understand the importance of evolutionary theory and comparative biology in the understanding of our own cognition. It also helps us see how the bodies of both humans and other animals are causally implicated in their cognition. This disregard for animal cognition, caused by dualism, has led to what Paul Churchland (2013) refers to as "linguaformal" models of knowledge and cognition. But since animals lack language, their cognition is thus declared by fiat, not by science, to have no relation to human cognition. If beliefs and desires, for example, are based on propositional, linguaformal representations, then animals cannot have beliefs and desires, which seems patently false (see Searle 1983; Patricia Churchland 2013; Panksepp and Biven 2012; among many others).

In his First Meditation (*Med.* 5, AT 7:64), Descartes says regarding truths that "on first discovering them it seems that I am not so much learning something new as remembering what I knew before." He also says of truths that

> we come to know them by the power of our native intelligence, without any sensory experience. All geometrical truths are of this sort—not just the most obvious ones, but all the others, however abtuse they may appear. Hence, according to Plato, Socrates asks a slave boy about the elements of geometry and thereby makes the boy able to dig out certain truths from his own mind which he had not previously recognized were there, thus attempting to establish the doctrine of reminiscence. Our knowledge of God is of this sort. (ibid.)

To Descartes, ideas were innate if they could be grasped without sensory experience. This seems misguided to me because he fails to distinguish types of sensory experience, from "directly relevant" to "enabling" sensory experience. Deprive an infant of *all* sensory experience, proper diet, affection, and so on, and they will learn ("recall") nothing, or will learn things perhaps deeply distorted, for example. A priori empirical reasoning is possible only in the presence of prior experience.

Many others who followed Plato also defended the idea of human nature as a function of universally shared a priori knowledge. One of the most influential was the Prussian philosopher Immanuel Kant (1724–1804). Kant's life and work are well known to most contemporary philosophers, and his philosophical program of thought is dense, well planned, and enormously insightful, such that it is difficult to imagine an intellectual of the Western world that has not been influenced by him in one way or another. Kant was obviously himself influenced by Descartes, though some philosophers see Kant as (nearly) bridging the gap between empiricism and rationalism. I don't agree with that assessment. To my mind, Kant was an innovative rationalist, but a rationalist nonetheless—we might say that he was to rationalism what Beethoven was to the classical tradition. Further, like other rationalists, Kant's philosophy suffers from a lack of understanding of culture, physiology, psychology, and linguistics. This is not anachronistic criticism. Kant's philosophy doesn't work in many crucial cases and those are the reasons why. For example, Kant championed the view that our *sensibility* (our perceptions and experiences of the world) is a set of powers of cognition fundamentally distinct from our *understanding* (our intellect or intelligence). Modern research (Gibbs 2005; Shapiro 2010; Panksepp and Biven 2012; and the present work) argues that no such separation is possible and that our intelligence and experiences are at most points along a continuum of selfhood, each an exemplar of dark matter.[7]

Kant was a hero of the Enlightenment, an era of human intellectual history almost single-handedly begun by Isaac Newton, with fundamental contributions by myriad others across Europe, especially Leibniz. Kant captures the spirit of the Enlightenment beautifully in a statement from his *Critique of Pure Reason*:

> Our age is the age of criticism, to which everything must submit. Religion through its holiness and legislation through its majesty commonly seek to exempt themselves from it. But in this way they excite a just suspicion against themselves, and cannot lay claim to that unfeigned respect that reason grants only to that which has been able to withstand its free and public examination. ([1903] 2007, xi)

These admirable sentiments have changed our patterns of thinking, from the Enlightenment, broadly speaking, to Kant's work as part of that *Kultur*. Though Kant has been the subject of innumerable works, it is Kant's thought directed to the understanding of the pedigree of dark matter that concern us here. Kant argued that unless our mind has *innate* a priori concepts, it cannot interpret the world it finds itself in. That is, the mind not only cannot be an Aristotelian blank slate, but it must come with pre-installed concepts that match our experiences of the world. Such concepts Kant referred to as "categories."

Interestingly, these categories did not nor could not lead in principle to an understanding of how things really are, "in themselves." Instead, they enable us to arrange and make sense of our perceptions. Apart from Kant's ontological commitments, I find this an appealing perspective. In fact it is easy to see the roots of American pragmatism in Kant's writing—in particular, in his views of realism and the limits of human knowledge. Kant's notion of a priori categories is perhaps best translated in my terms into the idea of an inborn ability of humans to generalize and learn by any means. Kant rightly observed in effect that without a learner, there is no learning. Thus humans must be born to be learners—individual humans have innate capacities to adjust to the world that we encounter (as all living creatures do). His view is that such learning is partially the application of highly specific categories to shape our perceptions. But statistical learners, even computer simulations, show that not all learning requires specific concepts.

Moreover, as we see later on, Kant's notions of the interrelationship between experience and intelligence fail to consider other mediators between the external world and the mind. First, like many before him (and most after him), Kant saw the mind rather than the individual as the locus of learning (which is why he had little to say about interactions between reasoning, emotions, and physiology). Second, Kant overlooked the richness of potential unconscious tacit knowledge, acquired during one's life—failing to recognize the vast amount of learning that is subliminal ("subceptive learning," Rogers [1961] 1995). Third, some of Kant's discussion of subjective vs. objective knowledge seems best recast in terms of emotion and perception—that is, complementary modes of relating to an object, which may be objective or subjective and which do not follow from properties of the object itself but from the role of experience-based dark matter in our relationship to the world in which we live and form a part. For example, take his illustration of the perception of a house, in comments on Leibniz (Kant B162):

> The I think must be able to accompany all my representations; for otherwise something would be represented in me that could not be thought at all, which is as much as to say that the representation would either be impossible or else at least would be nothing at all.

Thus at once Kant writes off dark matter of the mind as "nothing at all" because it is not—and often may not or even cannot—be the object of "I think," nor is it simply representational. Obviously, "knowing-how," therefore, does not follow under a Kantian conception of apperception. But neither does knowledge of grammatical rules, taste in clothes, likes in foods, prelinguistic experiences as a child, muscle memory and so on. Kant's view of apperception therefore seems acutely inadequate.

Kant was also influenced by Gottfried Leibniz (1646–1716), another major epistemological Platonist. Leibniz is responsible for the introduction of the novel notion that has been mentioned several times above, "apperceptions." In Leibniz's conception of this term, he refers to a perception that is registered by our consciousness, then stored away to build future knowledge—thus not terribly far from Kant's interpretation of his notion. William James ([1900] 2001, chap. 14) says of apperception that "it verily means nothing more than the act of taking a thing into the mind." All of our apperceptions, once registered are "drafted off . . . making connection with the other materials already" in our mind/memory. "The particular connections it [any impression] strikes into are determined by our past experiences and the 'associations' of the present sort of impression with them."

So my take (and reinterpretation) of the notion of apperception first introduced by Leibniz and later developed by Kant and James is that we experience things during our entire lives, from conception (not merely from birth) through death, through sickness, health, and even complete cognitive breakdowns, and these experiences enter our memories consciously such that we may reflect on them (Leibniz's and Kant's understanding of "apperception") while others enter our lives unconsciously (the sight of a particular color while we are ill, our mothers' reactions to loud noises while we're in the womb, and so forth). Although Leibniz referred to these latter experiences as "petite perceptions" and Kant said that they were of no account cognitively, I understand, to the contrary, that both types of experience are apperceptions. What binds all of these apperceptions together is long-term or episodic memory. The union of these apperceptions results in the notion of self we have (will, intelligence, and other components of the self are taken up in chap. 10). Thus Kant in my opinion does not resolve the rationalist vs. empiricist debate with regard to a priori knowledge but instead falls solidly on the side of rationalism.

Evaluating the philosophy to this point, I tend to agree with Unger (2014) that philosophy without science is largely "empty ideas." Speculations on the nature of human knowledge, reasoning, emotions, or cognition in some general sense without science lead nowhere. The great philosophers that precede us cannot be blamed for failing to use knowledge that was not available to them. But, now, possessing more knowledge, we can no longer continue to accept their writings as authoritative (or even informative in many cases).

Moving on in the Platonic tradition to other influential concepts of innate tacit knowledge, we come to ever-bolder theories. One of the most explicit and influential proposals mentioned already was that of Adolf Bastian (1826–1905), who claimed that there was universally shared tacit knowledge that formed the "psychic unity of mankind." Bastian's psychic unity thesis is in fact a set of hypotheses based on the idea that human cultures and consciousness derive from species-wide physiological mechanisms. Anticipating to some degree modern researchers in embodied cognition (Gibson 1966, 1979; Lakoff and Johnson 1980; Lakoff and Nuñez 2001; Shapiro 2010; Gibbs 2005; Skipper et al. 2009; etc.), Bastian argued that from our basic physiology emerges a set of "elementary ideas." If so, then all Homo sapiens' minds are constrained by the same inborn ideas, producing what some might call "human nature." These elementary ideas are shaped by local contexts and cultures, leading to culture-based "folk ideas" from the innate, universal, elementary ideas.

For these hypotheses, Bastian became an important historical figure in debates on human nature. His nineteenth-century proposals on this "psychic unity" emerged from his years of travel, from the mid- to late nineteenth century. During these travels and beyond, Bastian studied ethnology, culminating in a six-volume study, *The People of East Asia*, published in 1861. From collections made during his travels, Bastian made generous contributions to several museums in Germany. In addition, he cofounded the Ethnological Society of Berlin, one of the first in the world. The Berlin Museum of Folk Art was based largely on Bastian's own collections. While serving as curator of the latter museum, one of his assistants was the young Franz Boas, who was to become a founder of American anthropology.

Bastian's thesis of the psychic unity of mankind was influential in several domains. According to this hypothesis, the individual mind is embedded in a larger "social mind" or "social soul." Bastian argued therefore that to apprehend the psychic essence of Homo sapiens, we need to first analyze the local folk ideas in order to achieve an understanding of the elementary ideas from which these folk ideas emerge. According to Bastian, this deconstruction of folk ideas into elementary ideas required five analytical steps:

The first step is what Bastian labeled "fieldwork." His claim is the salutary one, that understanding human beings requires empirical research. It is not a matter for mere philosophical reflection. Cross-cultural ethnographic field research is crucial to assessing and fleshing out the psychic unity thesis.

The second step is to deduce from our fieldwork the "collective representations" of a given people. Field research in conjunction with the theory of elementary ideas provides the information we need to perform the deduction of these ideas in a given society from their collective representations. These representations—modifications from the universal elementary ideas—are the local manifestation of elementary ideas. They are what come to be represented as "folk ideas."

The third step in the understanding of elementary ideas is the analysis of the folk ideas arrived at from the first two steps. This is done by studying how collective representations break down into their constituent folk ideas. "Idea circles" are delineated when similar patterns of folk ideas are found in contiguous or proximate geographical areas (a concept somewhat similar to what linguists refer to as *linguistic areas*).

Following this, the fourth step is the deduction of elementary ideas, accomplished via the identification of similarities between individual folk ideas and patterns of folk ideas across regions that indicate underlying elementary ideas.

Finally, we come to Bastian's step five, the application of a "scientific psychology." This follows straightforwardly from the studies outlined above, as they us lead to an understanding of the psychic unity of our species, rooted in our underlying psychophysiological structure, leading to a scientific, cross-culturally grounded psychology.

By general consensus, Bastian succeeded in spite of himself. He was not thought of as a great writer. Additionally, many of his proposals were vague, and in spite of his aspirations to be taken as a serious scientist, his ideas were not always regarded as testable or clear enough to serve science. Nevertheless, they inspired subsequent researchers and intellectuals, and that is perhaps even better than generating merely testable ideas. His psychic unity theory became a touchstone.

One of those subsequent researchers was Sigmund Freud (1856–1939). As just about anyone will know, the "unconscious" was a central, novel, and crucial part of Freud's theory and clinical work in psychoanalysis. The unconscious mind—including tacit knowledge—arose partially in response to the phenomenon Freud referred to as "repression." As we experience certain negative events, states, or entities, according to Freud, we may "repress" these—that is, avoid conscious memory or reflection on them—rendering our psyche healthier in some respects. But the events of this negativity are always present in our minds, pushed "down" from consciousness to the unconscious portion of our minds. Freud argued that repressed, tacit memories could be recalled through hypnosis and, when so recalled, could enable a patient to be cured of the negative effects of this toxic tacit knowledge. These clinical aspirations were hailed as Freud's great contribution.

According to Freud the effects of repression and apperceptions on human psychology developed into more full-blown tacit ideas—for example, the "Oedipus complex" (for anyone needing a refresher, this is a purported, innate, universal desire by males to sexually possess their mother and destroy their father). Further developing his inventory of tacit knowledge and categories, Freud proposed to explain the hidden or unconscious and innate concepts, drives, and functioning of the human mind by proposing a tripartite division of the mind into *id*, *ego*, and *superego*. *Id* represents the completely unconscious portion of the mind, the source of all impulses and drives, seeking quick fixes to emotional needs, especially pleasure and gratification. In this sense, the id contains tacit, ranked values—that is, a mini-culture of its own. The *superego* labels that portion of the mind hypothesized by Freud as the "angel" to the id's "demon," as our morally absolutist enforcer of our conscience's sense of right without exception—another set of tacit, ranked values; another mini-culture. Finally, Freud proposed the *ego*—the rational component of the brain that manages and negotiates between the *Id* and the *superego*, the constructed self (chap. 10). Freud claimed that if we could bring the unconscious, repressed memories of our past to consciousness, then these repressed memories could be explicitly reconciled with our tripartite unconscious and our psychological health restored. Thus Freud's concept of *ego* in my terms is also dark matter, in the form of a set of values, cultural knowledge, and rules.

This brief glimpse of some of the basic tenets of psychoanalysis shows the fundamental importance of Bastian's psychic unity of mankind concept to Freud's work. Without this unity, a universal human psychic core, psychotherapy as conceived by Freud could not work. Freud was almost certainly aware of and influenced by Bastian's proposals. In the intervening years, Freud's own theory has been famously criticized by Grunbaum (1985) and Webster (1996), among others. There have even been suggestions that Freud invented some of the crucial data from nonexistent, "confidential" clinical cases (Webster 1996). Still, his theory remains one of the most influential in the history of the study of the mind. Perhaps no earlier theory made more lively use of the notion of innate, a priori dark matter. Together, Freud and Bastian also greatly influenced the work and theory of Freud's student Carl Jung, whose fame and importance came to rival those of his teacher.

Jung (1875–1961), another of the leading dark matter theorists in the Platonic tradition, was the founder of "analytical psychology" (Jung [1916] 2003). Fundamental to this form of therapy and the theory behind it was, again, Bastian's elementary ideas, which Jung reconceived as the "collective unconscious," that is, innate tacit information common to all humans. Like Freud, Jung saw the ability to come to grips with and manage our tacit knowledge, our various unconscious states, as crucial to the development of healthy psyches. He referred to this management or reconciliation of the two forms of unconscious states/levels as "individuation." Jung and Freud, along with Bastian, aspired to develop a science of the mind. Certainly, the two of them generated a number of intriguing hypotheses based on tacit experience and knowledge, which in turn have produced decades of research from around the world. Still, the lack of falsifiability of their claims, the imprecision of their terms, and the lack of cohesiveness between their theories and the rest of science were and are serious deficiencies. As we move to other researchers heavily influenced by Bastian, though, we come closer to science.

Perhaps the greatest early step toward a science of dark matter was the work of one of Bastian's early assistants. In the late 1800s as curator of Berlin's Museum of Folk Art, Bastian hired a young Franz Boas (1858–1942). Although Boas's research life eventually was dedicated to studying the richness and diversity of cultures around the world, the influence of Bastian is found in his belief that such studies would reveal "laws of cultural development," based on the uniformity of the foundations of all human cultures. Boas ([1940] 1982) argued that although all humans have emotions these are locally determined, so that it makes sense to say that culture determines emotion (Panksepp and Biven's [2012] research, on the other hand, seems to contradict Boas's claim, unless we see the latter as a proposal about the interpretation or categorization of emotions. This does seem to vary cross-culturally). The crucial observation is that Boas believed that diversity among cultures of the world were all built on a universal base, reminiscent in some ways of Chomsky's view of language. What separates Boas from Chomsky in this regard, however, was Boas's contact with pragmatism, through his colleague at Columbia, John Dewey (1859–1952). Boas was inescapably influenced by Bastian's notion of the "psychic unity of mankind." At the same time, Boas's pragmatism (in my interpretation) makes him a "straddler" of the border between nativism and nonnativism.

The final major figure influenced by Bastian who I want to discuss has been tremendously influential in the humanities, the social sciences, and in popular literature—namely, Joseph Campbell (1904–1987). This theorist and collector of myths from around the world also believed in the psychic unity of mankind, heavily influenced by Bastian. Joseph Campbell is another exemplar of the Platonic tradition. He also exemplified the stereotype of the "scholar" in Western culture. He loved books and ideas, the urbane urban life, and intellectual engagement, and he possessed an apparently encyclopedic knowledge of the major subjects of the humanities. He is rightly known and admired for his knowledge of and collection of myths from around the world. But his theory of myth, its place in the rationalist lineage, and its influence on the thinking of the general public is what draws our attention here.

Campbell's principal idea was that of the "monomyth," the idea that all human mythology derives from a single ur-story of our species. Campbell explicitly refers to Bastian in his writings and argues that the monomyth corresponds to Bastian's elementary ideas, whereas the local variations of the monomyth correspond to Bastian's folk ideas. One is tempted to quip that lots of stories can look similar if you edit out the dissimilarities—and this often seems to be what Campbell has done. On the other hand, Campbell is taken so seriously by so many worldwide that his ideas deserve more than such a dismissive assessment. For Campbell, just as with other legatees of Bastian and Plato, there was nothing vague about the idea of psychic unity. It was clear that it was the result of innate content, not merely innate abilities, as Locke, Hume, Berkeley, Sapir, and others would have it.

Campbell believed that myth functioned in the human psyche to enable humans to talk about the eternal source of life that existed before humans, before language, before culture. Thus most cultures have heroes, to enable them to talk of the eternal with less fear. These heroes undertake journeys to the netherworld, as Hercules to Hades and Jesus to hell. Like the rationalists in whose tradition he writes, Campbell preferred the dramatic hypothesis of a primordial species-wide myth to the idea that many people invent independently—the myth of beings who can do things we cannot, like Superman, Thor, vampires, and zombies (or ourselves in dreams). Unsurprisingly, the general public prefers Campbell's and Bastian's kind of dramatic, native-born ideas and mystery (regardless of scientific merit) to the less satisfying state of caution and silence, with their aggravating hedging of claims, deliberate vagueness, scientific disagreements, nonunanimity, and so on.

For example, as Campbell said:

> God is a metaphor for a mystery that absolutely transcends all human categories of thought, even the categories of being and nonbeing. Those are categories of thought. I mean it's as simple as that. So it depends on how much you want to think about it . . . half the people in the world are religious people who think that their metaphors are facts. Those are what we call theists. The other half are people who know that the metaphors are not facts. And so, they're lies. Those are the atheists. (2003, 135)

This statement is problematic in many places. But perhaps there is no better assessment of Campbell's monomyth theory to my mind than Robert Ellwood's (1999, x) comment that "a tendency to think in generic terms of people, races . . . is undoubtedly the profoundest flaw in mythological thinking." In my field research on dozens of indigenous languages and cultures of South America, for example, I have never observed anything like Campbell's monomyth or hero's journey. However, there were myths that I could perhaps have bent into such shapes had I so wished.

Campbell's views find resonance among many evangelical Christians. For example, missionary Don Richardson's 1981 book, *Eternity in Their Hearts*, or Pascal's "God-shaped vacuum" in the hearts of all men, are versions of the widespread Christian thesis that God has prepared every culture and every individual for the message of eternal life and salvation. Yet these "theories" and ideas, while colorful and obviously appealing to millions, are poorly supported by the facts. They seem more a reflection of their authors' personal histories than the results of careful, replicable field research.

It may seem deeply strange that anyone would propose that all humans are born with the same outline of an elaborate proto-story carried in their genes, in a species that has fewer genes than corn (Messing 2001). And yet there are far more elaborate nativist hypotheses on the market. Perhaps the most elaborate and in many ways the most fantastic story of all is Chomsky's universal grammar, or as we might refer to it in the present context, the monogrammar theory. We take up the "monogrammar" story again below, where it receives the longer discussion it deserves in the history of ideas on dark matter of the mind, in the context of understanding the nature of language. Right now, however, it is important to situate this work in the historical context.

Through Noam Chomsky (b1928), Platonic rationalism was revitalized and given a huge boost from the late 1950s. Chomsky has long supported the Platonic idea that we have access to knowledge without need for experience or sensory input. We can use reflection and intuition alone to "retrieve" truths from our minds because our minds are born with them (or born with access to them in some other plane, e.g., heaven or a previous existence, depending on the particular philosopher's theology). As Plato did, Chomsky seems to believe that much if not all of the most important knowledge that we humans draw upon to distinguish our character and lives from those of other animals is based upon our possession of inborn dark matter. This matter may be the result of evolution, physics, God, or whatever, but it is taken to be the inheritance of all Homo sapiens.

Chomsky's particular version of rationalism initially convinced so many across so many disciplines because it seemed to be richly supported empirically. Chomsky argued powerfully and brilliantly (see John Searle's [1972] summary of "Chomsky's revolution") that several facts about language lead us to the inexorable conclusion that humans are born with a tacit knowledge of many intricate details of grammar, a hypothesis that eventually came to be known as *universal grammar* (UG). Historically, Chomsky's work made the study of language central to epistemology, psychology, anthropology, philosophy of the mind, and computer science, among other fields. It is not an exaggeration to say that the fecundity of his ideas made him arguably the most important intellectual of his lifetime and one of the most influential intellectuals in history.

Chomsky's work since the 1950s has made the case for a plethora of principles, rules, transformations, constraints, and grammatical operations that are tacit and innate, part of the human genotype (though, unfortunately, none has held up from one of his theories to another). His views on tacit knowledge, however, clearly fit in the rationalist tradition:

> Linguistic competence is understood as concerned with the tacit knowledge of language structure, that is, knowledge that is commonly not conscious or available for spontaneous report, but necessarily implicit in what the (ideal) speaker-listener can say . . . It is in terms of such knowledge that one can produce and understand an infinite set of sentences, and that language can be spoke[n] of as 'creative,' as energeia. (Chomsky 1965, 19)

And also:

> The system of knowledge that has somehow developed in our minds has certain consequences, not others, it relates sound and meaning and assigns structural properties to physical events in certain ways, not others. (Chomsky 1986, 12)

He also makes it clear that this tacit knowledge is regularly drawn upon by children acquiring their native language and that this explains their rapid acquisition of their native language(s).[8]

Chomsky achieved arguably his most impressive result with the mathematical hierarchy of grammars now known as the *Chomsky hierarchy*. As his model of transformational grammar (a type of context-sensitive grammar) analysis began to conquer the intellectual world, he introduced a further, even more influential proposal. This was the idea that the ability of humans to learn their languages so quickly—perfect mastery before adolescence was claimed—and the fact that the structural principles of all human languages were apparently deeply similar, if not identical, could only be understood if the linguistics capacity were identical and innate in all humans. Along the way—in a(n in)famous 1959 review of B. F. Skinner's *Verbal Behavior*, a behaviorist account of language learning—he was perceived to have demolished the idea that children could learn languages merely by stimulus-response conditioning. Children's minds—not merely their behavior—were implicated in language learning, Chomsky reasonably argued, and among their mental attributes they must have some a priori "solution space"—if they are trying to learn language, how can they possibly know what it is they are trying to learn without innate a priori knowledge of where or how to find solutions? The UG that Chomsky eventually came to propose was quite detailed and structured. His idea of an innate UG became and still remains by far the most influential of all proposals on the tacit, dark matter of the mind.

Chomsky's case for innate tacit knowledge of language was made all the more persuasive to many because it was so technical, as one gleans from remarks by various interlocutors in the debate registered by Piatelli-Palmarini (1980). In an astoundingly brief time, Chomsky arguably came to dominate his field as no other scientist had ever done before, not even Einstein in physics or Darwin in evolutionary studies. After Chomsky, for a multitude of researchers, rationalism was taken as settled and proven. Empiricism, on the other hand, came to be regarded as passé. In spite of the prevailing zeitgeist, though, there is nevertheless a vital and contemporary empiricist tradition that offers a different explanation of dark matter.[9]

In later sections we discuss how concepts are acquired if they are not innate or in some other way a priori. My intention here was simply to provide a brief pedigree for the idea of innate tacit knowledge. It is worth remembering that for many decades, through the philosophy of Russell, Quine, James, Pierce, and others, as well as the psychology of B. F. Skinner, rationalism was not a particularly popular perspective. Empiricism to my mind has the virtue of rejecting the psychic unity hypothesis, which seems to tear and strain to fit the range of human cultures into its bed of Procrustes. To reiterate, psychic unity is unconvincing to me because it places the locus of human equality in what can only be described as shared concepts rather than biological heritage and capacities, and because it places the focus of human knowledge in the mind alone, rather than in the encultured individual.

More substantial insights into human nature—especially as this is revealed through the origins, nature, and use of dark matter—are better illuminated through science than philosophy. Philosophy is useful in stepping back to take a look at the issues from a higher perspective, perhaps, but the ancient philosophers were are best at suggesting questions rather than answering them. Like Paul Churchland (2013), Patricia Churchland (2013), and Unger (2014), among others, I believe that a science of the mind has to precede or at least accompany a philosophy of the mind/brain in order to make progress on the understanding of how people learn and think.

These writers are all part of an enormous, bifurcated tradition of the study of epistemology in Western literature. This forked road leads to (at least) two distinct views on the nature of tacit knowledge. From Plato to the present day, there are those whom we have reviewed above that advocate for *a priori* tacit knowledge, or *innate ideas*. The alternative we take up next is the view that all humans form tacit memories, knowledge, dispositions, and the like, from experience, not from universal ideas, a shared unconscious, or the same elementary ideas. All those that we have been discussing up to this point are Platonists in the sense that their theories were built on the foundation of universal, innate, tacit knowledge, regardless of the many other differences between them.

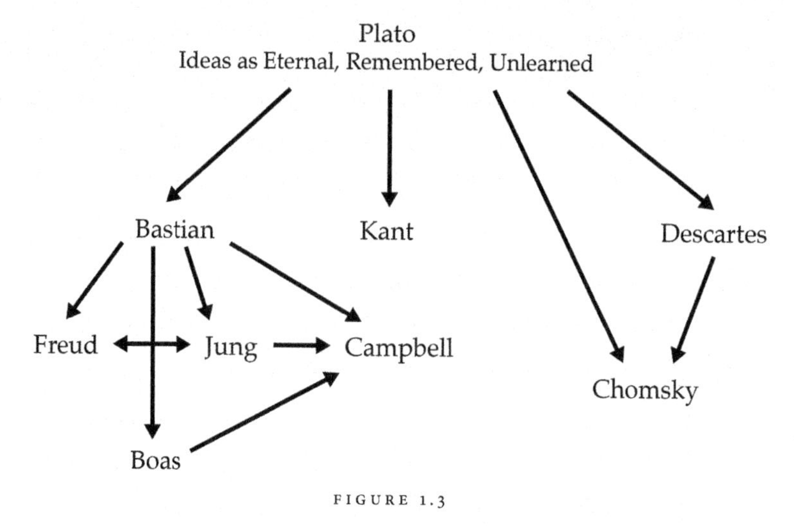

FIGURE 1.3

The clade of rationalism we have been discussing may be summarized as shown in figure 1.3. This chart is incomplete in many ways, obviously omitting a number of major thinkers; it merely highlights the writers focused on here. There were of course influences from others on Kant, Descartes, Chomsky, and so on, beyond Plato. But Plato's influence was foundational, as intended by the direct lines. Bastian's influence seems more direct on some writers, however, as I have attempted to indicate on the chart.

THE ARISTOTELIAN VIEW OF KNOWLEDGE

The same general considerations apply to the interpretation of the Aristotelian tradition, where arrows indicate intellectual influence along the parameters most relevant to our concerns relative to dark matter. Figure 1.4 (simplistically) illustrates that the roots of the empiricist tradition extend back to Plato's most famous student, Aristotle (384–322 BCE).

Aristotle initiates empiricism in works like his *Posterior Analytics*, where he states:

> So it emerges that neither can we possess them [premises of syllogistic reasoning] from birth, nor can they come to be in us if we are without knowledge of them to the extent of having no such developed state at all. Therefore we must possess a capacity of some sort, but not such as to rank higher in accuracy than these developed states. (2007d, 136)

In this first statement, Aristotle draws a stark dividing line between his understanding of innate a priori tacit knowledge (there is none) and Plato's. Both

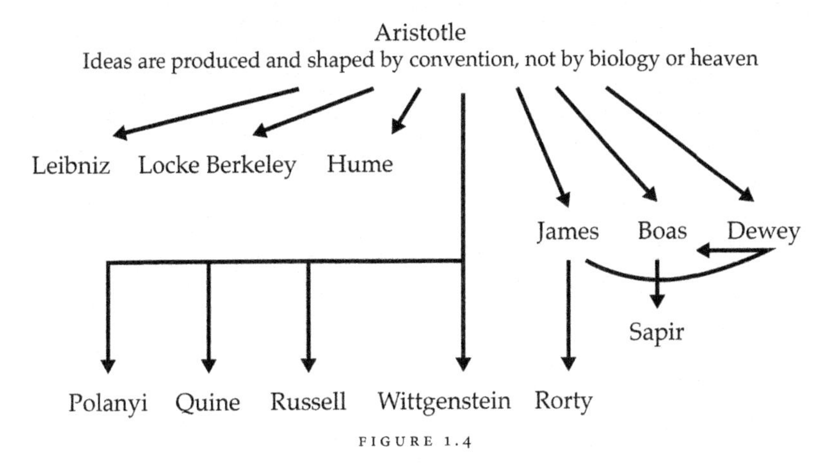

FIGURE 1.4

Aristotle and Plato attribute a great deal of importance to a priori knowledge. But while for Plato this prior knowledge is inborn, for Aristotle our ability to solve problems is based on knowledge acquired through experience, thus *a priori* to him means "knowledge acquired prior to encountering this problem." To be sure, we are able in Aristotle's view to acquire knowledge at all due to an innate *capacity* to learn, but such a capacity is not, he makes clear, to be confused with specific knowledge.

Aristotle thus breaks with his teacher and develops a concept of general ability in a flexible brain. He is evidently unconvinced by Plato's perspective in the *Meno*, though he does cite it occasionally in the deferential manner expected of a disciple. Aristotle sees a distinction between nature and convention, such that by nature Homo sapiens has a "social instinct" (in book 1 of his *Politics*, he states that "A social instinct is implanted in all men by nature"), though it is only by convention that man's specific societies and languages arise. Aristotle thus begins a new tradition. Though there are many passages in Aristotle (as in most ancient and many modern philosophers) that seem to be self-contradictory on superficial readings, the distinctions of Aristotle's philosophy, subtle or striking, launched thought on a distinct trajectory that, in the intervening 2,400 years, has led to a radically different approach to knowledge from his teacher's, especially as touching on what I am calling dark matter.

A couple of representative citations from *Metaphysics* are (2007a, 1):

> Now from memory experience is produced in men; for the several memories of the same thing produce finally the capacity for a single experience . . . science and art come to men *through* experience. [emphasis in the original]

And also:

> If then a man has the theory without the experience and recognizes the universal but does not know the individual included in this, he will often fail.

And:

> We conclude that these states of knowledge [the "skill of the craftsman" and the knowledge of the man of science] are neither innate in a determinate form, nor developed from other higher states of knowledge, but from sense perception.

And:

> So out of sense-perception comes to be what we call memory, and out of frequently repeated memories of the same thing develops experience; for a number of memories constitute a single experience.

Most controversially, Aristotle is responsible for the characterization of the mind as a "blank slate," or tabula rasa: "Mind is in a sense potentially whatever is thinkable, though actually it is nothing until it has thought" (2007b, 662). "And," he asks,

> how could we learn the elements of all things? Evidently we cannot start by knowing anything before. For as he who is learning geometry, though he may know other things before, knows none of the things with which the science deals about which he has to learn. (2007a, 511)

The contrast here with Plato's *Meno* could hardly be starker.

The successors of Aristotle proposed ideas at least as powerful and persuasive as the successors of Plato. There are several I want to discuss in what follows, found in the diagram above, because they seem most pertinent to my own perspective of human cognition and nature developed below. On the one hand, the three philosophers associated with the founding of empiricism—Locke, Berkeley, and Hume—are crucial. On the other, there is the anthropologist and linguist Edward Sapir, who never quite fits any set.[10]

John Locke (1632–1704) disagreed strongly with the rationalism and dualism of Descartes. In particular, he took issue with the Cartesian proposal that knowledge is comprised of those ideas which our minds conceive clearly and distinctively. Our ideas, according to Locke, to the contrary, are acquired by living and thinking. And our knowledge derives from the interaction of facts with our ideas, that is, in our determination of the fit between our ideas and the facts (as the philosopher John Searle [1983] would say, their "*mind→world*" fit). Locke rejected, following Aristotle, innate tacit knowledge.

Like Aristotle, Locke did not believe that the absence of knowledge on a tablet means that the tablet has no other properties. It has the capacity to receive and store information and more. Neither philosopher thought of the tabula rasa as devoid of capacity to be written on, not even of capacity to write upon itself. In my reading, they meant by tabula rasa not that there were no innate abilities, but that *there were no innate specific concepts*. Certainly the mind, in Locke's view, would possess nothing like the specific innate tacit knowledge of a universal grammar, Bastian's elementary ideas, Freud's unconsciousness, or Campbell's monomyth.

Locke has been criticized for, among other things, not having an explicit account of how, if our minds are blank at birth, humans can categorize or indeed learn anything. Where does the notion of "sameness" come from, the indictment goes, that enables a person to categorize *x* and y as belonging together but c as belonging to a different class? To my mind, this criticism lacks force. It seems to derive from a common confusion of knowledge with ability. Animals, certainly my dog, have the capacity to categorize. My dog recognizes the difference between humans and cats and dogs and trees, and between me and any other male. There is no need to propose a priori knowledge, though, for either my dog or me. We have bodies with visual systems, tactile systems, olfactory systems, auditory systems, and so forth, that are able to recognize physically those external properties that seem independent of specific objects such as physical similarities in colors, tastes, touch, and so on. (I will have more to say on this when I contrast embodied cognition with evolutionary psychology later on.) Thus this criticism of Locke does not present any serious problem. We simply must keep straight the distinction between innate capacities and innate knowledge.[11]

For Locke, the mind builds ideas from experience. The mind binds and shapes experiences within our consciousness. These experiences do not come preinterpreted or precatalogued or prewired. Locke agreed with Aristotle that the mind is a blank slate. Locke proclaimed the self to be a continuation of consciousness, built upon individual experiences and sensory experiences, rather than prewired or a matter of destiny.

Next there is George Berkeley (1685–1753), who went so far as to argue (as my friends the Pirahãs might have argued themselves, were they interested in Western philosophy) that abstract ideas are a fiction. Berkeley was a writer of iconoclastic brilliance. His criticisms of Newton's physics in *De Motu* ([1721] 1990) anticipated by nearly two hundred years the later arguments of Mach and Einstein. His work ([1709] 2011) *Essay towards a New Theory of Vision* was initially controversial but is now well accepted. Berkeley developed the philosophy of subjective idealism in which he argued that there are only spirits and ideas. Forgetting spirits for now, his view of ideas was that while there clearly are objects, they are objects only in our minds, and without a mind there would be no facts of the matter about them. There may be things outside our minds, but we can never know them because the mind is all that knows.

Berkeley's view is quite different from the eliminative materialism popular among many today. It certainly differs from the perspective I adopt here, in which there is no mind of any kind (except as an imprecise way of talking about brains), only bodies and the world (cultural, biological, ecological, etc.) in which they move.

The principal way, however, in which Berkeley advanced Aristotelian empiricism was in his denial of the ability to abstract away from immediate experience in order to discuss properties disembodied from such experience—that is, he criticizes the very notions of abstraction and generalization. As he put it ([1710] 1990, 10): "I deny that I can abstract one [property] from another, or conceive separately, those qualities which it is impossible should exist so separated; or that I can frame a general notion by abstracting from particulars in the manner aforesaid."

It is interesting that some cultures—for example, the Pirahã (D. Everett 2005a, 2008)—agree with Berkeley on the problematic nature of abstractions, for similar though not identical reasons. Whereas some before him naively denigrated generalization, Berkeley shows that such denial can in fact be an interesting and sophisticated position. Moreover, American pragmatists, especially William James, showed a similar, though weaker, distrust of generalizations.

Then there is David Hume (1711–1776), who argued against our facile interpretations of causality, based on what he intended to be a Newtonian-inspired theory of human nature. Hume's influence is pervasive in the history and practice of philosophy and science, on Darwin, Kant, and many, many others. Most modern researchers on the mind and most modern philosophers acknowledge a debt to Hume. The thread of his work continues through modern cognitive science. According to Fodor (2003, 234) Hume's 1739 *Treatise* is "the founding document of cognitive science."

The subtitle of Hume's famous *Treatise of Human Nature* (1739–40) is "Being an Attempt to Introduce the Experimental Method into Moral Subjects." His view was that ideas only make sense or exist through experience. In his discussion of experience, he elaborates conceptions of impressions and ideas that fit well with what we refer to here as apperceptions: "Ideas are the faint images of [sensations, desires, passions, and emotions] in thinking and reasoning" ([1739–40] 1978, book 1, parts 1, 4).

One of Hume's most important ideas relative to our current discussion is that our a priori reasoning cannot be the source of our ideas of cause and effect. This is because—absent experience—we can reason about causes without effects and effects without causes, whereas in our experience they must occur together. Causal inferences are thus a matter of experience, not a priori reasoning. In my terms, causes are registered in our apperceptions and forms of our dark matter, however the latter is instantiated.

The next name in the line of descent from Aristotle is William James (1842–1910). James's particular take on empiricism came to be known as *radical empiricism*. He claims that our experience necessarily underlies our debates and that only our experiences possess the sense of continuous structure (through memory). Therefore, there is no need for theoretical support for our own experiences.

Radical empiricism is the construction of one's understanding by means of one's singular experiences. In this James shows the influence of the transcendentalists Ralph Waldo Emerson (a friend of his father's) and Henry David Thoreau (Emerson's handyman). James's work is rich and, along with Charles Sanders Peirce, John Dewey, and others, led to American pragmatism (though Pratt [2002] argues that the Wampanoag Indians' influence on Roger Williams and through him Thomas Jefferson played a foundational role in this uniquely American contribution to philosophical thought). The influence of the pragmatists Peirce and James permeates not only philosophical thought, but also early linguistics and anthropology. Through Dewey, a colleague with an office at Columbia down the hall from Boas, and Peirce's work on signs (which influenced Saussure), pragmatist influence can be seen in the emphasis of Boas and Sapir on the particular, a preference for inductive and abductive reasoning over deduction, and various other ways.

From this tradition emerged perhaps the greatest linguistic anthropologist / anthropological linguist of all time, Edward Sapir (1884–1939). Sapir makes the case that human psyches and human cultures blend in subtle, complex, and always particular ways, such that generalizations regarding the nature and sources of tacit knowledge are often misguided. Instead—as a pragmatist might—Sapir's work shows that the particular and the variable, rather than the universal or invariable, are where the intellectual gold is to be mined.

Sapir's pioneering research, essays, books, and articles on North American Indian languages, on historical linguistics, phonology, descriptive linguistics, and psycholinguistics, as well as other subjects, are among the most important of the twentieth century. Unfortunately, personal, cultural, and historical circumstances—ironically, since the nexus of these factors was the focus of Sapir's studies—reduced his influence and prevented his ideas from having the effect on later generations of anthropologists and linguists they should have had.

Sapir arrived at a propitious moment in the history of North American anthropology and linguistics. Like Kant, Sapir was born in Prussia. He moved to the US with his parents as a young boy. Perhaps related to his multilingual background, he completed both a bachelor's and a master's degree in Germanic philology at Columbia University, initially on a prestigious Pulitzer scholarship (Columbia was at that time one of the few or only research universities to place no restrictions on the admission of Jewish students). He decided to pursue his PhD at Columbia, where two of the most important and influential intellectuals in American history had offices in the same building, John Dewey and Franz Boas. Boas accepted him as an advisee.

As we have seen, Boas believed in the psychic unity of mankind at least partially due to the influence of his former supervisor, Adolf Bastian. Boas's views were by no means as extreme as Bastian's, but his collection of texts from North American languages was partly motivated by a desire to see how elementary ideas were permuted by local cultures. Boas was also influenced by pragmatism, through his colleague Dewey, as well as earlier writers such as Peirce. Sapir eventually diverged from Boas, coming to believe that human differences were vital to understanding Homo sapiens, abandoning any vestige of Bastinian shared concepts. If he used any phrase similar to "psychic unity," it was in terms of universal capacities, not universal beliefs or knowledge. Cultures, for Sapir, could vary tremendously, and it was culture that made us who we are, by providing us with our core knowledge, concepts, values, and so on. For Sapir, culture was the epitome of dark matter, in that it was manifested by patterns that each individual implemented in his or her own thinking and activities, howbeit unconsciously for the most part. As Sapir put it, "*We act all the more securely from our unawareness of the patterns that control us*" (1985, 549; emphasis mine).

This applies to all of us. As I write this chapter, I work hard to put into written form what I believe are my own individual ideas. And yet I know that these ideas are available to me, interesting to me, and largely interpreted by me as a member of a specific culture who also possesses a specific individual life history. Or as Sapir might put it, "One is *always unconsciously finding what one is in unconscious subjection to*" (1985, 549; emphasis mine). Under the latter proposal we can fit Nazism, the Manson family murderers, the growth of scientific paradigms, the marketing strategies of clothing stores, the zeitgeist, and "national opinions."

Sapir thus diverged strongly from Boas as regards acquired vs. innate tacit knowledge. As Sapir developed his ideas on culture and psychology, he arrived at insights that are still at the cutting edge in the cognitive sciences. For example, in his posthumously edited *The Psychology of Culture*, he develops the concept of the individual and psychology as instrument, manifestation, and builder of culture:

> Any form of behavior, either explicit or implicit, overt or covert, which cannot be directly explained as physiologically necessary but can be interpreted in terms of the totality of meanings of a specific group, and which can be shown to be the result of a strictly historical process, is likely to be cultural in essence. (Sapir 1993, 37)

In this passage Sapir blends the tacit and the explicit, the individual and the social into a single definition (more like a description) of culture. His references to the tacit here grew from his experience with the intuition of native speakers, while writing his 1908 PhD dissertation under Boas on Takelma. Tony Tillohash, the Southern Paiute speaker, worked with him on the psychological reality of the phoneme, and this work further affected Sapir's sense of the connection between psychology and culture, both for what of this relationship could be seen overtly and what was not overt (see also Wallace 1970).

In 1931 Sapir was invited to join the faculty of Yale University with a partial charge to study the impact of culture on personality. But he faced his own problems with the dark matter of prejudice. One of four Jewish faculty out of a total of 569, he was denied admission to the Faculty Club, by anonymous vote, literally blackballed. The Faculty Club was where faculty voted on and debated administrative policy—more like a faculty senate in today's terminology.

Sapir's work is the best of the era, in any case. The unsurpassed quality of his research and writings on culture and the connection between culture and psychology ended with his death, none of his students developing these subjects with either the eloquence or insight of their teacher.

Moving on from Sapir to the central representative figure for this discussion of the Aristotelian tradition of dark matter, we come to Michael Polanyi (1891–1976), a pioneering contributor to ideas of tacit knowledge. Polanyi was an important contributor to physical chemistry, economics, and philosophy. Though the range, quality, and significance of his writings is impressive, his work on tacit knowledge is in focus here. Born in the Austro-Hungarian Empire, in 1926 Polanyi emigrated to Germany, where he was appointed professor of chemistry at the Kaiser Wilhelm Institute in Berlin. In 1933 he left the Reich of Hitler for the University of Manchester, where he was first a professor of chemistry for more than a decade, later becoming professor of social sciences. (The Polanyi home was evidently an intellectual incubator: one son, John Polanyi, won the Nobel Prize in chemistry in 1986; the other son, George, is a well-regarded British economist.)

Polanyi disagreed with his Manchester colleague Alan Turing when the latter claimed that the mind was a computer, its thinking largely reducible to rules. Polanyi's own view was much more nuanced and humanistic. He opposed the positivistic account of science and personal knowledge (which means that he rejected the popular concept that truth and human knowledge can be restated or discovered by algorithms). Rather Polanyi (building upon a perspective he first developed in his Gifford Lectures [1951–1952] at the University of Aberdeen and later published as his magnum opus, *Personal Knowledge*) argued that all knowledge is the result of personal judgments and commitments. And he argued that we quite reasonably believe more than we can prove. Likewise, and of particular relevance here, it thus follows that we know more than we can say.

Our discoveries, according to Polanyi, are driven by our values, commitments, and passions. Scientists choose the questions that are important to them from personal and social values. The context of our work, thoughts, and relationships to reality is our tacit knowledge. Polanyi viewed his discovery of the "structure of tacit knowing" as his most important discovery, across all the fields he worked in.

He was also one of the early advocates of "emergence"—the idea that not all human abilities, properties of the world, life forms, and so on, are reducible to lower-level properties that interact with higher-level features. His concerns with personal knowledge and emergence led him to study expert knowledge—for example, connoisseurship, musical ability, visual arts—rather than standard epistemology directly. We can see this concern reflected, however indirectly, in Michael Silverstein's (2003) modern work on "indexicals." Polanyi also exerted strong influence on the epistemological work of Thomas Kuhn. There is obvious resonance with Polanyi's work on expert knowledge as opposed to mechanistic processes in work by Freedman (2010) on causal inference and statistics (as well as works like Elster 2007). Some of Polanyi's most intriguing applications and developments of the notion of personal knowledge and tacit knowledge, however, are his writings on animal reasoning, animal personhood, and animals' tacit knowledge. This is consistent with his nondualistic approach to knowledge as well as the thesis of this study.

Other figures include major names such as Russell, Wittgenstein, Rorty, and Quine. I will end this section with a brief summary of Quine. If my purpose here were to develop a distinct perspective on empiricism, Quine would obviously be a central figure. However, Quine had little to say about tacit knowledge or dark matter in ways that directly impinge on this book. However, he did make forceful arguments for empiricism that have not yet been answered successfully, in my opinion.

His primary ideas relevant to our inquiry here include his naturalism, extensionalism, empiricism, naturalized epistemology, analyticity, holism, underdetermination, radical translation, indeterminacy and inscrutability (see Harman and Lepore 2014, 2ff). Without going into detail on these points, the upshot is that Quine believed that what we take to be nonnegotiable in our beliefs (called by some "analytic truths"), resulting from our lives, beliefs, and other forms of dark matter, are truths we are not prepared to disbelieve; for example, "a bachelor is an unmarried man" and "2 + 2 = 4." Other than this, they have no special claim to truth that falls outside of empirical acquisition or verification—for instance, as analytic sentences as opposed to synthetic sentences (Quine 1951). All that we know, we know via our senses and how these are interpreted—that is, apperceptions. There is no a priori knowledge. Moreover, communication across individuals and cultures is always underdetermined; either there will be multiple translations of a statement by one person into the dark matter of another, and it is hard to choose between them; or there is simply indeterminacy—there is no way to choose between translations. Across native speakers of the same language, this underdeterminacy/indeterminacy is one of the hallmarks (and overall salutary; see D. Everett, forthcoming) of human language and is key to understanding the evolution of human communication.

Quine's work thus is foundational for the present work, but because only in a few places did he mention indirectly our unknown knowns, he doesn't figure in our discussion again until the discussion of radical translation in chapter 8.

Summary

This chapter provided an overview of the pedigree of dark matter, tracing two broad traditions, the Platonic and the Aristotelian. The thesis of this book falls within the Aristotelian, empiricist line of descent. The first part of the chapter looked at kinds of knowledge and several proposed sources of that knowledge, from linguists, philosophers, psychologists, and anthropologists. The chapter also defined what is meant by the term *dark matter*, discussed its major divisions into unspoken and ineffable, and underscored how it is distinct from other types of knowledge, including tacit I. We looked at knowledge-how vs. knowledge-that, for example, arguing, along with others, that any dividing line between such knowledges is at best blurry.

In this regard, the chapter also considered the Pikean view of knowledge as "particle, wave, and field," illustrating its points with the linguistic knowledge of sound units, phonemes and allophones. Next the chapter discussed the Platonic tradition of innate knowledge, continuing from Plato to Chomsky and beyond, via many others. This section also looked closely at the influence of the thesis of the "psychic unity of mankind" first proposed by Adolf Bastian. Following this, we surveyed different works in the Aristotelian tradition of dark matter.

Having established the pedigree and some of the nuances of different ideas of the notion of dark matter, we are prepared to move to the next chapter for a discussion of one of the two major forces in the creation of dark matter: culture.

2

The Ranked-Value Theory of Culture

> The study of 'interpersonal relations' is the problem of the future. It demands that we study seriously and carefully not just what happens when A meets B—given that each is not only physiologically defined, but each also has memories, feelings, understandings, and so on about the symbols they can and must use in their interaction . . . In any [specific] situation when two people are talking, they create a cultural structure. Our task, as anthropologists, will be to determine what are the potential contents of the culture that results from these interpersonal relations in these situations.
>
> EDWARD SAPIR, *The Psychology of Culture*[1]

This chapter addresses the concept of culture from a linguistic perspective, drawing on optimality theory and tagmemics' slot:filler concept, among others. One question that it tries to answer, raised initially in the introduction, is, does each member of a culture, X, participate in a collective intention to "live according to the values of culture X"? Whatever the answer to this question (and the answer is, "occasionally"), the larger question this chapter looks at is, how is culture even possible? Is a society like a football team? Or perhaps like an orchestra? Or is culture simply the overlap in values, roles, and knowledge of individuals who live together and talk together? These questions try to get at the larger issue of how cultures hang together at all. In what sense might *e pluribus unum* describe culture? Since I claim that culture is an abstraction, it can only be found in the individual. It is the result of an emic gestalt in individual culture members, in a way to be made clear as the discussion progresses.

To get at these questions, I propose a model of culture in which the individual is the bearer of culture and the repository of knowledge, rather than the society as a whole. It examines the very idea of culture and its effect on the nature of national and local societies and individuals, via examples such as the height of the Dutch, the teacher in the classroom, businesses, and so on.

More specifically, the chapter lays out a theory of values, which is necessarily prior to a theory of culture (epistemically, not temporally). We then examine the value studies made famous by Kluckhohn's research group in Rimrock, New Mexico. We also consider Rokeach's theory of values and show how it can inform a theory of cultures.

The importance of the individual to the notion of culture leads to a discussion of culture as the core of cognition, requiring us to consider whether machines are capable of thinking (as McCarthy [1979], among others, argues) or not (as Dreyfus [1965, 1994] and the current discussion maintain). The argument is that machines cannot think because they lack culture and without culture there is no semantic understanding, no background, no dark matter in which thought can occur.

Other issues addressed in this chapter include "mining" dark matter from textual analysis, as well the emergence and nature of societal norms and conventions. The chapter concludes with a summary of its findings.

Individual vs. Social Knowledge

Dark matter is not simply another term for *culture*. But there is a clear connection between the individual, their community, and the formation of dark matter in individual minds. Like language, *culture* is an abstract noun, and we can never expect to find a "culture" or a "language" in the real world. Rather, what we find are people—people speaking and acting with one another. Their mutually shaping activities of "languaging" and "culturing" build similar values, concepts, social roles, linguistic constructions, and so on, in each person. Language is both action and knowledge simultaneously. And so it is with culture as well.

For example, consider the following interactions between two linguists:

A. "Colorless green ideas sleep furiously."
B. "They sure do."

The general population may have no idea what A's utterance means. But A and B know that this is a famous example sentence in Chomsky's early writings that is designed to show that a sentence can be grammatical yet meaningless. For the two linguists, A's sentence is an insider joke and B's response a humorous rejoinder. The function of the exchange might be largely phatic, simply to say, "Hey, we are both linguists." But additionally, B's reply shows that A's utterance is not in fact meaningless, because it indicates that whatever green ideas are, they sure sleep intensely.

Now consider the following. Persons C and D are watching the New England Patriots play the Miami Dolphins. The Patriots take the lead. C and D both yell "Yes!" and high-five one another. In this joint action, they show knowledge that there is a game of football, knowledge of how this game is scored, shared value ranking for the Patriots relative to the Dolphins, knowledge of what high fives are like and what they are for, knowledge that they are both rooting for the same team, and reinforcement of all of the above.

From such activities come knowledge-how, knowledge-that, community belonging, shared communication, and so forth—various forms of dark matter elicited, strengthened, and formed by very simply paired actions. These exemplify the role of culturing and languaging as dark matter and as forming dark matter. From these actions the individual creates from his or her etic experiences an emic gestalt of society.

In chapter 1 we traced the different historical sources of theories of dark matter—a priori and a posteriori tacit knowledges. The chapter made the case that there are two major traditions forming the major epistemological divide in approaches to the tacit. The Platonic tradition holds that humans are born knowing things. As we have seen, the principal contemporary proponent of this Platonist line is Chomsky, who has spent the last sixty years or so developing the theory of universal grammar, the idea that all humans are born with innate linguistic knowledge.

On the other side, there is the Aristotelian tradition in which knowledge is learned, facilitated by human capacities. This tradition ranges across the centuries to Sapir, the present work, and many others.

A LINGUISTIC TURN IN ANTHROPOLOGY?

I want to begin this section of the larger exposition with a claim that some might find startling or simply wrong: in the study of cultures and languages, the methods of linguistics are by and large superior to those of modern anthropology. Thus not only is linguistics a branch of anthropology, but it is the model branch for studies of ethnography. I offer no arguments for this now, though I have commented on the long tradition of linguistics as anthropology. But the proof is in the pudding, as we will see.

With this axiological preliminary, I plan to continue our discussion by developing a theory of culture based on the Aristotelian conception of tacit knowledge, arguing at the same time that tacit knowledge is too narrow as normally conceived and thus that it must be replaced by my idea of dark matter and is subsequently shaped by the dark matter of individuals thinking, speaking, and living together.

A historical incident—the famous Treaty of Medicine Lodge, signed in 1867 between the Arapaho, Kiowa, and Comanche peoples and the US Government at the Medicine Lodge River of Kansas—frames the first part of our discussion here. This treaty is worth examining not for what it contained, but for the readings it was given based on the dark matters of the opposing sides signing the treaty. After discussing various levels of implicit knowledge and perspectives on each side and the breakdown in communication that thereby resulted, we are prepared to review several influential definitions of culture entertained by anthropologists down through the years. We then take up the desiderata for a definition of culture and offer a novel proposal based primarily on knowledge structures, ranked values, and social roles. Again, culture is not a synonym for dark matter. But living culturally—in a community with shared culture in my definition—forms a good deal of every individual's dark matter.

From the discussion of culture, we explore various applications of this theory and how it elucidates and harmonizes concepts that some have taken to be orthogonal or contradictory to one another. We conclude the chapter with a discussion of the applications of the theory of culture to modern business as one example.

Key to the entire theory is the linguistic anthropological concept of emicization (Pike 1967), which describes the trajectory and point of all linguistic and cultural learning—the achievement of the perspective of the insider. Because dark matter is constructed by actions, observations, conversations, and the other components of a life history in a particular society, to understand it we must understand what this abstract notion of culture is that plays such a significant role in its emergence.

For more than a century, anthropologists have bickered about the definition of culture (Kuper 2000). Some have even argued that it should be replaced with a less global, less monolithic notion. But no anthropologist has seriously entertained the idea that it can be tossed out and replaced by nothing. The reason for this is that the members of a given family, community, society, nation and so on, clearly share some knowledge, some values, and relationships between their different sorts of knowledge and the rankings of their various values over time. They clearly talk alike. They clearly act in similar ways. They show disgust at similar things. And so on. In what follows I want to examine the evidence for culture and how it contributes to the dark matter of the mind, and then to consider one influential critic's arguments against it.

Let's say that you want to convince someone else that culture exists. What facts might you appeal to in order to support the existence of an entity called culture? Well, you might suggest social roles, or values, or knowledge transmission and learning, or skills and ways of being. Anything that can be explained by other means loses its force as an argument for culture.

I am a father, a teacher, an administrator, a husband, a shopper, a patient, and a counselor. Each one of these roles is arguably formed by the cultures to which I belong. Culture distinguishes and shapes social roles—even when they may seem universal, and thus might appear to be culture-independent.

For example, though it is true that there are Italian fathers and American fathers, the concept of "father" should not be conflated across cultures. It seems likely that between any two cultures, fathers will have overlapping but never identical roles. There are crucial differences between fathers of different cultures. Even American fathers or any fathers of ostensibly the same culture vary in the nature of their roles at different times.

For example, some societies may believe that (more accurately, "value the idea that") fathers should support their families. In such a society, it may be assumed that fathers have a responsibility to provide food, clothing, and shelter for their children. And—in Western societies, at least—both the society and many fathers themselves believe that it is good for fathers to help their children with schoolwork, heavy lifting, and tasks in general too difficult for children to do alone. Fathers of other generations may share exactly these beliefs and values. But these values are not identical across different cultures. For example, a Pirahã father will not pick up a child to comfort it if the child has injured itself, except in rare circumstances. He will expect the child to work hard and not complain on long treks through the jungle and will not offer assistance in many cases that the American father would. And his individual values emerge partly from the values of other members of his society.

But even in different generations within otherwise the same society, fathers may differ profoundly. For example, values shared by many of my father's generation included corporeal punishment, the expectation that women did the bulk or all of the housework, the belief that their wishes and orders would be carried out without question, and the attitude that their children were not deserving of respect or of a voice in family affairs. These fathers might regularly side with teachers against their own children in disputes. They considered the child and all its resources as mere extensions of themselves or their possessions. The fathers of the generation of my children, on the other hand, usually avoid corporeal punishment, see their family as a unit of equals, know they ought not to believe that their desires should be the only (or even the main) one heard, often help clean the house, would likely take their children's side in a school dispute, and so on. Being a father in the 1950s was therefore considerably different than being a father in the twenty-first century ("considerably" does not mean "entirely," of course). This is because the cultural role is constrained and defined by shifting cultural values.

If such an explanation is on the right track—if definable group values and role-expectations exist—then there is evidence that values are shared across individuals and therefore may partially define a group. This is in turn (part of) what it means for a group to be a culture: shared values. All cultural roles show similar diachronic, geographic, economic, and other shifts across time or across space or across populations. If we move from roles to beliefs or from beliefs to shared concepts, to shared phenotypes, shared food, shared music, and so on, we can find many examples of shared tacit knowledge that produce overlapping cultural groupings.

In part these shared mental items emerge because over the course of one's life, each of us accumulates experiences, lessons (both formal and informal), and relationships all being assimilated into our bodies and minds, partially via apperception. People who grow up in the same community (a relative term that in my usage can refer to family, village, nation, etc.) have similar experiences—climate, television, food, laws, and values (e.g., fat is wrong, honest is right, hardwork is godly). Their experiences are subjected to apperception and memory, both muscle and mental memory. Episodic and muscle memories hold our various experiences together as culture-guided apperception makes them our own. Arguably our "self"—or at least, our "sense of self"—is no more than this accumulation of memories and apperceptions.

This is then in broad brush how dark matter of the mind is formed. We want to go deeper, however, by exploring the roles of attachment and emicization in the emergence of dark matter and how this dark matter forms the bedrock of culture and individual psychology.

THE VERY IDEA OF CULTURE

There have been many definitions of culture offered in the anthropological literature over the years.[2] While we discuss several of these in what follows, I want to begin with my own definition:

> Culture is an abstract network shaping and connecting social roles, hierarchically structured knowledge domains, and ranked values. Culture is dynamic, shifting, reinterpreted moment by moment. Culture is found only in the bodies (the brain is part of the body) and behaviors of its members. Culture permeates the individual, the community, behaviors, and thinking.

From this definition, it follows that people may share cultural components without being part of an independently defined social group.

Although this definition emerges from a theory of culture, it also interrogates the construction of such a theory. What should, for example, a "theory of culture" do, exactly? Well, it should enable us to understand. It should remove or radically lessen our surprises regarding human social behavior and indexical values. It should predict or at least explain individuals' behaviors in specific cultural contexts. It should offer an understanding of the major institutions of a society. It should provide a guide and methodology for investigating specific cultures. To see what I mean, let's return to the example introduced earlier of the Medicine Lodge treaty.

In October 1867 a meeting took place at Medicine Lodge Creek, near present-day Wichita, Kansas, that led to one of the many treaties between the US government and the aboriginal peoples of North America. On the one side were representatives of the US's Indian Peace Commission. On the other there assembled members of the Comanche, Cheyenne, Arapaho, Kiowa, and Apache peoples. There were displays and speeches from both sides of the conference. The American military made a showy display with of colorful uniforms, howitzers, and other weaponry. The Indians gave a lesson in horsemanship that no US mounted soldier could match. Chief Ten Bears of the Yamparika Comanches made an impassioned speech. The treaty was signed. All agreed to it.

But the treaty was ineffective from the beginning. For once, at least, an official treaty with the Indians was invalidated not because of dishonesty on the part of the US government but because the signatories failed to realize that language—whether spoken or written in treaties—is merely the visible portion of an invisible universe of understanding that derives from the values, knowledge, and experiences—the cultures—of individual communities. Though people might read the same words in a treaty, as in all communication, our interpretations are slaves to our assumptions, based on background beliefs and knowledge that the literal meaning of the words rarely conveys.

In this case, the treaty called for the US government to provide food to the Indians so that they could feed their families through the winter months. The US Indian Agency was responsible for providing the food. The US Congress was responsible for ratifying the treaty that was signed. Each in turn depended on other cultural institutions, all with their own deadlines and priorities. The Indians could not have cared less about ratification and the like. But they should have, because when they arrived to collect their provisions, prior to ratification of the treaty, the pantry was bare. The Indians felt betrayed.

On the other side, the US government expected that the Indians, when they agreed to live in the reservations, would consider themselves bound to stay there in perpetuity and to forever abide by the "law." Perpetual obligations to anyone other than their own families were foreign to the Indians' values and understanding of the way the world worked. They could never have legitimately made the commitment expected of them. It made no sense. Although US officials could not have cared less about Indian interpretations rooted in their very different cultures, they should have. The Comanche chief Quanah Parker, present at this ill-fated gathering, at least learned from the experience. In his future dealings with whites, he learned to respect the importance of the dark matter of the unsaid. He subsequently inquired about every potential assumption that he thought whites might be making before signing future treaties (though no one outside a culture can ask all the right questions).

Treaties illustrate natural consequences of dark matter. But we need not look so high in the cultural hierarchy to find the effects of dark matter. There are plenty of examples in everyday, mundane transactions. For example, when you tell your friend, "Well, we're going to eat now," depending on your relationship, but only a little on your words, this could mean "It is time for you to go home therefore" or "It is time for you to wash up and sit with us." The guest's interpretation will be based on their relationship, their knowledge of their host's culture and personal expectations, their monitoring of the looks of other family members, and so on. It will not be based merely or even mainly on the words that a potential host speaks.

The point is that human language is not just a computer code. Fortran is not a language. Languages and human cognitive abilities draw on and emerge from cultures. Nothing in human language or human societies can be understood without getting at the dark matter of the background. This is what makes the study of human behavior more difficult than the so-called hard sciences—the variables of human interactions are not only nearly infinite, but the majority are hidden from direct observation. Understanding the nature and role of this dark matter in human behavior, language, and thinking is essential for comprehending, coexisting, and working together with fellow humans.

In spite of such obvious (though still superficial) evidence for culture, there are culture deniers. For example, in an essay addressing a question posed on the website Edge.org—"What scientific idea is ready for retirement?"—John Tooby (2014), a founder of so-called evolutionary psychology, argued that *culture* is a term that has not been found useful and should be abandoned. He argues that "'culture' and the related term 'learning' are . . . a pair of deeply-established, infectiously misleading, yet (seemingly) self-evidently true theories." He claims that "all 'culture' means is that some information states in one person's brain somehow cause, by mechanisms unexplained, 'similar' information states to be reconstructed in another's brain."

Tooby's concern is, frankly, hard to take seriously. For example, according to his caricaturization, screaming "Fire!" in a crowded theater would be an example of culture. Moreover, by his description, language would just *be* culture, equating two important, distinct concepts, since language is the primary means for getting ideas from one mind to another. Yet it presupposes ideas shared in both individuals for it to work at all. The contents of a movie would just be "culture" according to Tooby, because everyone watching the movie would share information if they understood the movie, regardless of when or where they watched it or what society they belonged to.[3] A poster of an equation becomes "culture" just in case the equation is solved or even understood roughly by a driver passing by, at any time, in any place, from any language.

Tooby further claims that a "science" of culture is no less gibberish than a science of how one building affects its neighbor building through architecture or power lines. He argues that instead of culture, specific phenomena—such as "ritual" or conversations—should be isolated and broken it into their specific components for study. Analysis of material into its constituent parts is of course often a worthwhile endeavor for scientists, depending on the level of analytic detail needed to answer their questions. But to claim that we can dispense with an overarching concept of culture in favor of its constituents is to repeat the errors of the Chomskyan paradigm in linguistics that ignores conversation, discourses, and texts in favor of the exclusive study of sentences—studying constituents apart from their distribution in larger units. Tooby claims that culture is no more helpful a concept than "protoplasm" (though comparing it to the "ether" would have been more powerful and humorous). "Culture and learning are black boxes, imputed with impossible properties, and masquerading as explanations . . . They are the La Brea tarpits of the social and behavioral sciences." And yet Tooby's remarks seem merely confused. The history of science shows us that failure to consider phenomena in the proper scale leads to misunderstanding.

If a famous scientist like the founder of evolutionary psychology is unconvinced of the usefulness of the term *culture*, then either it is indeed useless, he has read insufficiently, he has a vested interest in opposing the concept, or there is a need for a clearer understanding/definition of culture—or all of the above. I believe that there is indeed a need for a clearer or more accurate definition of culture. Yet even the inadequate definition that Tooby seems to use would not justify discarding the concept as useless, as Tooby urges upon us.

Some of the reasons for Tooby's apparent bemusement with the concept of culture are the following. First, he is understandably committed to his own theoretical model (which we consider in more detail in chap. 9). Evolutionary psychology (EP) rejects the idea that there is a force—culture—in which the mind is shaped and through which it learns things, for if this were true, it would render unnecessary EP's nativist theory of how cognition works, focused as it is on the mind instead of the individual, and on "what's in the mind," rather than "what's the mind in." Second, Tooby's bemusement arises from a disregard of more than a century of anthropological studies concluding that culture is a powerful explanan for human cognition, behavior, and social relationships, among many other things. Tooby's comments fail to refer to this literature. That is not the problem at all, in fact, especially not in a piece designed to be a short, pithy popular essay of the Edge type. More significant is that the remarks show ignorance and disdain for this literature, describing a concept that has not ever been more than a small aspect of culture (modes and types of information transfer). Nevertheless, Tooby can support his remarks with the accurate claim that there is no universally embraced definition of culture among anthropologists.

Culture indeed is appealed to as a way of making sense of many apparently unrelated phenomena. While some use culture to explain something as small as a conversation or even a word, it is also appropriately burdened to account for much larger entities. It has been appealed to in order to understand commonality within groups, even for entire nation states. For example, in *The Embarrassment of Riches: An Interpretation of Dutch Culture in the Golden Age* (Schama 1997), Schama discusses the Dutch motto, *Luctor et emergo*, "I struggle and emerge," and its role in both capturing and creating the Dutch people's sense of struggle to hold back and reclaim land from the sea—to build a nation where once there was only water and to provide a national rallying cry, a summary (to some) of the Dutch people's most salient cultural values.

Examined in this light, what are we to make of this phrase? Superficially, there is nothing profound about a state motto. Legislators have the authority declare any phrase they choose the state motto, any flower the state flower, any song the state song. This does not mean that such choice always reflects cultural values. Nor does legislative action automatically endow itself with a purpose or purchase in the society. On the other hand, Schama makes the case that not only does *Luctor et emergo* capture a fact about Dutch history and the personalities of some Dutch aristocrats, but that such phrases, once created, can in turn engender the very cultural value they name. The question Schama's book and many others raise is what the significance might be of intentionally shared ideas and values—through such means as advertising, public proclamations, slogans, aphorisms, flags, and so on—for effecting change within groups. When the group proclamation becomes a value of individuals, the social and the individual are shared or causally linked; this linkage leads to shared actions, and thus ultimately to shared values, getting us partway to what we mean when we claim that a people forms a specific culture (or that they "live culturally").

Thus while Tooby may be absolutely right that to have meaning, "culture" must be implemented in individual minds, this is no indictment of the concept. In fact, this requirement has long been insisted on by careful students of culture, such as Sapir. Yet unlike, say, Sapir, Tooby has no account of how individual minds—like ants in a colony or neurons in a brain or cells in a body—can form a larger entity emerging from multi-individual sets of knowledge, values, and roles. His own nativist views offer little insight into the unique "unconscious patterning of society" (to paraphrase Sapir) that establishes the "social set" to which individuals belong.

The idea of culture, after all, is just that certain patterns of being—eating, sleeping, thinking, posture, and so forth—have been *cultivated* and that minds arising from one such "field" will not be like minds cultivated in another "field." The Dutch individual will be unlike the Belgian, the British, the Japanese, or the Navajo, because of the way that his or her mind has been cultivated—because of the roles he or she plays in a particular value grouping, because of the ranking of values that her or she has come to share, and so on.

We must be clear, of course, that the idea of "cultivation" we are speaking of here is not merely of minds, but of entire individuals—their minds a way of talking about their bodies. From the earliest work on ethnography in the US, for example, Boas showed how cultures affect even body shape. And body shape is a good indication that it is not merely cognition that is effected and affected by culture. The uses, experiences, emotions, senses, and social engagements of our bodies forget the patterns of thought we call mind.

In this sense, denying the importance—or even the very existence—of culture as an object of study in favor of its constituents (knowledge, behaviors, beliefs, roles, etc.) would be like a linguist studying only phonemes rather than words, words rather than sentences, sentences rather than stories, or stories rather than conversations. Looking only at the parts and never at the whole leads to a defective understanding of all in this case. Saying that culture is just one person's mind affecting the mind of another person is like saying that language is just one person's voice affecting another person's ear, or that somehow the acoustics of speech transmission are the only proper object of study for linguistics. We should not forget that language is no less difficult to accurately define than culture. Yet we cannot do without the term *language*. People share a language, even though we know that no two people speak exactly alike. On the one hand, language is an abstraction, just as culture is. Moreover, language, like culture, is a property of individual speakers, not the group as a whole. Yet this does not make language, or culture, one whit less real as a shared possession of multiple people simultaneously. Anthropologists may have an even less sanguine view of culture as existing in the minds of individuals, but here exactly is the place where I believe linguistics offers a better model.

BUILDING CULTURES

Exploring this idea that understanding language can help us understand culture, consider how linguists account for the rise of languages, dialects, and all other local variants of speech. Part of their account is captured in linguistic truism that "you talk like who you talk with." And, I argue, this principle actually impinges upon all human behavior. We not only talk like who we talk with, but we also eat like who we eat with, think like those we think with, and so on. We take on a wide range of shared attributes; our associations shape how we live and behave and appear—our phenotype. Culture can affect our gestures and many other aspects of our talk. Boas (1912a, 1912b) takes up the issue of environment, culture, and bodily form. He provides extensive evidence that human body phenotypes are highly plastic and subject to nongenetic local environmental forces (whether dietary, climatological, or social). Had Boas lived later, he might have studied a very clear and dramatic case; namely, the body height of Dutch citizens before and after World War II. This example is worth a close look because it shows that bodies—like behaviors and beliefs—are cultural products and shapers simultaneously.

The curious case of the Netherlanders fascinates me. The Dutch went from among the shortest peoples of Europe to the tallest in the world in just over one century. One account simplistically links the growth in Dutch height with the change in political system (Olson 2014): "The Dutch growth spurt of the mid-19th century coincided with the establishment of the first liberal democracy. Before this time, the Netherlands had grown rich off its colonies but the wealth had stayed in the hands of the elite. After this time, the wealth began to trickle down to all levels of society, the average income went up and so did the height." Tempting as this single account may be, there were undoubtedly other factors involved, including gene flow and sexual selection between Dutch and other (mainly European) populations, that contribute to explain European body shape relative to the Dutch. But democracy, a new political change from strengthened and enforced cultural values, is a crucial component of the change in the average height of the Dutch, *even though the Dutch genotype has not changed significantly in the past two hundred years.* For example, consider figures 2.1 and 2.2. In 1825, US male median height was roughly ten centimeters (roughly four inches) taller than the average Dutch. In the 1850s, the median heights of most males in Europe and the USA were lowered. But then around 1900, they begin to rise again. Dutch male median height lagged behind that of most of the world until the late '50s and early '60s, when it began to rise at a faster rate than all other nations represented in the chart. By 1975 the Dutch were taller than Americans. Today, the median

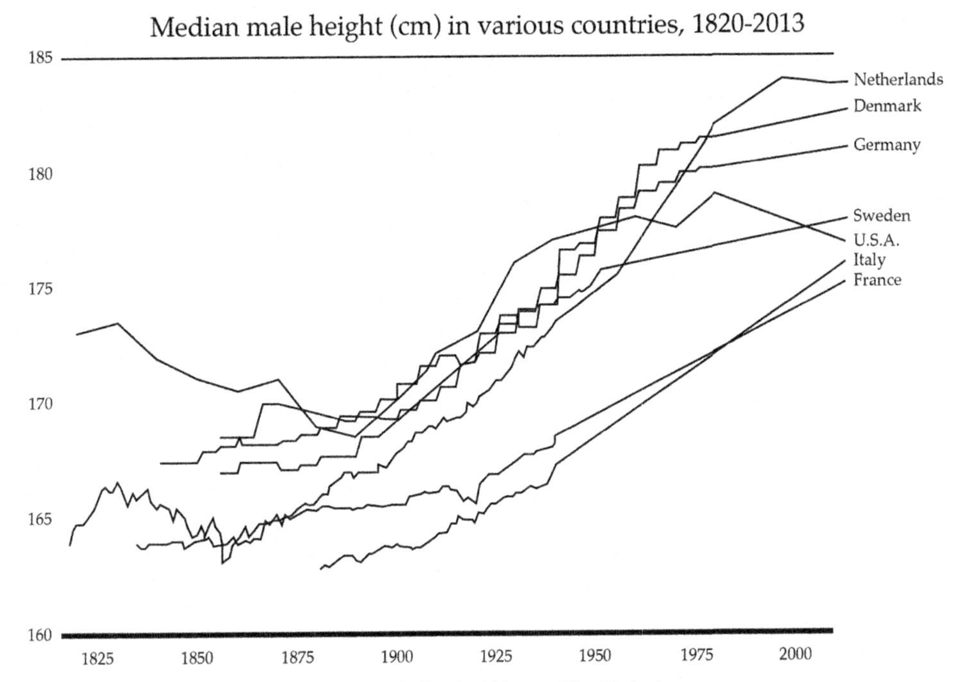

FIGURE 2.1

Dutch male height (183 cm, or roughly just above six feet) is approximately three inches more than the median American male height (177 cm, or roughly five ten). Thus an apparent biological change turns out to be largely a cultural phenomenon.

To see this culture-body connection even more clearly, consider figure 2.2. In this chart, the correlation between wealth and height emerges clearly (not forgetting that the primary determiner of height is the genome). As wealth grew, so did men (and women). This wasn't matched in the US, however, even though wealth also grew in the US (precise figures are unnecessary). What emerges from this is that Dutch genes are implicated in the Dutch height transformation, from below average to the tallest people in the world. And yet the genes had to await the right cultural conditions before they could be so dramatically expressed. Other cultural differences that contribute to height increases are: (i) economic (e.g., "white collar") background; (ii) size of family (more children, shorter children); (iii) literacy of the child's mother (literate mothers provide better diets); (iv) place of residence (residents of agricultural areas tend to be taller than those in industrial environments—better and more plentiful food); and so on (Khazan 2014). Obviously, these factors all

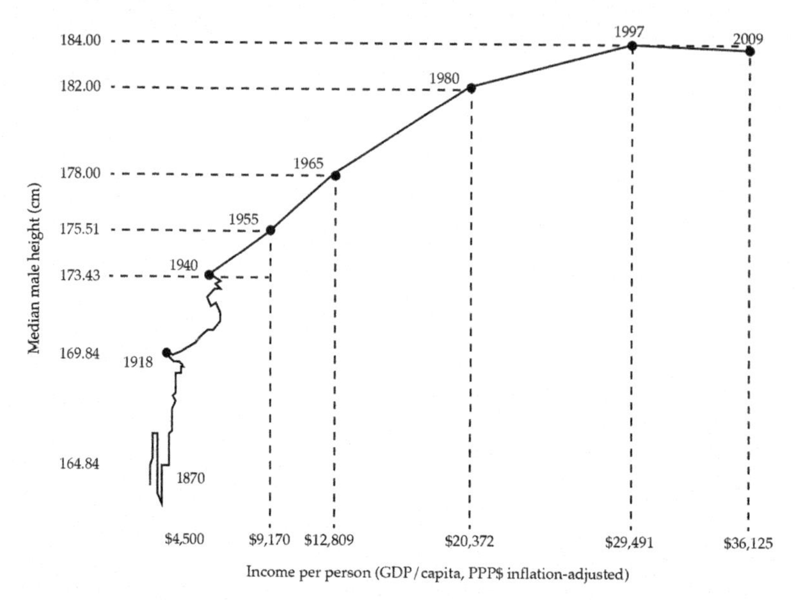

FIGURE 2.2

have to do with food access. But looked at from a broader angle, food access is clearly a function of values, knowledge, and social roles—that is, culture.

Just as with the Dutch, less-industrialized cultures show culture-body connections. For example, Pirahã phenotype is also subject to change. Facial features among the Pirahãs range impressionistically from slightly Negroid to East Asian to American Indian (to use terms from physical anthropology). Phenotypical differences between villages or families seem to have a biological basis (though no genetic tests have been conducted). This would be due in part to the fact Pirahã women have trysts with various non-Pirahã visitors (mainly river traders and their crews, but also government workers and contract employees on health assistance assignments, demarcating the Pirahã reservation, etc.). The genetic differences are also partly historical. One sizeable group of Pirahãs (perhaps thirty to forty)—usually found occupying a single village—are descendants of the Torá, a Chapakuran-speaking group that emigrated to the Maici-Marmelos rivers as long as two hundred years ago. Even today Brazilians refer to this group as Torá, though the Pirahãs refer to them as Pirahãs. They are culturally and linguistically fully integrated into the Pirahãs. Their facial features are somewhat different—broader noses; some with epicanthic folds; large foreheads—giving an overall impression of similarity to Cambodian features. This and other evidence show us that the Pirahã gene pool is not closed.[4] Yet body dimensions across all Pirahãs are constant. Men's waists are or were uniformly 83 centimeters (about 32.5 inches), their average height 157.5 centimeters (five two), and their average weight 55 kilos (about 121 pounds).

I learned about the uniformity in these measurements over the past several decades as I have taken Pirahã men, women, and children to stores in nearby towns to purchase Western clothes, when they came out of their villages for medical help. (The Pirahãs always asked that I purchase Brazilian clothes for them so that they would not attract unnecessary stares and comments.) Thus I learned that the measurements for men were nearly identical. Biology alone cannot account for this homogeneity of body form; culture is implicated as well. For example, Pirahãs raised since infancy outside the village are somewhat taller and much heavier than Pirahãs raised in their culture and communities. Even the body does not escape our earlier observation that studies of culture and human social behavior can be summed up in the slogan that "you talk like who you talk with" or "grow like who you grow with."

We unconsciously (sporadically—we may be conscious of this initially) learn the pronunciation, grammatical patterns, lexicon, and conversational styles of those we talk with the most. If you live in Southern California, for example, you are likely to say, "My car needs washing" or "My car needs to be washed." But in Pittsburgh, you are more likely to say, "My car needs washed" or "My car needs to be washed." There is a grammatical contrast between the two dialects in that one (Southern Californian) requires the present-participle form of the verb for the adjectival construction, whereas the Pittsburghese dialect requires the past-tense form of the participle. Both cultures converge in the *to be* construction. As another example, if you talk to people of my generation, you will likely say, "He bought it for you and me," whereas if you talked mainly with members of a more recent generation, you might say "He bought it for you and I."

The question is whether there is any more to the idea of culture than this sociolinguistic principle of "local mimicry." If not, we do not need a theory of culture per se, but only a theory of the role of mimicry in culture. And were this the case, Tooby's criticism might well be correct. Yet though there are good theories of mimicry (Boyd and Richerson 1988, 2005; Richerson and Boyd 2005; Arbib 2012; among others), culture is far more than mimicry.

Although imitation (joint influence to be more precise) may indeed be an entry point into culture—along with emotions, and basic survival needs of our species—the structures and values constitutive of culture take time to evolve. These structures and values emerge partially through conversational interactions, which are form-meaning exchanges, including the content of speech, perspectives on right and wrong actions or thoughts, acceptable levels of novelty of information or form of presentation, and levels and markers of conformity. This happens as people talk like who they talk with.

In other words, people grow to be alike. Raise two children together and they will be more alike than had they been raised apart. They will share values that children raised apart do not share, and they will (at least, early on) share knowledge structures that are more similar than had they been raised apart. The more people talk together, the more they talk alike. The more they eat together, the more they eat the same foods in the same way—the more they eat alike. The more they think together, the more they think alike. And so on.

CULTURE AS ROLES, VALUES, AND KNOWLEDGE

Culture has meant many things to different people over the years. This is telling. The diversity of views indicates that culture is not an easy concept to define—not a run-of-the-mill concrete noun like *book*, *relative*, *taco*, or *tomato*. Rather, it is an abstract noun like *love*, *ambition*, or *edification*. There is no (near-)universal consensus on what culture is. In fact, this lack of consensus has led some, as we have been discussing, to go so far as to declare the notion of culture dead and of no use to either science or simply a broader understanding of Homo sapiens. My own view is that culture has been hard to define not only because it is an abstraction, but also because it is a cover term for "sets of sets" of things united within a community. To better understand the definition of culture that was proposed earlier, it is helpful to first survey some of the better-known definitions from the anthropological literature. The question to keep in mind in light of our discussion to this point is whether such definitions can be improved.

Perhaps the best known of all definitions of culture is that offered by Edward Tylor ([1871] 1920, 1): "Culture, or civilization, taken in its broad, ethnographic sense, is that complex whole which includes knowledge, belief, art, morals, law, custom, and any other capabilities and habits acquired by man as a member of society."

Although Tylor's definition of culture is well known and still used by some, it is far from adequate. It isn't a bad point of departure to begin a discussion on the nature of culture, but it falls considerably short of a full definition. On the one hand, it is too coarse-grained. It doesn't allude to, for example, the relationships between kinds and structuring of knowledge that—almost as much as the knowledge itself—shapes a culture. It omits reference to roles that people play in their cultures and to the variety of relationships in which different roles may emerge. These are important research topics in contemporary anthropologically oriented research, so a definition of *culture* should make the concepts accessible/expected. Yet the most critical shortcoming of Tylor's definition is its lack of reference to either meaning, knowledge hierarchies, or values. Culture is how we structure our knowledge, how we impute meaning to the world, how we value things and people in our environment, among other things. It is more than the epistemological union of multiple individuals.

So perhaps we need to consider another classic definition of culture. Talcott Parsons (1970, 8) offers an alternative: "Culture consists in those patterns relative to behavior and the products of human action which may be inherited, that is, passed on from generation to generation independently of the biological genes." Well, of course culture must be heritable. But this is far from all it is. For example, a Filipino-American boy could acquire some patterns from his American father and pass these down to his children, but continue to mainly interact with Filipinos. By Parsons's definition, the boy would be culturally American and Filipino, yet the set of things inherited by the boy from his father too small a set for anyone to treat him like or mistake him for an American. There has to be more than merely inherited knowledge. So, like Tylor's definition, Parsons's is not incorrect, but neither is it necessary or sufficient. We have just seen that it is insufficient. But it is also not necessary that information be inherited. For example, one people could view another people as "the enemy" during wartime and share this attitude and knowledge of the other people. But the attitude might not (hopefully would not) be passed down to a subsequent generation after the particular war. The knowledge was cultural knowledge, but it was not necessary that it be inherited to count as such.

Another leading anthropologist to offer a definition of culture is Ward Goodenough (1981, 167): "A society's culture consists of whatever it is one has to know or believe in order to operate in a manner acceptable to its members." Again, one can say, "Yes, of course," but still be unsatisfied with this definition for the same reasons as previous definitions—its lack of reference to structure, roles, hierarchy of knowledge, and so on.

Like Goldilocks, we can try other definitions. One is from Claude Levi-Strauss, who offered the following: "Man is a biological being as well as a social individual. Among the responses which he gives to external stimuli, some are the full product of his nature, and others of his condition . . . But it is not always easy to distinguish between the two . . . Culture is neither simply juxtaposed to nor simply superposed over life. In a way, culture substitutes itself to life, in another way culture uses and transforms life to realise a synthesis of a higher order" ([1949] 1969, 4). In Levi-Strauss's definition, "nature" and "condition" refer to biology and culture or genotype vs. phenotype, respectively. Culture, as Levi-Strauss puts it "substitutes" for life—that is, it buffers us from our biology as it offers us a way to live that is not merely biological (see also Descola 2013). It changes our lives from those of mere animals living in proximity, to a society, with values and the other components of mental and social life. Lovely as the sentiments are, and however true, Levi-Strauss here merely describes aspects of culture; he does not provide a definition. At the same time, he offers insights that should be captured by a successful theory of culture.

Moving on to the next bowl of porridge, Clifford Geertz (1973, 5) offers a definition of culture that surpasses the previous ones, in my opinion, because it focuses on meaning: "Man is an animal suspended in webs of significance he himself has spun . . . I take culture to be those webs, and the analysis of it to be therefore not an experimental science in search of law but an interpretative one in search of meaning. . . . Once human behavior is seen as . . . symbolic action—the question . . . to ask [of actions] is what their import is." Geertz's definition of culture as "webs of meaning" is at least partly correct. Surely, as I argue below, meaning is a vital component of human life that culture furnishes. But meaning is not all there is to culture. Culture also establishes constraints, a variety of socially vital roles, and values, as well as conventions, symbols, and many if not all of the *forms*—grammatical, lexical, societal—used to encode and decode meaning.

In spite of its shortcomings, however, Geertz's definition is particularly interesting to me because it is the first propose that culture is a way of seeing meaning in the world. Yet it places too much emphasis on the symbolic. It ignores, for example, the "grammar of culture" that Pike proposed and that in some form *must* be part of our understanding of culture. Even grammar itself, although it manipulates symbols, is not itself purely symbolic. Yet without a grammar, we cannot share the meanings of the social and physical world around us that constitute Geertz's notion of culture.

Finally, let's examine Marvin Harris's (2001, 47ff) definition of culture: "Culture . . . refers to the learned repertory of thoughts and actions exhibited by the members of social groups—repertories transmissible independently of genetic heredity from one generation to the next." This is another great start. But like the others, Harris's definition falls short. It seems to make culture little more than our overt, or "exhibited," behavior. But it is much, much more than this. In fact, taken literally, Harris's concept of culture lacks an adequate role for dark matter (not unintentionally). Yet in my view, culture is defined *primarily* through dark matter.

Another feature absent from previous definitions of culture is culture's dynamicity, its refusal to remain static and fixed, its variability even within individual members of the culture, where members' values, roles, and symbols fluctuate frequently.

In light of the lack of fit of previous definitions of culture to our program here, let's revisit the definition of culture that was proposed above:

> Culture is an abstract network shaping and connecting social roles, hierarchically structured knowledge domains, and ranked values. Culture is dynamic, shifting, reinterpreted moment by moment. Culture is found only in the bodies (the brain is part of the body) and behaviors of its members. Culture permeates the individual, the community, behaviors, and thinking.

From this definition, it follows that culture is a relative concept and that people can share culture to a greater or lesser degree, even if they are not members of a well-defined social group or community or geographical area. There are various components to this definition, each subject to a range of (mis)interpretations, so let me unpack it before trying to apply it.

By "abstract network" I mean that culture is a postulate to help explain what it is that people share when we say that they are members of a particular society. Culture resides exclusively in individuals, though its artifacts, rituals, tools, and so on, provide concrete evidence for it. And no two individuals are exactly alike culturally. There are only idiosyncracies. My next-door neighbor and I may share many similarities, but we are never identical. Yet we are part of a network—people who have some overlapping values or social roles or knowledge structure or all three.

The more our values, roles, and knowledge structures overlap, the more connections we share and, therefore, the stronger our connection in a cultural network. Thus we can form a generational network, a CEO network, a rap-lovers network, a "Western culture" network, and industrialized society network, and even a Homo sapiens network, so long as we share values, knowledge structures (not merely knowledge), or roles, however slight the overlap.

This is recognized by many laypeople when they claim that "people are all alike." We *do* all share some values. Likewise, the other extreme—represented by cultural relativists—is also right when it claims that no two cultures are alike; no two cultures (or even individuals) share all the same values, all the same social roles, or all the same knowledge structures.

Values are the assignment of adjectives of morality for the most part (more clarification of values will come directly) to specific actions, entities, thoughts, tools, people, and so on. They are also statements about how things should or should not be. To say, "He is a good man" expresses a value. This can be broken down into finer-grained values such as "He treats his children well" or "He likes stray animals" or "He gave me a ride home" or "He is polite." Values are also seen in the tools we choose—a bat instead of a gun for home defense, or a machete instead of a hoe for digging vegetables in the garden. They are seen in the use of our time. Value sets are vast and varied. (We will return to them below.)

"Hierarchical knowledge structures" refers to the idea that human knowledge, at least—perhaps this also applies to other animals—is not an unordered set of ideas or skills. What we know is broken down in various ways according to context. All is structured in relation to all. And this hierarchy inescapably produces a gestalt output.

Consider playing the guitar. I say "I know how to play the guitar." I do not describe what I know as "I know where to find the frets as I need them." Take my knowledge of a specific song, such as the ability to play the peerless classic "Louie Louie." If you ask me if I know it, I will say yes. If you ask me to perform it, I will. And in this exchange we treat it as a unitary item. But if I were to teach you how to play it, it would be of little use to say, "Here is 'Louie Louie'" and just run through the entire song. I first need to explain its component parts: "Start with a G chord. Hit it three times. Then play a C chord, likewise striking it three times. Finally, form a D-minor chord, also hitting it three times. Then repeat as many times as you like, in 4/4 time, singing a number of lyrics (which don't matter because no one knows them)." Then I might show you the melody, the lead guitar part, the drum part, and so on. And each of these subparts is also broken down into component parts. All arranged hierarchically, just like the stories and sentences we utter. The parts as we first hear them are etic experiences. As we internalize them and learn their role in the whole, the parts and the whole together fall within our emic understanding. And only then can we understand that to go beyond the basic performance of a song to "playing it with soul," as we internalize the song and arrange it among our values, knowledge, roles, and so on.

Moreover, the song is explicable ultimately by appeal to dark matter. We first learn the chords, guitar lead, and words in small bits. Then we learn to play them together, very gradually, without thinking of them consciously. These small parts—as Gestalt psychologists would have recognized via their ideas developed more than one hundred years ago—are not the song. The song as a whole is not dark matter knowledge. But its component parts become dark matter *after* they are learned. They were not originally tacit knowledge. But then they undergo the transmogrification from overt to covert knowledge, illustrating the effects of emicization (taking on the insider's perspective of guitar playing) on access to knowledge.

"Social roles" describes actions as conforming to a particular node one occupies in an abstract network. Any grouping of people will be defined by its values, the knowledge structures it devolves from and develops, and the expected duties of each of its members by virtue of their membership classification. Whether the group is a college department faculty (each is expected to specialize, and their specialization is one of their roles—e.g., syntactician, Amazonianist, chairperson, administrator, and so on).

Department chairs in the North American academic system, China, or the United Kingdom will differ in many of their values, administrative knowledge, and more, but in their roles (independent of what they are called), they share some aspects of administrative knowledge and values. Moreover, when there exist larger homogenizing forces across communities, roles and knowledge will grow even more alike. An example of such forces in modern academics is international accrediting bodies, such as the AACSB (Association to Advance Collegiate Schools of Business) or EQUIS (European Quality Improvement System) for business schools. They grant accreditation to schools for sharing and implementing the agencies' values.

No two individuals will share all values, identical value rankings, the same set of acknowledged social roles, and so on. And each individual will regularly update or modify their knowledge, the concepts of social roles, their values, and their rankings. This dynamicity is another reason we say that culture is an abstraction—it is a generalization with considerable smoothing.

In order to better understand the model of culture I am suggesting, let's consider how it might solve a problem proposed by Marvin Harris (2006, 23ff):

> Much evidence exists that the cultural information stored in the brain contains contradictory information. For example, in a study of how Americans conceptualize the family, Janet Keller (1992:61–2) recorded these competing "schema":
>
> > Family members should strive for the good of the whole group
> > but
> > The good of the individual takes precedence over the good of the whole group.
> >
> > Family is permanent
> > but
> > Family is always in transition.
> >
> > . . .

Family is nurturant
but
Family is smothering.

From such "contradictions," Harris concludes: "Indeed, from my cultural-materialist perspective, the emphasis on the proposition that ideas guide behavior, but not the reverse, is the mother-error of contemporary anthropological theories" (22).

Although my remarks here are not intended as a criticism of Harris's overall theory, they do intend to show that his arguments here have no force. Everyone appears to hold mutually inconsistent values. But the inconsistency is not always real. For example, consider the last contradiction above, "Family is nurturant/family is smothering."

Most people would talk about their family in a similar way but not as a contradiction, saying instead something like, "Families are often nurturing/should be nurturing, but unfortunately they are often smothering." "Nurturing" and "smothering" express value judgments. But because they are values, in my model we need to know how they are ranked. And because all ranked values are violable, such that lower values can be violated by higher-ranked values, it is possible for higher-ranked values to mask the effects of lower-ranked values, as we see in the rankings following this paragraph. Getting back to the terms at hand, the idea of nurturing includes widespread positive judgments in society as a label for a set of more finely grained values, such as {love, assistance, sharing of wisdom, independence, imposition of limits, trust, financial support, judgment, etc.}. Ironically, smothering can follow from the same values, differently ranked (and these are just two of many possible rankings that could derive these two concepts, meaning that we expect similar concepts to be different at the level of dark matter even when producing superficially similar results):

NURTURING (rough pass): TRUST >> FINANCIAL SUPPORT >> LOVE >> ASSISTANCE >> SHARING OF WISDOM >> IMPOSITION OF LIMITS >> JUDGMENT, ETC.

SMOTHERING: IMPOSITION OF LIMITS >> LOVE >> JUDGMENT >> SHARING OF WISDOM >> FINANCIAL SUPPORT >> TRUST, ETC.

The >> follows from the use of this symbol in optimality theory (Prince and Smolensky [1993] 2004) and means that any value to the right of the double arrow can be violated in order to obey a value to the left. Thus a smothering parent may love and trust their offspring, as a nurturing parent does. But love and trust may be ranked differently in each parental behavioral set, leading to very different types of families or parenting. Moreover, as our definition of culture makes clear, different times or circumstances can have their own rankings, such that the same parent can be smothering for one request and nurturing for another from the same child.

There are many different rankings possible for any set of values. And rankings can vary from individual to individual and even from situation to situation for the same individual or group. A family will contain nurturing and smothering values simultaneously, with the difference being in how the values are ranked for a given family member relative to a particular act (work or play) or person (son or daughter, for example) involved—your mother may be smothering regarding your love life and nurturing regarding your choice of profession, for example. Rankings and roles lead to seeming contradictions but, contra Harris, many ideas, once broken down into their component, violable value hierarchies are not contradictory at all. (Of course, some ideas may turn out to be mutually contradictory; e.g., "The earth is flat, but I think I can sail around it," but we cannot even conclude this without careful analysis.) It turns out that people are rational and that the notion that ideas control much of behavior makes a good deal of sense. *Assume rationality and work backward—from behavior and a theory of culture and the individual—to understand others.*

The same goes for all the other examples Harris uses. For example, he lists considerations in selection of a location for defecation for Hindu farmers in India:

> A spot must be found not too far from the house.
> The spot must provide protection against being seen.
> It must offer an opportunity to see anyone approaching.
> It should be near a source of water for washing.
> It should be upwind of unpleasant odors.
> It must not be in a field with growing crops. (Harris 2006, 24)

He then observes, "Fulfilling all of these rules on a small farm leads to behavior that violates the rule of fecal avoidance, as evidence by the elevated incidence of hookworm."

Again, Harris cites this seeming set of contradictions as evidence against an idea-over-behavior understanding of culture. His own view of culture is also given as: "a culture is the socially learned ways of living found in human societies and that it embraces all aspects of social life, including both thought and behavior" (ibid., 19).

But neither his criticism nor his definition of culture get us to the conceptual place I think we need to arrive at. First, the criticism doesn't follow. Assume that the Hindu farmer's desiderata for dump-taking are in fact ranked, violable values. Assume that "avoid fecal matter" is not the highest ranked of the values (outranked perhaps by "stay close to home"). Then the ranked-values approach accounts easily for the behavior, sans contradiction, in spite of what Harris claims. It also simultaneously shows his definition of culture to be underarticulated and thus inadequate. This is not to claim that all behavior is value- or idea-driven; for example, a baby's grasping may be gene-driven. But most behaviors probably are value-driven when we take on the concept of violable constraint ranking. The same transition from *inviolable*, contradictory rules and constraints to violable, ranked constraints ("linguistic values," from one perspective) is what has made optimality theory (OT) so influential among linguists—it solves the apparently intractable. Neither linguistic ranking nor the violable value ranking I am proposing here are original to the humanities. Ultimately they derive from "Hopfield Nets/Networks" (Hopfield 1982), which in turn emerged from the desire to understand why some physical states are achieved rather than others (why does molten glass dry "smooth as glass," for example).

The theory of culture defended here, however, is not simply an application of OT *tout court*. For example, OT assumes that linguistics constraints are universal—that every person is born with the same set of constraints and they simply have to learn how their particular language ranks the universal set. To capture this, the theory contains a GEN(erate) function to ensure that all languages have the same constraints. The theory developed here, on the other hand, has no place for universal values, and therefore it has no need of this function. The anthropologist must discover the values and their rankings of each individual in each society.

VALUES EXPLAINED

With this brief introduction to my violable value-ranking model of culture, we can consider in more detail just what values are. Values, as I have stated, are fundamental to culture in various ways. They shape cultural forms, group intentions, meanings, aspirations, conventions, and so on. Before discussing their role in more depth, however, it is important to understand what a value is.

There are several kinds of values, and each has an important role in society and culture, though there are a couple of broad sources. The first kind of value to mention are *terminal values*. Terminal values include "leading a comfortable life," "having a sense of accomplishment" (see Rokeach 1973, 160ff), "freedom," "security," and so on—they are what society and the individual sanction as laudable goals. *Instrumental values*, on the other hand, are how we think it is best to achieve our goals; for example, "ambition," "cleanliness," "honesty," "politeness," "self-control."

There are also what I refer to as *biological values*. These are likely shared in one form or another by all humans, though that is an empirical hypothesis. These include things like self-preservation, not going hungry, being warm, and staying healthy. To the degree that these are universal, they are *less* interesting for the present discussion. But they are definitely part of our dark matter—more connected with emotions and bodily functions than higher cognition, perhaps. Their satisfaction is essential to our well-being, however, and they can override all other values in some circumstances. So it is well to remember them. At the same time, cultures rank and interpret these differently, so comparison is not trivial. What one considers healthy, another may consider disgusting. What one considers pain to be avoided, another may understand as pain to be sought.

Rokeach spends much time discussing another set of values, what he labels "immanent values"—values that hold universally by virtue of their very nature: "Thou shalt not kill," "Thou shalt not commit adultery," "Treat the weaker fairly," and so forth. It is far from clear to me, however, whether such values actually exist. Therefore there is little more to be said about them here. They fall mainly in the domain of axiology and are orthogonal to our current objectives of understanding how dark matter arises and comes to guide our actions and thoughts.

This is enough to get our discussion of values, culture, and dark matter under way. But for values and value ranking to be of utility in the anthropological enterprise, we need to say how they might be studied—the *methodological requirements* for a study of values. The first methodological question that arises is how one goes about identifying the values of individuals and groups. One obvious way is to follow the method employed in the famous Rimrock study, headed by Clyde Kluckhohn and his wife, Florence.

In 1948 the Laboratory of Social Relations at Harvard University planned and undertook groundbreaking research entitled the "Comparative Study of Values in Five Cultures," or more familiarly known among those close to the research as the "values study." The site of the study was one of the most austerely beautiful regions of the United States—Rimrock, New Mexico—where Clyde Kluckhohn, the nominal leader of this project, had conducted research among the Navajos since 1936. The values study emerged from his (and others') "value orientation theory." This Harvard research group launched a pioneering series of studies of human values, completing the values study in spite of Kluckhohn's sudden death in 1960 at age fifty-five.

In the Harvard study, the research team examined five separate cultures in the Rimrock area, cultures of particular interest because they shared the same environment but with significantly different ways of life. It was hypothesized that since the material environment was identifical, differences in the ways of life would derive exclusively from cultural differences, in particular to differing value systems. Linguists call differences in the same environment *contrast*—as /p/ and /b/ contrast in the same environment of *pat* and *bat* (spelling notwithstanding, looking only at the phonemes). The five cultures were Navajo; Zuni; Mexican-American; Texan and Oklahoman farmers and ranchers; and Mormon. Each of these represents a separate community and culture.

Kluckhohn prepared a statement on the project that included the following: "There is general agreement among thoughtful people today that the problem of 'values' is of crucial importance, both practically and from the point of view of scientific theory" (Vogt and Albert 1966, 1).

Following the publication of Vogt and Albert 1966, however, the study of values in anthropology eventually petered out. According to D'Andrade's (1995, 13ff) assessment, "The results [of the Harvard Rimrock study] were generally agreed to be disappointing. The major problem seems to have involved the *identification* of values. If a universal framework was used, like Florence Kluckhohn's universal framework for the analysis of values, specific cultural values were left undescribed and unanalyzed. But no procedures had been developed to determine specific values." D'Andrade's criticisms are cogent and appropriate. And I suggest an answer directly.

The problem, as D'Andrade describes it, is reminiscent of the criticism of Chomsky's research program in linguistics. For example, there is no good treatment in that theory for either (i) the nonoverlap of supposedly universal principles in either abstract or concrete structures of specific languages (e.g., no two "passives" work exactly alike) or (ii) there are no overt universals at any interesting level (see N. Evans and Levinson 2009).[5] Universalist theories of socially affected cognition and behavior or linguistics often run aground on the reefs of the details. Boas ([1911] 1991) was one of the first to point out in modern times that we ignore diversity at our scientific peril.

Gallagher (2001, 1–2) summarizes the Rimrock study in the following way:

> They hypothesized that ". . . there are a limited number of common human problems for which all societies at all times must find some solution . . . How a group is predisposed to understand, give meaning to, and solve these common problems is an outward manifestation of its innermost values, its window on the world: its value orientation." The five common human problems, posed as questions, that provided the most useful "value orientations" in creating a cultural typology were:

Orientations	Possible Dimensions		
Time	Past	Present	Future
Activity	Doing	Becoming	Being
Relations	Individual	Collateral	Lineal
Person-Nature	Humans dominant	Harmony with	Nature Dominant
Human Nature	Good	Mixed	Evil

FIGURE 2.3

> What is the temporal focus of life? (Time orientation)
>
> What is the modality of human activity? (Activity orientation)
>
> What is the modality of a person's relationship to others in the group? (Relations orientation)
>
> What is the relationship of people to nature? (Person-nature orientation)
>
> What is the character of innate human nature? (Human nature orientation)
>
> . . . From their research they deduced that societies would respond in one of three ways to each of the five questions or orientations (figure 1 [figure 2.3 in this volume]).

This passage at once illustrates the innovative and very important nature of this research, as well as its fatal flaw: universalism. Assuming that all people share the same values or that similarly labeled values are the same values is unwarranted. The paragraph that follows immediately below is interesting in that it shows that the Harvard group had grasped the idea of value rankings, though not the concept of violable values.

> In 1961, Kluckhohn and Strodtbeck published their theory and findings in their book, Variations in Value Orientations, in which they proposed that the rank-order of preference—from most to least—gave the society its cultural character. The different patterns of rankings allowed one culture to be distinguished from other cultures. It was this rank-order of preferences, they argued, that was the foundation for the more-visible cultural values, beliefs, norms, and actions—and even heroes, rituals, songs, etc.—of the society. They also proposed that, although a society may have a general preference that is dominant, there is a great deal of diversity within cultures and all cultures will express all possible dimensions at some time or through some individuals. Carter (1990) added to these propositions with his finding that cultures could share the same rank order of dimensions, but differ substantially if there was relative difference of preference for each of the dimensions. (Gallagher 2001, 3)

Numerous researchers (see D'Andrade 2008 for a detailed criticism and review) have criticized the Kluckhohns' study for conflating values into an artificial set of "orientations." Rokeach's body of work addresses, in my opinion, many of those shortcomings, however, and could serve as a useful basis for a typological comparison of values. On the other hand, that said, there is no reason to suppose that all values—nor even most, apart from biological values—will be universal. (And in fact, perhaps not even all biological values are universal given the enormous variation in their rankings and interpretation cross-culturally.) The Rimrock study failed, if this is even the correct judgment, because of a failure to sufficiently articulate its theory of value ranking and, most important, for its unwarranted assumption that values are universal. Yet the study has rightly been seen as pioneering. It is an extremely important milestone. In fact, it may have been due to this study that even philosophers began to recognize the significance of values and their rankings.

What seems common to both Harris's inadequately developed notion of cultural contradictions and the Rimrock study, in retrospect, is that both treated values like lists of inviolable constraints (this is also true of newer studies such as Graeber 2001). In other words, if two values conflicted (say, honesty and accomplishment) a quandary arose—how to understand why one value was respected while another was not. People understood, of course, that not all values received equal priority. Davidson even refers to "value ranking," in the following:

> Now suppose that our judge owns the house coveted by B and C and that she has decided to sell it to one of them. We can imagine that there are . . . distinct steps in her reasoning, insofar as it involves the desires of B and C. First, she determines what she can of their preferences, their *value rankings*, perhaps on an interval scale. Second, she compares these preferences, her judgment or judgments being of the kind just mentioned. (2004, 60; emphasis mine)

Conceptually, at least, values are ranked fairly easily, as can be seen in the following example. Assume that we are comparing the values of the inhabitants of two cities—say, Paris and Houston. Let us further assume that Parisians and Houstonians value "good food," however they define *good* and *food* locally. And let us suppose that both of them value being in good shape. Now, for the sake of discussion of this simplified value system, we propose the following rankings:

Parisians: GOOD SHAPE >> GOOD FOOD
Houstonians: GOOD FOOD >> GOOD SHAPE

It seems fair to say that the different rankings of just these two values could produce different body shapes (add to that a finer analysis of what each group considers to be "good food"—fried chicken and mashed potatoes vs. coq au vin, etc.—and the differences grow), in spite of the fact that it is true to say that the two cities have the same values. In this case, it is not the values but their relative ranking that makes the difference. Thus it is essential not only to have some idea of what a group's values are, but—as the Rimrock study presciently observed—also the prioritization of the values (and the rankings themselves are also dynamic, changing according to situation, subgroup, etc.). And since I do not accept the idea of a "universal value set," my claim is that we must first discover values and their rankings in each group separately.

How do we come up with reasonable hypotheses of a people's values and their rankings? Questionnaires and interviews are not enough, though they can provide some useful data. We crucially require participant-observer records accompanying the former. We need to study a variety of texts, conversations, and interactions. How are values portrayed, described, lived out, discussed, reacted to? Actions and words often are in conflict. People can imagine what their values are, while their actions may reveal different values, different rankings, or finer nuancing.

Consider interviewing two populations. Both claim that to be vegetarian. The first group never eats meat. The second group eats meat in the home of friends who do not know they are vegetarians. Thus although they share the value of "avoid meat, eat vegetables," the first group seems more consistent. Is the second group hypocritical? In a theory of inviolable values, yes. But in the real world of violable values, no. Here is how we might characterize the difference:

Group 1: AVOID MEAT >> PLEASE FRIENDS IN THEIR HOME
Group 2: PLEASE FRIENDS IN THEIR HOME >> AVOID MEAT

Notice that the question of cultural relativity does arise in this context, but not in a naive way. Societies must have values, along with structuring of roles and knowledge. And these values and structures can differ in ranking, identity, complexity, number, and so on. But there will always be some degree of overlap, due to the ecology of adaptation—we all have to get by, make a living, and follow Shannon's (1949) basic model of communication (see also D. Everett 2012a; Sterelny 2014). We never reach what Davidson (1973) referred to as incommensurable "conceptual schemes." This overlap between humans need not be ascribed to innate knowledge, but to external organizational pressures, as well as innate emotional structures, our physiology, and so on.

Nevertheless, just as there are many aspects of theoretical linguistics that are vital still for understanding languages—especially of the Platonic formal variety (Katz 1972), construction grammars (Goldberg 1995, 2006; Croft 2001), and role and reference grammar (Van Valin and LaPolla 1997)—the problems with Kluckhohn's program do not reduce the utility of values in understanding culture.

Let's consider again the pioneering work of Milton Rokeach. Rokeach defines the notions of value and value-system as follows:

> A value is an enduring belief that a specific mode of conduct or end-state of existence is personally or socially preferable to an opposite or converse mode of conduct or end-state of existence . . . "A happy life is preferable to a sad life," for example would fit this whereas "A happy life is preferable to fried eggs," would not." And also, "A value system is an enduring organization of beliefs concerning preferable modes of conduct or end-states of existence, along with a continuum of relative importance. (1973, 5ff)

Using this conceptualization of values, I suggest a methodology for the study of values.

A METHODOLOGY FOR CULTURE STUDY

There is a minimum threshold of research to incorporate an understanding of values into a particular ethnography, which includes several steps:

1. Through observation of behavior (we must begin with what people actually do and what they avoid), identify potential etic values. How are resources (money, time, fuel, food, etc.) used? How are activities and ends related? And so on.
2. Identify professed values: What do people say that they value, in Rokeach's sense of values?
3. Posit emic, tacit values using the standard distributionalist methodology of linguistics (Pike 1967, etc.). For example, if stray dogs are treated with violence while domesticated dogs are treated lovingly, then we see different values of "dog" and distinct manifestations of "lovingly" (presumably it is different toward family members and friends.)
4. Test emic values: Conduct experiments to see how people behave relative to the predictions of the proposed value system.
5. Propose potential value systems via experiments, interviews, and observations. (How are values ranked relative to one another in systems such as food, religion, philosophy, relationships, employment, language, etc.? How do values shape or affect different roles of an individual in society?)

6. Test value *systems*: For example, how do values interlock and vary within and across communities such as food, government, religion, and so on? Are potentially useful predictions made? Do they hold up? This could have been done, for example, in the Rimrock study; and if it had been, they might have seen contrasts and similarities across systems.
7. Test value rankings.
8. Look for individual variation in value inventories, value systems, and value rankings.
9. Write up findings across value systems.

Here the write-up would follow the basic principles of the definition of culture suggested earlier. Emic values become visible and differentiated from etic values when we ask questions such as, "What values do native speakers pay attention to?" "How are different values distributed among the population?" "How do different etic values get grouped together by native speakers as the same emic values according to circumstances?"

Once we have undertaken such initial studies, we can attempt to craft an ethnographic description of the target culture. This will be a very similar effort to the description of a new language. Though each language or society may be relatively independent of other languages or societies, there may be causally significant interaction affecting pairs of languages via contact between societies at a macrolevel or by individual bilinguals/biculturals. Both kinds of structures are relatively stable, outlasting the lives of the individual members of the society or the speakers of a language. For example, a language may have dialects distributed in space or distributed in terms of social levels, time, style, idiosyncracies (idiolect). There are subgroupings and interlocking groupings (where members belong simultaneously to different groups, yet share values and rankings to some extent), roughly identifiable along the lines of linguistic isoglosses.[6]

It is with the work of the linguist Kenneth Pike (1967) that notions of culture begin to capture the dynamics and structure missing from other definitions, because Pike's research is informed by his research in language and linguistics. Using his understanding of language, Pike argued that a society is "a structured group of individuals sharing in . . . behavior [whatever social behavior one chooses to examine]."

This notion of structure is further clarified by Pike and Pike (1976) as a "grammar of society" and, in my terms, a "grammar of culture." In this sense, a culture is partially a C(ulture)-grammar. Like any grammar, a C-grammar can be revealed and tested only by a solid methodology and rigorous testing of hypotheses. A C-grammar is like a linguistic grammar—a link between the forms instantiating it and the meanings derived from the resultant forms and their distributions.

D'Andrade (1995) correctly observes that earlier treatments of cultures as grammars were too simplistic to be of much use. But the claim I am making here is that culture is only partially constructed grammatically, or alternatively, that a culture contains many grammars produced and interpreted by the culture's values. Whether the entire culture can be said to be a grammar is not the point of the discussion here. But it nonetheless seems clear that every member of a culture has a variety of ever-changing roles that are sanctioned by, produced by, valued by, and understood by the culture in question.

Society and culture are of course more than merely grammars—but they are connected and constructed in grammar-like ways and especially in their local contexts, groupings, and actions. A Bostonian investment banker and an Amazonian hunter find their place (or if preferred, "make their place") by occupying a series of grammatical nodes in society. These nodes are rarely invented. One cannot be a professional musician without an entire technology, social role, and payment structure produced by society over time. And the structures and roles and fillers of the sociocultural grammatical system into which we are born themselves emerge from the ranked values and supporting beliefs of the culture. In this sense, if we take culture as beliefs, knowledge, and values, and society as the roles and structural relationships between them, with members of society as "slot fillers," then we begin to develop an idea of culture as meaning projecting society as structure/grammar, from which individual "meaning" and purpose emerge in conjunction with the small vocabulary of individual psychologies.

How might this work out in practice? A C-grammar must begin with the individual. We can thus think of all the individuals of a society as "fillers" for C-grammar slots. In this regard, consider a classroom experience. First, a grammar must have slots and fillers. The fillers are easy—these are the students and the professor. The slots are "speaker," "talk," and "hearers." The semantic roles of each of the slot:filler pairs are "lecturer," "lecture," and "class," assembled in a simplified form like:

College Class
Speaker:Lecturer + Talk:Lecture + Hearer:Class
Fillers: professional specialist, student, types of lecture

This unit of a C-grammar is equivalent to a sentence, though it will have immediate constituents, as shown in figure 2.4. There is a cultural structure, a college class, that has at least three components: a predicate (what is being done culturally, i.e., a lecture), an actor (the lecturer), and recipients of the

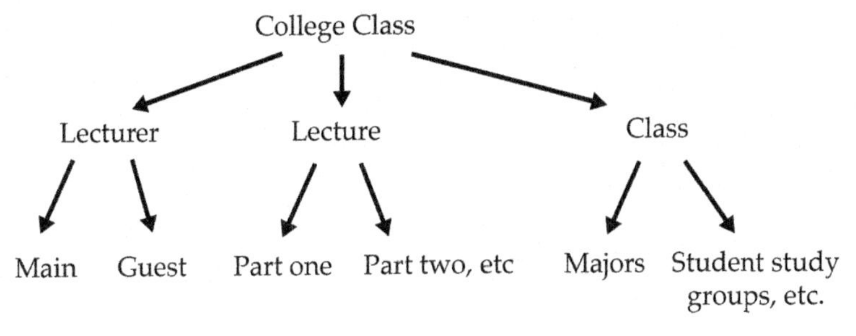

FIGURE 2.4

lecture (the class). These are the immediate constituents of the gestalt (i.e., emicized) cultural event. They are themselves broken down into constituents. So a lecture may be split between a guest speaker and the main lecturer or teacher, each contributing a portion; or between the teacher and YouTube or a film, and so on. The lecture will itself be organized into different sections, in order to facilitate more effective communication. And the students will fall into subconstituents of the class—different study groups, different majors (when relevant to the composition of the class or targets of the lecture), and so on.

Taking Pike's analysis of society as a type of grammar seriously, a society will have a range of forms to express its values or meanings. We can analyze these forms as generated from a dynamic combinatorial grammar of roles, fillers, and slots, as above. Or we might instead take forms of the grammar to emerge from a more static inventory of cultural "constructions," along the lines of construction grammar in linguistic theory (see Croft 2001; Goldberg 1995, 2006; among others), or as an OT grammar. The choice is not critical at this stage, though there could be significant empirical differences as the ideas are explored. I have chosen a simple model of the C-grammar, though of course others could be imagined—for example, constraint-based grammars. This remains as a future project.

Different constructions or grammatical outputs will be found in different societies based on different cultures. For example, a culture that values sports may have units called sports teams, with fillers that ultimately come down to individuals called athletes. There could be an entire constituent—say, the University of Michigan football team. Such a unit would be a C(ultural)-syntagmeme. This syntagmeme will be composed of smaller units, or "immediate constituents," such as defensive team, offensive team, special teams, bench, or coaching squad.

A token of a C-tagmeme will be a group on the field at a particular time. This formal unit will have constituents and a meaning, namely, the function of the team on the field. Each member of the team will also have a meaning—his or her role on the team. According to Pike, society will be structured into many such teams with crisscrossing membership. This is not new, of course, since Simon (1962) argued more than fifty years ago that information is most efficiently organized hierarchically.

But what about simpler societies? In Pike's sense, a society will be simpler if it manifests fewer C-syntagmemes and less hierarchy. Consider, for example, an Amazonian society such as the Pirahãs. That society will be manifested by its individuals and "parsed" into its immediate constituents: village communities, families, women, men, children, adolescents, and so forth. Another group might instead be parsed into more structured kinship hierarchies: families, clans, lineages, or more professional specializations, and so on.

To act together, a society must in some way share the intention that our individual actions produce a result of the group. Voting is arguably such an action. Participating in a classroom lecture is another. These are all *predicates* in the grammar of culture, in which each person occupies a role, alone or jointly, a slot in the C-sentence or C-discourse. For example, in the diagram above, the lecturer is the C-subject. The students are the C-object. The lecture is the C-predicate, a result of a group intention. In the linguistic act of the lecture, the topic or theme is what is being taught. But in the social organization, the students are the object, not the subject matter. We are describing their social roles in this moment in time to a particular teacher.

When participants are from different cultures, as in the earlier Medicine Lodge example, they often assume understanding of roles, structure, and meaning of the joint act they are engaging in. But they rarely realize that each participant possesses a separate cultural semantics for interpreting the activity. In my view of the entire situation, this is what happened: The Comanche interpreted the predicate of the Medicine Lodge event as immediately-in-effect, conditionless promises, considering all actors as equal plenipotentiaries. The American negotiators saw themselves as subordinates to Congress, the Indians as a group that should accede to a greater authority, and their joint act of treaty signing as entering into a conditional, time-delayed initial offer. They also saw the Indians as inferior beings whose opinions and understanding mattered less.

Pirahã society lacks sports teams, specialized professions, and so on. As an egalitarian, unspecialized (professionally), smaller society of intimates, it will be parsed into fewer constituents. Still, we can look into its constituents. Roughly (very much so), Pirahã can be described by the diagram in figure 2.5. Once again, the chart of Pirahã society is not intended to be complete. But it does cover a lot of what is important and, in my understanding, the principal

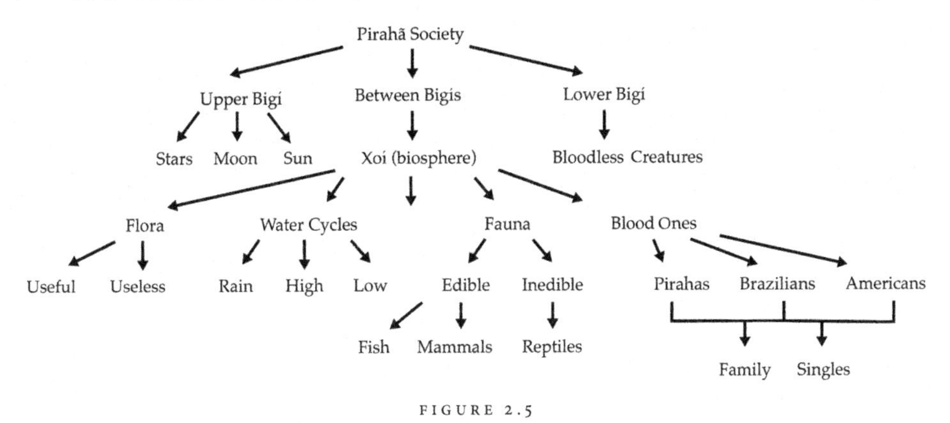

FIGURE 2.5

constituents of Pirahã society as perceived by the Pirahãs. Their society incorporates setting, humans, humanoid creatures (*kaoáíbógí*), different groupings of flora and fauna, nuclear families, and the like. So the biosphere is the *xoí*. This word means "the environment in which we live, move, and have our being." When one is going to the jungle, one says, "I am going to the *xoí*." When a parent wants a child to sit still in the canoe, they will say, "Move not into the *xoí*." When the Pirahãs talk about moving to another river or far downriver or upriver, or far to the interior, they describe all of these as "moving to a different *xoí*."

Bigí refers to a natural barrier. The sky, in Pirahã, is *bigí*, as is the ground. This is a vital constituent of Pirahã social organization, with different types of entities defined by their position relative to the barriers.

The next immediate constituent of Pirahã society are its villages. Every village is known to all Pirahãs at all times, even though they can be more than a hundred miles apart and even though their composition is constantly shifting. The river is the next immediate constituent of Pirahã society. More than half of the Pirahãs' diet comes from the river. All Pirahãs always live at the river's edge, wherever that is in the course of a rainy or dry season. River animals are crucially important to them in ways that jungle animals are not. But the river also organizes their village living, such that in the dry season they are on the beach and in the rainy season in the forest. However, at no time are they more than a hundred yards or so from the bank or edge of the river. In the river there are fish—of paramount importance to Pirahãs—and other animals (e.g., caymans, snakes, manatees, porpoises, stingrays) that they do not eat, or at least, very rarely eat. The fish will be divided into constituents corresponding to the Pirahãs's idea of species (as will other animals). The plants around them will be either useless to them (a few kinds of bushes, ferns, etc.)

or useful. Of the latter, some will be edible (e.g., Brazil nuts and fruits), while others will not be (e.g., vine for baskets, bark for bowstrings, wood for bows). The nuclear family will be manifested by parents, children, and anyone else occupying the same hut—grandparents on occasion, spouses of children, and so on. This is rather fluid. Some single people in the village live alone. There are a few single parents. This diagram, then, is just a way of illustrating how the Pirahãs break down the important components responsible for their values and day-to-day life as a society.[7]

Cultural values, conventions, and so on, will serve as the interpretative "linking rules" (see Van Valin and LaPolla 1997 for a discussion of form:meaning linking rules in language) from the C-grammar to the C-semantics, linking rules (connecting meaning to form)—how people understand themselves and their social and physical worlds (and how those worlds are integrated into their culture; see, for example, Kohn 2013).

Taking a field perspective of Pirahã C-grammar, the structure of a particular society comprises a set of relationships in a network (White and Johansen 2006). Individuals belong to a network of criss-crossing hierarchies and relationships. Each individual thus is located by coordinates in simultaneously co-existing matrices. Pike (1967, 644) says that "as language utterances in sequence form a hierarchy of parts, whereas the language units as an abstract system constitute a network of intersecting hierarchies, so society likewise has the activities of its individuals sequentially but hierarchically structures while the relationships between those individuals themselves are integrated in a total network."

Also important to understanding culture and society is the linguistic notion of "complementary distribution." Individuals can have distinct identities or a single identity in multiple modes of manifestation (think Clark Kent vs. Superman). In a strictly distributionalist view of the individual, Siamese twins would represent what structuralists might refer to as a "portmanteau being," where the unit boundary is less clear. A culture is an amorphous mass in some ways. It is like an enormous discourse. It isn't likely useful to propose a single-tree diagram to map it out in any useful way. On the other hand, its constituent components, like the sentences of a story, may be individually diagrammed in order to represent their major organizational units and principles. Thus the construction-based language of thought is also the language of language and the language of society—a mode of human organization. Still, it is worth considering more deeply the linguistic model of cultural analysis (see also Searle 2010 and cf. Hacking 2000).

For example, consider an analysis of leadership in a given society, along the lines of *emic* leadership vs. *etic* leadership—members of a society see a leader has having characteristics even when her or she does not, while ignoring other characteristics because of the "emic unit" of president, fireman, nurse, and so on. Changing the *emics* of roles—*nurse = female* to *nurse = either male or female*—results from changes in the nature and later the perception of the role.

In the study of a cultural grammar, we need to know how our theory should help us understand the constituents of a culture. One example might be the study of the leadership role of president. We discover, let us say, that a president signs documents, makes speeches, talks to people, and gives orders. But so does just about every modern professional. So these are etic activities initially in our analysis. Yet if we watch many different individuals engage in etically similar (or even identical) activities, the question we need to answer is, what meaning underlies the actions that distinguish the role in question (e.g., the US Presidency) from other roles? Some questions that we might ask as we try to understand the insider, emic perspective are: When does the president do these things? What does the president sign? What are the contexts and contents of the president's speeches? Who are the addressees of the president's letters and speeches? What social authority underlies the president's activities?

The president signs laws passed by Congress. The president speaks on matters of local, state, federal, and global significance. His or her words influence others because of the office he or she holds. The president addresses—qua president—individuals, groups, and organizations. Holders of other roles have no societal sanction to do as the president does and the meanings of their actions, even when similar, are different. OK so far, but how does a specific role and set of behaviors arise in the ranked-value, grammatical view of culture? To see this, let's consider a harder problem.

This problem was suggested to me by Dr. Eugenie Stapert (pers. comm.) from her ongoing research. How do members of a speech community distinguish a native speaker from a nonnative speaker? Though there is no single sufficient condition, categories that could distinguish etic and emic perspectives could include the following: discourse and use of gestures; constructions and lexemes; conversation turn-taking conventions; semantic fields; prosodic and gestural guides to the pragmatic organization of speech; body orientation; pronunciation; use of jargon and slang; or clothes. (Stivers et al. 2009). Etically a speaker may manifest these features. But emically, one only counts as a native speaker if one manifests these features in *meaningful* ways—using the appropriate etics in native contexts with native intentions, to show emic grasp. This requires participant observation and distributional analysis, along the lines of linguistic fieldwork (Sakel and Everett 2012).

The definition of culture offered earlier should help us understand the multitude of manifestations of communities living culturally: patterns of behavior, social roles, institutions, tools, museums, and so on. It follows from my definition, "culture-with-a-small-*c*" underlies "Culture-with-a-capital-*C*." An example that stands out in this regard is my favorite museum, the majestic Rijksmuseum in Amsterdam, the Netherlands.

The Rijksmuseum was founded in 1800 in the capital of the Netherlands, the Hague. Eight years later it was moved to Holland's most beautiful and important city, Amsterdam. The state museum, like many others around the world, was founded to reflect the values (shown in particular histories illustrated, artists selected, accomplishments highlighted, etc.) and the wealth (another value) of the nation.

Cultural values ooze from its walls. Each of the one million or so objects owned by the museum reflects a choice motivated by a value of some subgroup of Dutch culture. And of that million, each of the eight thousand or so objects actually selected for display owes its place in the public space of the Rijksmuseum to a choice that at least in some cases reflects the value perceived by the selection subgroup as reflecting the values of the entire community of the Netherlands.

The art library in the Rijksmuseum, the display of the country's naval history, the Rembrandts and Vermeers and Van Goghs—all of these selections help create Culture-with-a-capital-*C*, (or what some might refer to as *Kultur*). The expression of the values of a people. The literature that becomes popular likewise. And music, and so forth.

But in examining the components of culture or Culture, we must not forget that culture's "bits" can be shared across a diversity of networks. The American rock 'n' roll or rap played on a Chinese radio station may reflect a cultural network of teens around the world rather than merely a component of American culture. In fact, I remember hearing teens on the BBC years ago express the sentiment that they feel closer to people who like the music they like than to people who live in the same country, share a religion, or speak the same language.

If we examine art within the Rijksmuseum, we see further examples of the enveloping cultural network. The most famous work of art in the Rijksmuseum is Rembrandt van Rijn's imposing painting *The Night Watch* (or more accurately, *The Company of Captain Frans Banning Cocq and Lieutenant Willem van Ruytenburch Preparing to March Out*). This painting dates to the golden age of the Netherlands, and was completed in 1642. The painting is—like so many others—imbued with symbolism relevant to its home community's values, knowledge, and roles. In the painting we see values of Dutch culture of that time, in the very idea of a "watch" (vigilance against invasion and crime); the ideas of virtue and noblesse oblige of the wealthy; and the larger need for vigilance in all areas of one's life, the circumspection of the "good Christian."

Rembrandt's work was enthusiastically received, but his reputation also declined precipitously in his lifetime. The rise in popularity of painters such as Anthony van Dyck, whose work contrasted so strongly in color and light with Rembrandt's (his were much brighter, much more upbeat than Rembrandt's) adversely affected the reception of Rembrandt's work, as the newer work reflected changing values and standards of beauty. Since Rembrandt's death, of course, he has been recognized as one of (if not the) greatest of all Dutch masters, based on more recent changes values. And yet those who attend to the connection between culture and dark matter realize that change in any artist's reception or reputation reflects the shifting values of those who behold the art rather than their art alone. In fact, the reader or looker or hearer's culture is the most significant factor in success of the artist (or inventor, or CEO, etc.). The artist who can best read or predict those values and their shifting rankings among their audience—and adapt to them—will enjoy the lion's share of success.

THINKING IS CULTURE-BASED

Neuroscientist Gary Marcus, in an op-ed piece in the *New York Times* (Marcus 2015), argues that everyone should "Face it, Your Brain Is a Computer." He spends the first part of his brief piece exposing weaknesses in arguments that the brain is not a computer. He summarizes a few of the arguments against brains being computers in the following:

1. Some people claim that "Brains are parallel, computers are serial," therefore, brains are not computers. But Marcus correctly responds to the effect that "No, that's wrong. Computers can be serial or parallel."
2. He points out that others say things like, "Brains are analog, computers are digital." But analog devices—for example, a vinyl record—work on a "smooth continuum," whereas the digital is discrete; therefore brains are not digital computers. But again Marcus responds correctly that this analog argument also fails. Computers can be either analog or digital or both. In fact, many early computers were analog.
3. Another argument raised against brains being digital computers is: "Brains generate emotions, computers do not." Marcus tries to answer this criti-

> cism as well, by claiming that emotions are simply information transfer and so computers will be able to do this in the future. What they already do is just like this.

Here, however, Marcus is incorrect. Emotions are not simply the transfer of information. They are largely the transfer of hormones (and if hormones are merely information, every organ in the body is a computer). And emotions are partially constrained and wholly defined by culture. Computers lack hormones and culture, so they cannot have emotions, period. Again, we could call hormones "information," but that would immeasurably weaken the claim that brains are computers. In fact, I would go so far as to say that until a computer can urinate, be scared spitless, or need a change of underpants, it has no emotions. Being scared spitless is not merely transferring information—it is *experiencing fear*. It is being in a state that a computer cannot be in.

But why does Marcus even want to join the hyperbolic salesmen of artificial intelligence and claim that brains are computers? He states that "the real payoff in subscribing to the idea of a brain as a computer would come from using that idea to profitably guide research." He argues that this is particularly useful when examining the special computer known as a "field programmable gate array." Why does he think that research on computers profits from being compared to brains? Practically, the benefit doesn't emerge clearly. Theoretically, on the other hand, Marcus is a subscriber to the (neurologically mistaken) view of the brain as organized into distinct "modules" (an idea that I find as convincing as the notion that I have a "boating module" of my brain).

He therefore claims that a *field programmable gate array* could be like the brain, or vice versa, so we should investigate this. A field programmable gate array (FPGA) is mainly an array of logical units that may be programmed via wiring the units together in different configurations, such that they are able to perform all sorts of functions. There are many advances in this technology and FPGAs have taken on a wide array of functions. Their special quality is their ability to reconfigure themselves such that they can be said to have a certain amount of flexibility. On the other hand, the use of FPGAs as models of the brain escapes none of the criticisms of Dreyfus (1994) and others, and computers simply transfer and receive and remember information. They do not have, for instance, apperceptions, incorrect memories, justifications for mistakes, emotional overrides of computational functions, or culture-based languages.

Moreover, if information is what makes the brain a computer and if we are willing to interpret the wet stuff, blood flow, electric currents of the brain, and so on, as information, then the kidney or the heart or the penis is also a digital computer. There simply is no evidence from biology or behavior that the brain is like a digital computer except in ways so superficial as to render a rock a digital computer (it is always in a single digital state).

In sum, the overblown claims that still persist that the brain is a computer are at once too strong and too weak. They are too strong because if the brain is a computer, so are the body as a whole, the liver, and the kidneys. The claims are too weak because they omit crucial components of the brain, such as its organ-like exchanges with the rest of the body, as well as apperceptions, muscle memory, and especially, cultural perspectives that can shape both the brain and the body. They also focus on the syntax of brain operations and omit semantic operations. In fact, the semantic problem of brains—pointed out in Dreyfus (1965, 1994) , Searle (1980b), and others—has never been solved (and is unlikely to be until computers can acquire culture; see D. Everett 2015).

Our perceptions and the full range of our thinking is shaped significantly by our cultural network. This observation leads me to a consideration of claims for thinking in which the dualism of Descartes and the mind as computer of Turing form the core of ideas of the cognitive—my first criticism of the research program of artificial intelligence. So long as we refer to AI as *artificial* intelligence, I agree that it is an interesting and extremely important research program. Unfortunately, many of its proponents (Newell and Simon 1958; McCarthy 1979; Marcus 2015) want to drop this qualifier.

Following on Simon's (1962, 1990, 1991, 1996) and Newell and Simon (1958), long-term research program on understanding and modeling human problem-solving, the authors discuss the automatization of (at least parts of) the process of scientific discovery. They make three fundamental assumptions at the outset. These assumptions are worth considering in some detail because they illustrate the contrast between the noncultural reasoning of the artificial intelligence community vs. my thesis that intelligent agents are cultural agents. To my way of thinking, comparing artificial intelligence to natural intelligence is conceptually similar to comparing the flight of a Boeing 747 with the flight of a bumblebee. There will of course be general physical principles of flight that will apply to both. Yet there is no interesting sense in which the Boeing 747 teaches us how the bumblebee flies qua bumblebee or how its flight "feels," nor how its flight is shaped by those it is flying with. Nagel (1974) makes a case for some of what I have in mind with his arguments that consciousness is both specific to a particular organism and gestalt in its emergence from smaller parts.

Langley and his coauthors (1987, 8) list three principal assumptions that guide their investigation. First:

> The human brain is an information-processing system whose memories hold interrelated symbol structures and whose sensory and motor connections receive encoded symbol structures from the outside via sensory organs and send encoded symbols to motor organs.

Second,

> The brain solves problems by creating a symbolic representation of the problem (called the problem space) that is capable of expressing initial, intermediate, and final problem solutions as well as the whole range of concepts employed in the solution process.

Third,

> The search for a problem solution is not carried on by random trial and error but is selective. It is guided in the direction of a goal situation (or symbolic expressions describing a goal) by rules of thumb called heuristics.

What these authors are trying to understand, of course, is the *dark matter* underlying scientific discovery. Describing the process does not imply that there is either a normative theory of scientific discovery for humans, nor that the heuristics of scientific discovery can be transformed into algorithms.[8]

This work was important, and Simon's own pioneering understanding of human problem solving won him the Nobel Prize in Economics in 1978. Nevertheless, the points above are problematic. For example, one serious question is raised by what the authors intend by the phrase "problem solution," and the relevance of their focus of scientific discovery to human cognition more generally. Not much, in fact. One could respond that their decision of what interests them hardly counts against their account. At the same time, the unlikelihood that their solutions will scale up to human reasoning render them too specialized for any general account of human psychology. The examples they select, for example, are tailored to solutions via symbolic processes. They nowhere address the wider range of problems human must solve, including some of the most important ones of life, such as "How should I live," "Who should I marry," "What should I believe," "How can I learn to square dance," "Who should I vote for," and so on. These deeper problems involve physiology, culture, emotions, and a myriad of other nonsymbolic aspects of human cognition. To be sure, many of these other problems have large informational-symbolic components, but what makes them hard and vital are precisely their non-informational components. And there are larger problems. For example, even the discovery procedures Simon et al. are interested in are cultural constructs. Therefore, I do not find strong AI to have become more relevant to the understanding of human thinking than it has ever been, in the more than three decades since the devastating criticisms of Dreyfus (1965, 1994) and Searle (1980b).

Since the earliest days of artificial intelligence, eminent proponents of the idea that brains are computers have proposed, often quite emotionally, that of course machines can think. McCarthy (1979, 1) says the following, for example: "To ascribe certain *beliefs*, *knowledge*, *free will*, *intentions*, *consciousness*, *abilities*, or *wants* to a machine or computer program [emphasis in original] is legitimate when such an ascription expresses *the same information* [emphasis mine] about the machine that it expresses about a person."

Asking the question of why anyone would be interested in ascribing mental qualities to machines, McCarthy (1979, 5ff) offers the following reasons (which I have paraphrased slightly):

1. We may not be able to directly observe the inner state of a machine so, as we do for people whose inner states we also cannot see, we take a shortcut and simply attribute beliefs to predict what the computer will do next.
2. It is easier to ascribe beliefs to a computer program—independent of the machine that is running it—that can predict the program's behavior, than to try to fully understand all the details of the program's interactions with its environment.
3. Ascribing beliefs may allow generalizations that merely simulating the program's behavior would not.
4. The belief and goal structures we ascribe to a program may be easier to understand than the details of the program are.
5. The belief and goal structures are likely to be closer to the objectives of the program designer than the program's listing is.
6. Comparing programs is perhaps better expressed via belief-attribution than merely comparing listings.

But these are each built on both a faulty understanding of beliefs and a faulty understanding of dark matter more generally—at least, if we are on the right track in this book about dark matter. Moreover, this type of personification of computers is too powerful—it can be extended in humorous, but no less valid, ways to circumstances no one would ascribe beliefs to: thermostats, toes, plants, rocks teetering on the edge of a precipice. In fact, there are many cultures—the Pirahãs and Wari's, for example—in which beliefs are regularly ascribed to animals, to clouds, to trees, and so on, as both a convenient way of talking (like McCarthy) and a potentially religious or pantheistic allusion to deity in all objects.

Beliefs are intentional states (Searle 1983) that occur when bodies (including brains) are directed toward something, from an idea to a plant. Beliefs are formed by the individual as he or she engages in languaging and culturing, becoming culturally articulated components of individual dark matter.

People can talk about some of their beliefs and the qualia of their beliefs. The beliefs of humans can be logical or illogical, consistent or inconsistent, with other beliefs. They are partially ordered, as values are. For example, in the statements "I believe in science" vs. "I believe in God," the ranking can be either GOD >> SCIENCE or SCIENCE >> GOD. If it were the case that programs could be meaningfully described as McCarthy claims, they would have to be able to rank their own beliefs, not take a shortcut by having someone else program their beliefs.

To the degree that programs can be described as McCarthy claims, they do not *rank* their "beliefs," and they lack "values"—unless, of course, certain states describable as "beliefs" or "values" have been either (i) directly programmed by a human programmer or (ii) indirectly programmed (e.g., algorithms that allow for multiple responses or "beliefs"). Moreover, by omitting any role for culture, biology, or psychology in computers, belief-ascription is nothing more than a metaphor in a way that it is not for humans. Belief ascription in humans is not merely an epistemological convenience, or façon de parler; it is retrievable psychologically, culturally, and biologically from the individual and their society without need to assume an external designer.

Dreyfus (1965) rightly rejects such claims for computers, based on several theses (from Haugeland's [1998] summary): (i) intelligence is *essentially* embodied; (ii) intelligent beings are *essentially* situated in the world; (iii) the world of intelligent beings is *essentially* human (thus I believe that Dreyfus has culture in mind when he says "world").

Before closing this chapter, there are a couple of issues that need to be examined to further clarify the concept of culture being developed. Again, the claim is not that dark matter is culture but that culturing produces dark matter, which includes "idioculture" as one of its components; that is, the value rankings, knowledge structures, and social roles manifested or possessed uniquely by a given individual, just as a "language" is found only in individual idiolects and only used by societies via individuals.

One of those issues in need of understanding is the role and emergence of tools. How are we to characterize tools culturally—things that are used to aid individual cultural members in different tasks? Tools are dripping with dark matter and culture. I conceive of them as *congealed culture*. Examples include physical tools—shovels, paintings, hats, pens, plates, food—but nonphysical tools are also crucial. D. Everett (2012a) makes the case that a principal non-physical tool is language (and its components). Culture itself is a tool.

The tool-like nature of language can be seen easily in its texts. Texts (discourses, stories, etc.) are used to exhort, to explain, to describe, and so on, and each text is embedded in a context of dark matter. Texts, including books, are of course unlike physical tools in the sense that as linguistic devices, they could in principle tell us something about the dark matter from which they partially emerge, though generally very little is conveyed. And the reason for that is clear: *we talk about what we assume our interlocutor does not know* (but has the necessary background knowledge to understand). And dark matter, which we do not always even know that we know, is simply overlooked or presupposed.

Language as a tool is also seen in the *forms* of texts. Consider in this regard once again the list of so-called contradictory principles that Harris provided above with regard to the Hindu principle "Avoid fecal matter."

> A spot must be found not too far from the house.
> The spot must provide protection against being seen.
> It must offer an opportunity to see anyone approaching.
> It should be near a source of water for washing.
> It should be upwind of unpleasant odors.
> It must not be in a field with growing crops.

The first line uses the indefinite article *a*. In the second line the definite article *the* is used. From that point onward, *spot* is pronominalized as *it*. This is because of English conventions for topic tracking (Givón 1983) through a discourse. The indefinite article indicates that the noun it modifies is new information. The definite shows us that it is shared information. The pronoun reveals that it is topical. As the single word is referenced and re-referenced throughout the discourse, its changing role and relationship to shared knowledge is marked with specific grammatical devices. This is shared but unspoken, and largely ineffable knowledge to the nonspecialist.

How does the understanding of culture promoted here compare to the wider understanding of culture in a society as a whole? It is common, for example, to hear about "American culture," "Western values," or even "pan-human values." According to the theory of dark matter and culture developed above, these are perfectly sensible ideas, so long as we interpret them to mean "overlapping values, rankings, roles, and knowledge," rather than a complete homogeneity of (any notion of) culture throughout a given population. From laws to pronunciation, from architecture to music, to sexual positions and body shape, to the action of individual humans as members of communities ("likers of Beethoven," "eaters of haggis," and on and on) in conjunction with an individual's apperceptions and episodic memory—all are the products of overlapping dark matters.

A similar question arises as to whether it makes sense to speak of "national values," and if so, how these might arise. Again, such values are values shared by a significant number of members of a nation, due to similar experiences. For example, during childhood, people of my generation had three television stations to choose from. There were few if any fast-food chains. The food available in supermarkets across the US was fairly constant. Thus a person growing up in Indiana would have very similar experiences with cultural products as a person growing up in Southern California (as my wife and I have discovered by comparing stories). This has changed dramatically over the last fifty years, but still there are a number of things shared in common—uploading videos to YouTube, reading news online, using smartphones and apps, and so on. Such experiences trigger the formation of similar dark matters across large numbers of individuals. They trigger them even at the level of an entire country, continent, or the world of television viewers, Internet access, and so on.

Just so, values can produce in an individual or in a community a sense of *mission*—for example, the Boers, the Zionists, the American frontiersmen and settlers in *Manifest Destiny*. The Third Reich. This sense of mission and purpose is what many businesses are after today as the use of the term *culture* has been adopted by companies as "what we are all about." Businesses commonly dedicate web pages, documents, lectures, meetings, and so on, to establishing a sense of culture (which is often little more than an unranked list of values and occasionally goals—as though these were separate from values—with no discussion of knowledge or roles).[9] Though these business-co-opted ideas of culture differ significantly from some academic understandings, they are nevertheless not terribly wide of the mark, even if they represent an unarticulated subset of the idea of culture that I am urging here.

When we look at individual businesses, the ethnography of commerce more generally, the anthropology of money, or any other area of business from a cultural perspective, there are various possible approaches we see, from diachronic to synchronic, from symbols to actions, patterns, practices, and so on. But the fundamental question in the study of business from my perspective is just the same as the fundamental cultural question for any domain: What are the emic units, the etic units, and process of emicization?

While people can be expected to share values regardless of where they are found, this should not be overstated. It is of course true that as biological entities, we paint our lives from a palette of overlapping "colors" (e.g., the similarity of the problems we face or our biological resources). Yet if I am correct, our differences are far more profound than our similarities. The dark matter that includes the ineffable, the unspoken, the not-often spoken, and the hard-to-say is a product of a specific place, time, individual psychology, apperceptions, memory, and culture. In this sense, my theory agrees with Edward T. Hall's "silent language." As he put it so perspicaciously: "What is most difficult to accept is the fact that our own cultural patterns are literally unique, and therefore they are *not* universal" (Hall [1959] 1973, viii; emphasis in the original).

Although there are most certainly general principles of human behavior and the formation of dark matter, the combinations of individual apperceptions with exposure to mere subsets of larger value, knowledge, and role networks means that no two people will be exactly alike in any way. And certainly no two cultures will be.

Other examples of dark matter are useful and easy to find. So consider the important things we rarely say. How do you stop at an intersection? How do you aspirate a consonant? Why do we recreate the spatial configuration of events by their position relative to the anterior portion of our bodies rather than global directions such as north and south? When I went to the Amazon to begin field research, I wanted to work on texts that would simultaneously teach me about the Pirahãs' language and culture. I did not realize at the time that there is in fact no other kind of text. All texts reveal culture and language. Still, some aspects of culture were more obvious to me as a beginner so I began with arrow making. It turns out that this is knowledge all male Pirahãs share; knowledge they rarely if ever talk about; but knowledge they obviously can talk about. In fact, the first text I ever collected (see below) was on arrow making. As my years among them went by, I became fascinated by this topic-knowledge transmission without language.

There are better examples of knowledge that is unspoken, though. Nonhuman animals present superior examples in some ways. These animals have beliefs, desires, and emotions, and learn complicated behaviors and ways of interacting with the world. Yet they lack language altogether and so, by definition, cannot talk about their knowledge. Almost all nonhuman animal knowledge is therefore dark matter. Most people wave their hands at these fascinating phenomena, sweeping them all under the label of "instincts" rather than knowledge.

Dogs, humans, and other animals go through an attachment period, are driven by emotions, learn tricks, learn to obey a range of commands, come to sense ownership/relationship/belonging to certain items in its environment, and so on. My 140-pound Fila Brasileiro, for example, barks when even slight changes are added to the environment—a stack of books in a strange place, cushions from the sofa piled for cleaning, a new car in the driveway, and so on. While my dog cannot "tell" me about this in English, through her barking and body posture she communicates relatively well, though many of her actual feelings remain ineffable. Her dark matter in this sense has both "communicable" (via actions and barking) and ineffable components, just as humans' does.

Humans have non-informative labels for many of the things we know ineffably. Similarly to the way that we label some animal knowledge as "instinct," we often refer to ineffability of human abilities as "talent." For example, there are creative writing courses at many universities. But no one can teach you to write *quality* lyrics, novels, creative nonfiction, music, visual art, and so on. You can improve through instruction. But you cannot "arrive" solely by instruction. Someone might be able to teach you how to write in a style *like* Tom Wolfe or to compose *like* Bach, but you cannot be taught how to write *as well as* Wolfe, compose *as well as* Bach, paint *as well as* Van Gogh.

There are several reasons for the ineffability of talent. First, talent is to some degree a social ascription. When we say that "She has got *it*," that *it* is a social appeal in a limited temporal context, along the lines of the formation of a band in the Springsteen epigram at the beginning of this book. Second, talent sits near the middle of the knowledge-that and knowledge-how continuum. We can talk about style and analyze art, but there is still what Dreyfus called "know-how." Third, talent is not a cause but an *effect*. Writing a novel is an illocutionary act, but producing a novel that is considered brilliant requires a perlocutionary effect. The same goes for dancing well. Or any other ability. The society must sanction them and *agree that you are good at them*. Talent is often associated with seeing things differently than others of one's culture, of being "on the cutting edge." One way to conceive of this in my terms is that to have talent (or being a crackpot) or creativity is to think at the edge of the emic: being an outlier of—though still within—one's own cultural system.

Another form of ineffable tacit knowledge is how we gesture to accompany spoken speech. We can describe (see chap. 7 for an in-depth discussion) gestures scientifically and study them. We can distinguish etic from emic gestures in particular cultures. But speakers themselves cannot tell us why they make certain gestures, when they make them relative to the speech stream, why some are used repeatedly throughout a story or conversation while others may emerge only once.

Exposure to overt knowledge in other contexts related to dark matter in our more familiar contexts becomes enormously informative if we can relate the two (which is nontrivial). For example, I really only began to understand my native English grammar when I first studied Spanish in sixth grade. When any person begins to probe for alternatives to their customary behavior, the diffuse light of their intellectual torch shines back on what they left behind, however momentarily—their path as they walked just before "here" and farther before the "there" ahead. Just as *hablar* taught me that "to speak" is a verb, so *tortilla* helped me to understand the function of "bread." In the many years I lived among peoples of the Brazilian Amazon, especially among the Pirahãs, I saw the norms I had left behind in ever-brightening bas-relief. As my relationships changed to include a wife who loves animals, animals entered my beliefs, desires, values, and thoughts in the heretofore unexperienced role of friends and companions. This intrusion of animals has illuminated my relationships and thinking about humans. With each person I learn to like, love, or dislike, I learn more about those who already populate my past and present.

The desire to understand the various levels of human knowledge—from the explicit to the deeply embedded muscle memory of our flesh—is hardly new with me. This is *the problem* of human existences. One can talk about this knowledge as a psychologist, as a shaman, as a philosopher or a jurist, an anthropologist or a linguist, a minister, an imam, a chef, or a novelist. The crucial step is to recognize one's vantage point, to the degree that this is possible in a rain forest carpeted with trees that block out one's narrow horizons, or on a busy street valleyed between skyscrapers, or an open field from which no other reference point is available for miles.

On the other hand, we each begin from the same vantage point—our self. But what is this self I look out from upon the foreground and background of my visual field? How does this self come to be? In a delicious irony, determining the nature of our self-centered vantage point can answer the most profound questions of the "other," as the other can answer the most profound questions about me. Anthropology begins simultaneously with me and them.

The knowledge that makes us who we are is part body and part culture-mind, where "mind" is a way of talking about neurological development and "culture" is way of talking about the societal contribution to that neuromental development—procedures our brains go to first out of habit, in turn determined to large extent by the society in which the brain is found. For each member of a society, there will be a group with which he or she identifies. In fact, there will be several—some overlapping, some without any point of connection other than the societal member in question. Each human is a social nexus.

There is a Brazilian expression, *Se der bolo eu tiro meu corpo daí!* "If it gives cake, I take my body out!" I love this because literally interpreted, it is impossibly opaque in English. The first time I heard it, it was completely opaque to me. The meaning derives from the perceived complexity of cake baking, especially among those whose literacy skills make it difficult to follow recipes closely, rendering the final product lottery-like. But even for those who can follow recipes easily, a cake is a very complex item. The same flour that makes chapatis and gravy can make German chocolate cake. This humans-as-cakes metaphor, or humans-as-any-complex-mix concept, seems more useful to me than the "man in a can" view of the nativists.[10] Sure, parts of us are innate. But the real action is in the variation of combinations and connections formed by our movements through the social world that make us each unique in nontrivial ways. This mixing, though, cannot be seen directly. The invisible matter of our brains is crucial in the composition of our selves and natures.

Human psychology and human culture construct one another, as Sapir argued more than eighty years ago. On the other hand, the work here differs significantly from Sapir's in that it addresses not only the biological and cognitive "platforms" (see D. Everett 2012a) upon which language, culture, and interactions are built, but also the ways in which our selves and the very construct of human nature emerge from this interaction over the generations.

NORMS AND CONVENTIONS

By-products of culture that play out in texts and out of texts are norms and conventions (simplifying, norms are conventions with moral expectations). Considering norms, for example: What are these, and how are they enforced in any culture? Norms are not merely statistical regularities (Brennan et al. 2013), though even the statistical regularities of behavior need to be sorted out, because some of them will reflect the culture (such as Pirahã posture) but others will not (such as Pirahã canoe use vs. canoe building—they do not build the kinds of canoes that Brazilians make, even though these are the principal canoes that they use). Canoes reflect Pirahã economy and relations with the outside world. But limited (i.e., only a couple of hours per day) canoe accessibility for men is not a norm. It is a function of a shortage. On the other hand, there are canoe norms—who can use a canoe, which canoe, where, when and how, and so on.

The idea that our own values, knowledge structures, and social roles enable us to interpret the world has been around for a while. Davidson (2004) and Searle (1980a, 1980b) refer to this cultural knowledge as "the background" (though this term has a more technical definition in Searle's work). Clifford Geertz, however, is the one whose work is most associated with the view of culture that I have in mind. For Geertz, culture is our way of imputing meanings to the world, our means to interpret our experiences.

We have been asking from various perspectives how the hermeneutic role of culture illumines the idea that there is a dark matter of the mind. To take yet another tack in answering this question, it is useful to address one more aspect of culture that has not been referred to yet in this exposition, namely, *conventions*. Conventions are signs, behaviors, reactions, and so on, that emerge from dark matter. They are identifiable behaviors of groups, not individuals alone, and are useful tools to extract meaning, predict behavior, reduce decisional complexity, and make us "feel at home." They are the cues that give us comfort, the idea that we know what is being done and what we should do next. One example of a convention is phatic language.

We pass each other in the hall. I say, "Hi. How are you?" If you start to tell me about your hernia or that you had a bad night's sleep, you do not understand the conventionality of phatic language. The purpose of conventional phatic statements like "Hi. How are you?" is not to elicit information. They take the syntactic forms of questions on occasion but are culturally the equivalent of "grooming." Yes, they do take the form of questions and so some people will affect that they should answer them as such. But the form of phatic language as statements, questions, and so on, is an example of homopraxis—two acts that share a common form. In some cases, "How are you?" is intended to be an actual question. In others it is not. In those cases, the phatic cases, the point is to recognize you, similar in some ways to grooming in primate social groups. Some cultures, such as the Pirahãs', lack phatic language. It is a matter of cultural convention. And in the case of American queuing, the queue is also a norm—it is considered immoral by some to fail to queue, and there are informal sanctions for failure to do so (e.g., "Get to the back of the line!" "Who do they think they are?" "I cannot serve you, sir, because you have cut in line.")

Still other conventions include queueing. In an American store, for example, no matter how crowded, most people will, without being instructed, form a queue in front of the cash register. In some countries, without rigorous enforcement, such queuing will not occur—everyone will crowd around the cash register hoping to get waited on first. Queuing is thus a convention of some cultures but not others. And, as with all conventions, when we experience another culture, we will always be bothered by the absence of our culture's conventions in the other culture. The reason is that conventions make life easier by requiring fewer decisions, by bringing a sense of the familiar to the foreign.

Formulaic expressions—again, such as "Gezundheit" following another's sneeze; using irony to defuse serious or sad situations; and expected responses to jokes (laughter), meals ("That was excellent"), and so on—are conventions. Though, due to dark matter, cultural members may see conventions as "natural," they are in fact cultural.

Among the leaders in reflection on the nature of conventions are Lewis (2002) and Millikan (1998). According to Millikan, there are requirements for something to be a convention that are often overlooked or misunderstood in other work, even work as influential as Lewis (2002). (And, closely related is work on "norms," which, like conventions, also follow from ranked values, knowledge structures, and roles. See Brennan et al. 2013.) Millikan's list of the components of convention includes (1)–(8) below:

1. *Reproduction.* Structures of languages are reproduced, not by everyone, but recognizable by all. Language structures are thus one example of the conventional. The phatic language just mentioned is an example of this. But so is grammar. To say, for example, "the good ol' red, white, and blue," rather than "the good ol' white, blue, and red," or even "good the blue, red, white, ol'" is conventional.
2. *Weight of precedent.* Conventions have little tendency to appear in the absence of precedent. This is another manifestation of the actuation problem that we mentioned earlier. Saying "'ta dum" after a joke, or rolling one's eyes after an inane statement, are conventions. But they do not begin to be conventions until they have been so established by precedent, by a first exchange.
3. *Coordination conventions.* Some conventions help us coordinate activities, thoughts, interpretations, and so forth. Queuing is one such example. Others including raising one's hand in a meeting to "have the floor," and so on.
4. *Conventions do not imply regular conformity.* Sometimes in meetings people talk out of turn, even though everyone knows that it is expected to raise your hand and be recognized by the leader of the meeting first. Conventions are themselves violable constraints.
5. *People want, expect, and seek conventions.* In a waiting room, you place your hat on a chair to signal conventionally that you are reserving that chair for someone and that no one else should sit in it. Since conventions are neither inviolable nor hold any special legal status, you would not be arrested if you picked up someone else's hat, handed it to her, and sat in that chair. After all, who or what gives her the right to save a seat? Convention does. There are two conventions at play here. The first is "first come, first served." If you see an empty seat and sit in it, someone is not "allowed" to come along later and try to move you out of it. But in the case of the seat-saving convention, you have in effect "first come, two served."

We allow this behavior and in effect depend on it, expect it, and look for it, because it brings order to circumstances that would otherwise be unorderly. Thus conventions can, in the right circumstances, develop into norms.

6. *Conventions are not prescriptive rules.* There is, of course, no rule written anywhere or memorized that says, "If you arrive first, you may save as many seats in a row of seats as you like." And in fact, some people may sit in seats you are trying to reserve because there is no precise rule that tells you how many seats you can reserve at a time. However, in general, when we see someone behaving—even at the extremes—in accordance with recognized conventions, we avoid conflict and let them have their way via convention.
7. *Conventions crisscross.* I refer to conventions that crisscross as *homopraxes* (deriving from *homo-*; *homo*nyms, *homo*phones, etc.). Homopraxes can be teased apart only by emic analysis; two conventions that have the same etic form but different meanings—such as raising your hand in a classroom and offering a Nazi salute—can be distinguished only by emic understanding, dark matter.
8. *Conventions produce perlocutionary effects.* Imagine someone telling a story of a murdered loved one. In the US the expected perlocutionary effect is to express sadness and solidarity with the storyteller. But in other societies, such as among the Pirahãs, the expected response would be laughter, a convention that shows that one should not take things too seriously, that one has not lost face, and so on.

That is one view of culture and its symbiosis with dark matter. In the next chapter, I explore the ontogeny of culture and dark matter. How in particular does the child negotiate the transition from alien to insider?

The Cultural vs. Natural Basis of Roles

The roles that contribute to the formation of culture are not always obvious, nor do they always map well across cultures. For example, consider the *kagi* role among the Pirahãs. I first heard this word when a dog trotted into one part of the village from another with a small squirrel monkey holding on to its fur, riding it. The Pirahãs said that the monkey was "*kagi apoo.*" I knew that *apoo* meant "on" or "on top of," but I wasn't at all sure what *kagi* meant. I next heard this exact expression in a description of beans and rice, with the beans on top of the rice—the beans were *kagi apoo*. Then once when Pirahã men were talking about sex, they described the man on top of the woman as *kagi apoo*.

This all seemed quite mysterious. I then learned that my family is my *kagi*.

One brother with another brother is with his *kagi*. What could this possibly mean? Well, it turned out to mean "expected associate." Rice is the expected associate of beans. Wife is the expected associate of husband or woman of man. One brother is the expected associate of the other. When a monkey mounts a dog for a ride, the dog has become its "expected associate." Thus "expected associate" is contextually determined, and it refers to a specific relationship in a particular situation, not to a single entity. To understand this word, I needed to understand the relationships that the Pirahãs understood and to understand the expected partnerships as the Pirahãs understand them. Only in this way can one understand why the same expression is used for the act of coitus, a monkey on a dog, and beans on rice, or how *kagi* can refer to brother, rice, wife, dog, and so on.

The upshot for our construction of a new approach to the nature of culture and its relationship to individual psychology is that social roles emerge from particular cultural relationships, perspectives, and knowledge structures. The Pirahã notion of "expected associate" is more abstract perhaps than the English word *partner*, though even the latter word can itself look quite abstract indeed to someone learning English for the first time (e.g., one who works with you, someone who lives with you, a casual acquaintance of a cowboy). Participant observation in a society and culture is indispensable to an understanding of social roles, as it is for values and knowledge structures.

CONCEPTUAL TOOLS IN THE CREATION OF CULTURE

Many anthropologists over the years, such as Levi-Strauss ([1949] 1969), Leslie White (1949), Chagnon (1984), Descola (2013), Marvin Harris (2006), among others, have attempted to develop theories of culture that go far beyond the notions of values, social roles, and hierarchical knowledge that I appeal to here. They have used organizing principles such as animism, totemism, universal mythology, and principles of analogy to suggest universal principles of culture. These principles might be universal, according to the hypothesis, because they are innate or because they are ancient, when there was only one or a few cultures. Or they could exist because of concepts shared in culture areas. Such principles are hypothesized to constrain our normal cognitive abilities, our reason, our emotions, and so on, such that we converge on these universals of culture.

Descola (2013), for example, offers a lovely-to-read, erudite, and highly empirical tour of people groups of the Americas, Asia, Australia, and so on, to offer one such sweeping scheme of generalizations. I want to consider some of his ideas here, in order to explain what I like and what I do not like about them. By this point, I suspect that what follows will not be terribly surprising to anyone, since the theory of dark matter taking shape here is largely unsympathetic to notions of universality that range beyond basic physiology—that is, universal systems of knowledge, other than those that are learned via experience and which are functionally so useful as to lead to convergence as natural solutions.

Descola's work is an attempt at a grand synthesis in the Bastian tradition, so far as I can tell. He attempts to classify all peoples of the world in terms of the four-way ontological distinction between animism, totemism, analogy, and naturalism. In this system, animism represents relations based on differences in physical manifestation but similarities in "interiorities"—our "spirits" or mental life or "souls," and so on. From this perspective, a tree and a man are very different on their exteriors, but both are inhabited by God or gods or souls. Their interiors are thus roughly the same. Totemism, on the other hand, is a way for addressing the world as similar in physical form and interior life, such that a wolf and a man are *like* each other internally and externally in some ways. Analogy, according to Descola (2013, 201ff), is "a mode of identification that divides up the whole collection of existing beings in to a multiplicity of essences, forms, and substances separated by small distinctions and sometimes arranged on a graduated scale so that it becomes possible to recompose the system of initial contrasts into a dense network of analogies that link together the intrinsic properties of the entities that are distinguished." We say, for example, that "in this respect, John is like a leaf," focusing on the comparison of specific traits in one creature and another.

Finally, there is to Descola the important ontological category of naturalism. Westerners are naturalists, in the sense that we see nature in opposition, as a background to ourselves, in contradistinction to ourselves and nature as parts of a single world, in which we move, reside, exchange places, and appear as different manifestations of thought without really being distinct entities.

From these four ontologies, Descola creates an interesting hermeneutics of the world, offering accounts of differences, similarities, behaviors, and mental lives across cultures. However, although Descola's work is obviously important, innovative, and far-reaching, if the theory of this book is on the right track, grand unifying schemes like his—based upon human knowledge that is specific and shared among all peoples—are on the wrong track. What is needed is a different kind of unifying scheme, based on the idea that there are no universal ideas except those that can be arrived at via languaging, culturing, and having human bodies.

Summary

This chapter developed a theory of culture, inspired in part by linguistics research, drawing on optimality theory, tagmemics' slot:filler concept, among others. A main question that the chapter attempted to answer was "Does each member of a culture, X, participate in a collective intention to 'live according to the values of culture X'?" The answer was that it is unnecessary to characterize all culture as shared, group intentions, but as confluence of shared backgrounds—exposure to similar value rankings, knowledge hierarchies, and social roles. Thus we answer the question "How is culture even possible?" by imitation, language, and—most importantly—emicization, the formation of subjective gestalts, via apperception, from objective experiences. A society is therefore not like a football team nor like an orchestra. Rather, the cultural cohesiveness of a society arises from overlapping values, roles, and knowledge of individuals that live together, eat together, think together, language together, and culture together. The result is that culture is the epitome of *e pluribus unum*. Culture is an emic gestalt in individuals.

The chapter developed a model of culture in which the individual is the bearer of culture and the repository of knowledge, rather than the society as a whole. It also presented a theory of values, which is epistemologically prior to a theory of culture.

The discussion of the importance of the individual to the notion of culture led us to an understanding of culture as the core of cognition, leading simultaneously to the conclusion that machines are incapable of thinking. The argument offered was that without culture, there can be no semantic understanding, no background, no dark matter in which thought can occur.

3

The Ontogenesis and Construction of Dark Matter

> The repeated claims of cultural psychologists and anthropologists, as well as evolutionary biologists, for decades to recognize contextual/cultural variation and systematically introduce it into attachment theory have been largely ignored.
>
> HEIDI KELLER, in *Different Faces of Attachment: Cultural Variations on a Universal Human Need*

In the previous chapter we laid out a theory of culture in which groups of people share values, knowledge structures, beliefs, and understandings of and participation in social roles and experiences—"culturing." As these various mind-body phenomena are incorporated into the apperceptions, memory, and muscle habits of individuals we can say that they "share a culture," though we know that no two people will share exactly the same value rankings, beliefs, and so on.

In this chapter, the goal is to understand how dark matter is acquired. Of course, cultural learning is a large area of study, and my own thinking has profited tremendously from work that has gone before.[1] In related work, there is a vast psychological literature on child development, much of which necessarily engages societal learning. And Bruner's body of work is pioneering, exemplary, and foundational.[2] There is also the work of discourse and language in cultural learning, at least indirectly.[3] And these works only scratch the surface of the literature.

However, my goal here is to describe in detail my own field research on cultural learning and then to extrapolate from that lessons that I believe to be most relevant for understanding of how humans construct/acquire the dark matter of their minds.

There are reasons why computers are unlikely to ever be able to think or talk, except in artificial ways. The main reason is the conglomeration of factors that make human thinking what it is: consciousness, emotions, apperceptions, dark matter, cognitive plasticity, culture, society, and physiology. Without these, there is no human thinking by robots.

What makes humans unique from other animals and machines is not computational ability. Other animals have tremendous problem-solving, navigational, emotional, and other cognitive ranges, as well as consciousness. Studies of canine cognition, for example, reveal that (unsurprising for any dog owner) dogs can think and reason across a variety of tasks.[4] In fact, one reason that humans and dogs have developed such a close relationship over the millennia is that each possesses mental and physical abilities that benefit and complement the other's. Canines and humans are similar in some ways emotionally; their reactions, needs, and so on, are based on very similar subcortical emotional centers (Panksepp and Biven 2012), thus their mutual contributions to each other's well-being produces strong attachment between them. Dogs can even become part of a culture in the sense of our discussion above and throughout this study, based on social roles, attachments, and their meanings (companion, ally, aid, etc.) to their associated humans. In fact, among the Pirahãs, dogs are referred to as *kagí*, the same word used for *spouse* or *family*. (This term actually means "partner" or "normally associated entity," so it can even refer to rice with beans, as explained in chap. 2.) The Pirahã expression *kagí ʔapoó*—"partner on top of"—can mean several things, including rice on beans, sexual intercourse, a pet monkey riding a pet dog, and so on. Here are what I consider to be the necessary requirements to develop dark matter:

1. A body—for muscle memory, tastes, sights, sounds, and the like.
2. "Culturing"—engaging in the normal array of practices of a community guided by similar dark matter. An enormous component of dark matter is knowledge derived from culturing, languaging, and otherwise behaving as a member of a particular community, engaged with other members of the community. Anything a dog knows, believes, tastes, values, and so on, is dark matter by definition, since it is in principle unspeakable for the dog (which is not to say always ineffable, because tail wagging, barking, and many other canine behaviors are forms of communication). Human dark matter, varying by culture, includes things they believe and that they don't know they believe (at least until stimulated to discuss it)—for example, that the arch can support a bridge, that the capricious extra payment by an employer to one colleague but not another is unfair, that a person can travel a thousand miles an hour without protection (on a rotating earth), or that it is disgusting to reassimilate body fluids. Dark matter also includes reasons for the uniformity of behavior.
3. A flexible brain—to make new associations, to innovate, to tie specific sets of experiences together, to be able to unify apperceptions, to learn by subception, to improvise, to find humor, to invent humor, and so on.
4. An emotional brain—emotions are necessary conditions for culture/dark matter, though not sufficient conditions by any means. Mr. Spock's Vulcan culture could not exist. Animals bond with humans and humans bond

with other humans, animals, carpets, guitars, food types, and the like, because of the interaction of emotions, cognition, and physiology. Without such emotional attachments and motivations, culturing and dark matter would not emerge; nor would its by-products, such as semantics.

5. Semantics—the study of the structures, qualia, and sources of meaning, which crucially distinguishes humans from machines in the use of language. But semantics is simply a by-product of dark matter, itself a by-product in part of acting and belonging in a community.
6. Human intelligence—it is not clear that possession of a rich dark matter is quantitatively different from animals' abilities, knowledge, and so on. In fact, I think that the Cartesian idea that only humans think or use tacit knowledge is a form of species-centrism, compounded by centuries of looking at cognition dualistically.

Attachment Theory

The creation of the individual-culture connection is known as *attachment*—the initial experience of relationship formation by infants. Years ago, a philosopher friend told me that he believed that you should give your children your undivided attention for the first two years of their lives. After that, their personalities and relationship to you are formed, and so the time you need to spend with them will diminish dramatically.

I suspect that this isn't far wrong. During their first couple of years of life, children, like other mammals, are forming crucial bonds with their primary caregivers and other humans and entities intruding into their senses—such as siblings, grandparents, pets, things around the house—as they construct their identities as individuals in a cultural network. Of course, although attachment happens in the first months or couple of years of life, interactions, observations, and so on, continue throughout one's life. Some of these additional foundational pillars for the formation of dark matter are the following.

While engaging with his or her community, a human learns both positive and negative lessons. Humans learn how to do things. We learn what unacceptable behavior is. We develop skill in tasks, including the use and source of their own bodies (in fishing, dancing, playing games, riding elevators, eating, sex, defecation, etc.). When a child transgresses an expectation, it may be scolded, slapped, or ostracized for a period, depending on the culture and other factors. When a child helps a fallen sibling, it may be complimented or given a candy or some other type of reward. Actions that conform to the values of the people we interact with will garner different responses than actions that violate those values.

We learn as we interact with others that our bodies are perceived as ugly, plain, beautiful, fat, skinny, strong, nicely colored, and so on. We will learn that we are perhaps weaker, stronger, faster, slower, taller, shorter, poorer, richer, whiter, darker. Our self-image will be formed by the perceptions of others' perceptions of us and our perceptions of others. This is the sociocultural influence on the psychology of the individual.

In our actions, physiological reactions, and so forth, we learn our limitations relative to the environment (e.g., withstanding temperatures and their fluctuations, killing or eating animals, planting fields, the relative value of different crops). Through rituals, culturing, conventions, practices, and other categories of activities, people learn meanings, illocutions, perlocutionary expectations, how to think, what level of effort to invest in different types of tasks, the nature of concepts such as "duty," "freedom," and "sacrifice." We are not merely self-locomoting computers on legs. We are affected physiologically, emotionally, and mentally with every step we take through our environment. These steps build our concepts, from our self-image to our philosophy of science. No learning without doing; specifically, no learning without cultural doing.

But once we have mastered a language, we also enter the world of language-based learning. Socialization and culturing pick up speed and enter a new dimension. We are socialized in school at all levels and encultured by fairy tales, discourses, television shows, music lyrics, textbooks, mathematics, and all else that comes our way. This process begins with the topics or themes of our stories. What do our stories presuppose? (The latter is the shared dark matter of the entire group.) How do they introduce new information? How do these stories highlight important information? What information is important? How do they structure arguments? Where and on what matters are arguments and argumentation appropriate or inappropriate? The myriad ways we *interact with, interpret, and remember* our physical, ecological, social, cultural, and linguistic environments produce our dark matter.

It is unnecessary to assume prewired knowledge in a theory of attachment. (At least, I will not do so here.) There are certain abilities that are required, however, in order for an infant to construct its identity (learning its position in the social hierarchy, relevant roles of others in relation to itself, value learning and ranking, preliminary learning and hierarchical arrangement of knowledge, ability to interpret its own experiences, etc.). Childhood, as Alison Gopnik (2010, 4ff) observes, "is not even just something that all human beings share." But it is, she argues, "what makes all human beings human." Gopnik also accurately describes infants as "profoundly alien." We do not know as much about the newborn as it knows about us. It has a several-months' head start, since its learning begins in the womb (Paul 2011). The importance of childhood derives from the fact that human beings enjoy a great survival advantage via their ability to escape from the constraints of purely biological evolution. And "new research shows that babies and young children know and learn more about the world than we could ever have imagined" (Gopnik, Meltzoff, and Kuhl 2001, viii).

Children are born as outsiders. This is why they are aliens. They have no emic knowledge other than what they might have learned in the womb (perhaps nothing emically). They must quickly construct an identity as insiders for themselves as they undergo emicization for the first (and for most, only) time in their lives. The first step in their journey from the strange to the familiar, from observer to knower, is attachment.

Attachment (Bowlby 1969; Otto and Keller 2014) is the construction of the primary bond between infant and caregivers, the first step in emicization. Dark matter has already been forming before birth, but through emicization via attachment, language learning, cultural learning, and so on, the child becomes able to better categorize, store, and arrange its apperceptions, thus constructing itself, its environment, and its society (other examples of related cultural learning are Deloache 1997 and 2000).

Attachment cannot be understood well—or even presented as a useful idea—unless it is studied in a variety of cross-cultural contexts. Although early studies focused on mothers and infants, the work summarized in Otto and Keller (2014) shows that there are various important, different manifestations of attachment from the standardly assumed Western model of mothers and infants—dyadic relations—to more varied, network attachment patterns as found in many traditionally agrarian communities, hunter-gatherer societies, or more close-knit, smaller village societies around the world. As Otto and Keller (2014, 3ff) note, "The evolutionary/ethological foundation does not justify the assumption that attachment has the same shape, emerges the same way, and has the same consequences across cultures." In other words, the variation in attachment of children to caregivers not only devolves from variation in cultures but also reinforces and further separates different cultures from one another. Attachment relations serve a variety of functions and take on a variety of forms.

The importance of attachment studies is twofold. First, they add to our understanding of the range of attachment possibilities selected by our species. Second, they help us to at least begin to tease apart the relative contributions of nature vs. nurture in the shaping of human development. But rather than merely survey the literature on attachment, what I want to do below is to describe concentric circles of attachment in Pirahã and to discuss the significance of this attachment sequence for the acquisition of dark matter. It is useful to offer this (surface) description of how attachment works in Pirahã culture, in order to provide a case study of how people build their identities—personally and culturally—in specific settings. As the description proceeds, I also comment on how values, norms, practices, and conventions emerge from such attachment, returning to all of these issues at different points in later chapters.

This section can be taken also as a sustained discussion and description of emicization in a particular society, the transmogrification of the child from alien to native. One thing that does set humans apart from other species is our ability to accumulate knowledge by means of language. If a chimp, for example, learns that a particular plant is poisonous, its offspring might learn that valuable lesson if they observe their parent avoiding the plant (or getting sick and dying from it). But European chimps (in zoos) do not know about poisonous plants in "the old country" (Africa) because of lore passed down through their grandparents. Chimps do not build tools whose basic design is elaborated with additions and improvements in each succeeding generation. Culture can be shared transgenerationally, via the transmission and elaboration of knowledge.

Through language, each generation learns not only from the generation before them but also potentially from all the generations that have ever lived. Language is not the only tool by which we construct knowledge, values, ways of behaving, and so on. But it is the most important, even as it itself is shaped by culture (D. Everett 2012a).

Beyond language, there are other tools used in constructing a cultural identity. For example, imitation is crucial. Thus, Boyd and Richerson (2005) argue that imitation plays a significant role in the learning and transmission of culture. And as actions that we imitate change across time, imitation alone can carry some cumulative cultural knowledge (e.g., the transmission of a bow and arrow's design changes down through the history of a specific culture through mere imitation of the latest and best design). But imitation is not quite enough for cultural transmission. One cannot learn the specialized hierarchical knowledge (e.g., the theory of relativity) to support many social roles merely by imitating another's actions. Language is crucial for the construction and transmission of culture that goes beyond the day-to-day. Still, our quotidian life is dripping with cultural knowledge and values that can be transmitted by imitation alone; and for the child learning its culture, these are crucial to the initial formation of "self."

PIRAHÃ EMICIZATION AND ATTACHMENT

The attachment I describe here works as a series of "concentric circles," from caregivers to family to society as a whole. At each stage of development, the child is attaching to particular sets of individuals, each more inclusive than the other—mother, parents, family, village, larger Pirahã population.[5]

Hunter-gatherer attachment begins differently, often radically so, from Western practices. For example, I once watched a young woman named Xioitaóhoagí set off from her village with some other women to harvest sweet manioc from her family garden.[6] She walked out with only the clothes on her back, machete, and a basket woven from palm leaves. What made Xioitaóhoagí stand out from the other women walking with her, all laughing and talking loudly, surrounded by numerous small, malnourished dogs, was that she was heavily pregnant.

Later that afternoon, the women returned from the gardens. Xioitaóhoagí passed by my study hut carrying about forty to sixty pounds of manioc roots packed in her basket, secured by a tumpline across her forehead. Like other Pirahãs, her arms were crossed across her chest. I started to return to my work when I suddenly realized that she was carrying something else. Looking again, I saw a newborn baby in her arms and that her stomach was much smaller. She had given birth and then continued on with her work. The baby was born on the ground by the side of the field. After the birth, the mother continued harvesting and trekked back with the other women, carrying two external loads instead of the single load she had intended to carry when the day began.

Not all Pirahã births happen at the side of a field or in the jungle. At low water, women often wade into the Maici river up to their waist, crouch down, and deliver the baby underwater. Births also occur on the raised sleeping platform in the woman's hut. The first question that arises for me, then, is how the Pirahã way of birthing affects—if at all—the Pirahã mother-child relationship.

It is important to understand that Pirahã birth is dangerous and hard for both the mother and newborn. It happens without pain medication, without a comfortable bed, without the assistance of a physician or midwife, on the jungle floor or crouched in a tropical river with piranhas, electric eels, anaconda, caimans, and so on. Pirahã mothers are usually unassisted in childbirth, the solitary effort of the mother to bring forth life. The husband is rarely present. This form of birth leaves the mother with a greater sense of the unbuffered immediacy of her biological connection to her child. A Pirahã woman will have a different understanding of the physical and psychological act of giving birth than the suburban American woman. Yet, beyond the stoicism of this or that particular Pirahã woman, what does her different attitude toward birth mean for Pirahã culture?

Well, it means several things. First, the immediately postpartum and subsequent experiences of the newborn—as well as the experience, observations, and expectations about birth of other children—are also important in-group identity formation. As the Yanomami leader Davi Kopenawa remarked on his own birth in his insightful autobiography, *The Falling Sky*, "I fell on the ground from the vagina of a Yanomami woman."

Second, it means that Pirahã infants' personal characters are formed partially in response to the different material and cultural circumstances. An infant can be cared for by others for brief periods during the day, but will by and large be cared for and spend most of its hours with its mother for the next three to five years. Third, it means that Pirahã mothers and children will attach differently than, say, American mothers and children, due to the distinct personal, unmediated (by teachers, doctors, and other professional caregivers) nature of their relationship.

It is also worth commenting on the marital status of the mother in relation to identity formation and attachment. Unlike some Western mothers, the Pirahã woman is not concerned with stigmas associated with her marital status or age at birth. No one condemns her for children out of wedlock—the concept doesn't even exist, aside from the economic challenges of a woman alone. And even in the latter case, she will always have family of some sort to depend on. Attachment of the child only optionally (though usually) includes the biological father.

The child spends brief periods with other Pirahãs, relatives, or simply others of the same village its mother is in. The child easily distinguishes Pirahãs from non-Pirahãs early on because Pirahãs have a particular smell (of smoke, fish diet, etc.), look (brown skin, black hair, little body hair, etc.), feel (calloused, sinewy), and talk (Pirahã). The average Pirahã infant will willingly go to almost any other Pirahã who stretches their arms out to it, while rejecting the embrace of non-Pirahãs violently.

The Pirahãs occasionally go hungry. They have few material possessions. They have rare access to medical help. They have almost no connection to the economy of the national culture. Moreover, infant mortality among the Pirahãs is very high; perhaps 60 to 75 percent of all children die before the age of ten (though this is improving due to regular visits of the National Health Agency of Brazil, FUNASA). In these ways, the Pirahã mothers' circumstances seem similar to those described for Alto do Cruzeiro mothers by N. Scheper-Hughes (2013).

Yet their responses are quite different. According to Scheper-Hughes, the Alto do Cruzeiro mothers show a degree of aloofness to their infants, whose "spirits" may leave them unpredictably. These mothers rely on a strong sense of religion to enable them to cope with the tragedy of losing their young at a higher rate than the wealthy Brazilians they see on television or work for in one way or another. However, unlike the Alto do Cruzeiro mothers, Pirahã mothers show no aloofness to their infants, showering their infants with affection and attention, sobbing long and loud when they die. What might account for this contrast between mothers in materially similar circumstances? I think that there are two potential explanations. First, the Pirahãs, unlike the Alto do Cruzeiro mothers, have no concept of poverty nor a desire for alternative material conditions (e.g., the lives of the relatively well-off or even of Americans like me, since nothing I have seems to attract them or interest them a great deal). This absence of a concept of poverty means that the Pirahãs' experience is not filtered through any perception of how much better life should or could be if only they had more material resources. As far as they are concerned, they live in ideal circumstances and this is just how life is. Once when I asked a Pirahã man why he thought I had come to their villages on the Maici, he responded without hesitation, "Because the Maici is a good place." Taking a Pirahã to visit another group I worked with, the Banawá, he opined, after only a few minutes, "They are ugly. Their water is ugly. Their jungle is ugly." Second, the Pirahãs have no religion. They believe that you live life as it comes, with no thought of God protecting, killing, or otherwise affecting anything of their daily lives, much less the health of their infants.

The Pirahãs are their only world. This explains why infant attachment is to Pirahãs only. Most infants will turn and scream if a foreigner—especially a bearded foreign male like me—tries to take them. Commenting on the occasional baby who will accept my extended hands, their parents usually say, "My child is unafraid."[7] These parents appear to believe that their babies show courage by coming to me. Yet the deeper value for their parents is that children *not* come to me. The parents who appeared proud earlier that their children came to me might be overheard later telling the child to beware of me because I might take them away to another jungle, *'merica*. Caregivers will often intentionally scare their small charges by acting like they are going to throw them to me or hand them to me when I am leaving in my boat. This type of caregiver behavior dramatically increases fear of me, and most parents laugh out loud when their children scream at my approach—a valuable lesson from their perspective is being learned (i.e., non-Pirahãs can be dangerous—a lesson based on history). Simultaneously, this sharp delineation between Pirahãs and "others" strengthens attachment of Pirahã children to Pirahã adults.

Pirahã children are raised to be physically and emotionally reliant only on other Pirahãs, to avoid imitation of or admiration of foreigners or their ways. The bond between Pirahãs is partially built around the homogeneity of their sensory experiences, though none of this is ever stated. This is all dark matter, most of it of the ineffable variety—how can you put into words the totality of apperceptions structured by episodic memory that makes you "you"? Thus, although Pirahãs do not normally talk about the distant future, they do at times, when asked, say something along the lines that their children will be like other Pirahãs.

As is the case with other cultures' infants, Pirahã infants are cared for around the clock. Mothers, nursing or otherwise, are not inseparable from their infants, however. Occasionally one mother may nurse another mother's baby, allowing the latter to spend more time gathering or engaged in other activities. This depends on the supply of food, (i.e., how well fed both mothers are), both mothers' health, and the relationship between the mothers (e.g., neighbors in the village, kinship, and so on—with mothers' sisters being the most common surrogate milkgivers, but not the only). Also, older siblings of the infant often care for it between feedings, but even when carrying it about (such as to proudly show it off to the anthropological linguist), the mother is never far away. Others may also carry the infant, but rarely out of earshot of the mother.

Infants are regularly talked to, but without special "baby syntax" or "baby phonology." So far as I have been able to tell, on the other hand, mothers (and, to a lesser degree, fathers) often use hum speech (D. Everett 1985, 2005a, 2008) with their babies. They also often speak to them in a high-pitched voice, full of laughter and punctuated with kissing, tickling, and playing.

A strong, somewhat paradoxical attachment/identification practice is the nursing of nonhuman mammals. That is to say, Pirahã mothers not only nurse their own and other mothers' infants, but they also nurse other mammals, as in figure 3.1. In fact, I have seen Pirahã women nurse dogs, monkeys, peccaries (as in the photo), and other animals (even the smaller, tree-dwelling anteaters—*Tamanduá mirim*). Pirahã men joke that women will nurse anything except piranhas (then laugh very loudly and raucously). The Pirahãs are aware—based on comments from river traders, government employees and others—that this is an unusual practice (though it is not unique to the Pirahãs), but they continue it for a couple of reasons.

FIGURE 3.1

The first is that all Pirahãs love animals, and the Pirahã women in particular like to raise young animals. They enjoy playing with them, training them, and so on. But the second—paradoxically to Western thinking, perhaps—is that they raise these pets, nursing them as needed, in order to eat them when they reach adulthood. This doesn't prevent a close, caring relationship while the animal is moving inexorably toward esculent adulthood. The Pirahã name these animals, raise them with much affection, and take them almost everywhere they go. Multiple mothers may nurse the mammal, or it may be nursed by only one, depending on many factors, such as perceived ownership (if one woman or her husband makes a strong claim to possession of the animal, other women are less likely to nurse it); who killed or captured the animal's parent; who has the most breast milk (e.g., a woman just beginning to wean a toddler but who has no new infant); and so on.

Thus children learn values about animals, build animal experience into their all-important "apperceptional set," and integrate these experiences into their emerging selves even as their selves are merged into these experiences. The sharing of human breast milk with animals is observed by children keenly, who seem to find it entertaining. Infants occasionally nurse alongside or immediately following animals. Despite the likelihood that the taste of the animal lingers on the woman's breast, they display no overt reaction. But this taste adds to the child's formative apperceptions and is not shared with members of "buffered" societies—for example, American or European—as Pirahã children blend these experiences into their emergent selves.

All of this builds the child's connection to the Pirahã community and to nature—the nursing of other mammals adds a highly peculiar Pirahã sensory experience to the child's development, both conceptually and physically. For example, one remarkable feature of Pirahã children is their almost complete lack of fear or repugnance of animals, even dangerous ones (e.g., harpy eagles and weasels, which are also raised among them), when in the village. They learn the behavior of many jungle animals from direct observation (even ones that are captured, but not raised, and killed soon after capture, such as caimans). As Pirahã children age, they share not only their mother's milk with animals but solid foods as well, usually sharing their plates with dogs and occasionally other animals, both contentedly eating from the same mound of fish and manioc. Values are created and ranked about community, animals, dirt, mother, and so on. Culture encompasses the entire ecology.

Through their relationship to birthing, animals, and nursing, the Pirahãs establish an immediate and lifelong learning, living connection with nature and their community from the moment they are born. This is a connection that distinguishes them sharply from Brazilians and the rapidly assimilating Parintintin and Tenharim Kawahiv-speaking groups whose own reservations abut the Pirahãs'. Animal relationships and knowledge underscore the distinctiveness of their community from the river traders, explorers, missionaries, pilots, and others who visit their village from time to time. In recent books, Descola (2013) and Kohn (2013) have argued strongly that an understanding of human cultures is only possible as we understand them in their surrounding environment and ecology—that flora, fauna, topography, and so on, play a nontrivial role in human roles, values, and knowledge structures.

These points of group attachment are strengthened during the children's maturation through other natural experiences of community life as the children learn their language, the configuration of their village and to sleep on the ground or on rough, uneven wooden platforms made from branches or saplings. As with other children of traditional societies, Pirahã young people experience the biological aspects of life with far less buffering than Western children. They remember these experiences, consciously or unconsciously, even though these apperceptions are not linguistic.

Pirahã children observe their parents' physical activities in ways that children from more buffered societies do not (though often similar to the surrounding cultures just mentioned). They regularly see and hear their parents and other members of the village engage in sex (though Pirahã adults are modest by most standards, there is still only so much privacy available in a world without walls and locked doors), eliminate bodily waste, bathe, die, suffer severe pain without medication, and so on.[8] They know that their parents are like them. A small toddler will walk up to its mother while she is talking, making a basket, or spinning cotton and pull her breast out of the top of her dress (Pirahã women use only one dress design for all), and nurse—its mother's body is its own in this respect. This access to the mother's body is a form of entitlement and strong attachment. However, it is transitory, leading to the huge shock that is produced by the onset of weaning.

At about four to five years of age—or much sooner, if the mother gives birth to a new infant—the confident, satiated toddler loses access to its mother's milk. The cutoff is sudden and unexpected, and exposes the toddler to hunger, work, independence, and an end to the sense of ownership of its mother's body. The transition is always unpleasant for the toddler, who begins to scream and cry most of the night and day, sounding to the unaccustomed ear as though they are suffering horrible pain of some sort (though as one gets to know Pirahã crying patterns, it becomes easy enough to pick out the signs of anger and petulance in the crying).[9]

During the day, one sees children throwing tantrums as a protest against being hungry, cut off from the mother's milk, and losing the privilege of its mother's arms to a newborn infant. I have seen young children writhing in the dirt screaming, pounding their faces with their fists, deliberately throwing themselves full force on the ground, not infrequently close to or even in the fire (serious burns have occurred), spitting, and carrying on as though they were in the throes of epilepsy. The reaction by the entire village is almost always the same—complete indifference. The children are ignored even though they carry on for hours and occasionally hurt themselves. They are ignored even though they are in the hot sun, not drinking, seemingly using all of their available energy to the point of exhaustion. They are ignored even though they are pitching their fit in the main path of the village, forcing everyone to step over or around them. The thrashing little discontent will usually tire, stop pitching fits, and become much more stable within a few weeks. Any long-term psychological effects of this non-ritualistic rite of passage are invisible to the external observer. The attachment to the mother has been weakened. In its place now begins the accelerated growth in attachment to the Pirahãs as a group.

This accelerated group identification comes from what I call the "hard edge" learning phase for weaned toddlers. The newly weaned child goes from a soft life of no hunger, no work, and pampering to a life in which it must begin to walk more (to the field, to the river, into the jungle, etc.), rather than be carried most of the time; to take on duties, especially carrying small (and always appropriate for its size) loads of firewood, legumes, fish, and so on; to go fishing with older children; to watching and carrying its younger siblings, and so on. They play without adults watching their every move as well—for example, paddling canoes alone on the Maici River. At all stages of life, but especially in this transition to a new independence, the Pirahã child faces risks with little supervision. Children run carrying sharp knives, walk near the fire, reach out to touch living, dangerous animals, and otherwise engage in many activities that some Westerners would consider unsafe. They get bitten, burned, cut, banged up, lost, stung, and hurt in numerous ways during this stage. But in my experience, Pirahã children build from these experiences selves of confidence, grace, and pragmatic knowledge. Such risks and their consequences are crucial for learning and living in Pirahã culture.

One might think (incorrectly) that Pirahã are children innately programmed to carry knives, walk near fires, and be more graceful than Western children. Raising my own children (two girls and a boy) among the Pirahãs and later seeing my grandchildren play in the village, the contrast between the quietness, lack of clumsiness, and common sense and awareness of dangerous things in the environment between the Pirahãs and my own offspring (and certainly myself) was stark. American children (and many adults) run screaming loudly, pounding the ground with their feet. They fall, get stung by wasps and other insects, bump their heads, fall in the river, fall in the canoe, can't sit still, and on and on. Pirahã children show poise and elegance as they move, are relatively quiet, rarely trip and fall, rarely bump their heads or get stung or hurt, compared to American or European children. Pirahãs believe that the contrast between their abilities and ours distinguish us as peoples. These contrasts in apperceptibly shaped carriage and skills strengthen their sense of group identity.

As an example of learning skills early on, consider the use of a bow and arrow. One day while talking to a Pirahã man, I felt a sharp poke on my upper back. The man started laughing as I turned to see toddler, still wobbly on his feet, picking up his blunt-tipped, six-inch-long arrow from the ground where it had fallen after striking me. He was shooting at a mosquito on my shoulder. The child returned my gaze with a serious expression before turning to take aim at a leaf on the ground. The man explained that all Pirahã males learn to handle a bow and arrow the same way—trial and error from childhood. By the time they are adolescents, they are good enough to hit just about anything they shoot at.[10]

Sexual behavior is another behavior distinguishing Pirahãs from most middle-class Westerners early on. A young Pirahã girl of about five years came up to me once many years ago as I was working and made crude sexual gestures, holding her genitalia and thrusting them at me repeatedly, laughing hysterically the whole time. The people who saw this behavior gave no sign that they were bothered. Just child behavior, like picking your nose or farting. Not worth commenting about.

But the lesson is not that a child acted in a way that a Western adult might find vulgar. Rather, the lesson, as I looked into this, is that Pirahã children learn a lot more about sex early on, by observation, than most American children. Moreover, their acquisition of carnal knowledge early on is not limited to observation. A man once introduced me to a nine- or ten-year-old girl and presented her as his wife. "But just to play," he quickly added. Pirahã young people begin to engage sexually, though apparently not in full intercourse, from early on. Touching and being touched seem to be common for Pirahã boys and girls from about seven years of age on. They are all sexually active by puberty, with older men and women frequently initiating younger girls and boys, respectively. There is no evidence that the children then or as adults find this pedophilia the least bit traumatic.

To summarize, much of cultural identification and attachment is achieved by nonlinguistic imitation and learning. Such knowledge is almost exclusively tacit and largely ineffable. Thus tacit knowledge of a community can only—or at least best—be ascertained by the old-fashioned methods of participant observation, note taking, hermeneutics, and conversations, and the interpretations of a variety of behaviors, looking for the links between them, whether linguistic or below the threshold of consciousness. This epitomizes dark matter. Such learning is not limited to culture broadly speaking, however. It is seen in language learning as well.

LANGUAGE AND ATTACHMENT

Piaget (1926) and Vygotsky (1978) champion different, apparently incompatible, perspectives on the connection of language and society—such as egocentric language vs. language as socialization—and sociocultural development. Piaget's egocentricism appears incompatible with Vygotsky's views, at least as Piaget characterizes them. However, I believe that the two may in fact be reconciled if we interpret egocentricism as the "formation of identity," at once a deeply personal psychological process that is nonpathological only during socialization in a specific culture. Thus a child learns language to form itself as an autonomous psychological being, but this autonomy makes sense only in comparison and contrast to others—that is, in a social environment.

Pirahã language acquisition, though I have not studied it experimentally, follows the broad outlines of language acquisition in other cultures. The child begins to learn its language and culture from the womb. Culturally, the fetus learns its mother's biological rhythms, her diet, her pitch range (modulo the wet medium the sound waves must cross), and so on. Linguistically, it is exposed to the mother's prosody (tone, stress, and intonation) and other features of (at least) the mother's speech. As soon as the child is born, it is exposed to the clearer and louder linguistic cacophony of its native community.

In the case of a Pirahã child, it will be almost immediately exposed to five channels of speech (D. Everett 1985, 2008): hum speech, yell speech, musical speech, whistle speech, and consonant-vowel speech, each of which plays a different but important role in Pirahã culture. Mothers and other caregivers do not speak "baby talk" or "motherese" to babies. However, many mothers use hum speech more frequently with babies than other channels. Pirahã children thus learn the importance and use of the prosodic complexity of Pirahã at the very outset of their lives, from the womb. And this prosodic complexity is highly distinctive in Pirahã, setting them apart from any other known group of Brazil. Also, they are exposed as infants to sounds that occur in no other language of Brazil (one of these sounds occurs in no other language of the world)—a voiced apico-alveolar laminal double flap (a form of [l]) and a voiced bilabial trill (D. Everett 1982).

In addition to sound features of their language, however, Pirahã children must master the structure and meanings of words as well as Pirahã grammar, the range of acceptable story topics, the way stories are told, structures of conversations, and so forth.

Consider first stories. Pirahã children learn the topics that are appropriate in their culture for talking about and discussing—just as American children, German children, and Sesotho children do. Pirahã children will not learn any talk of creation, God, the end of the world, oral literature about the forest, and so on. They will learn that talk about nature is as they have experienced it—hunting, fishing, gathering, unexplained sights and sounds—are the most common topics.

They will learn about their words and that verbs can take up to 65,000 possible forms (D. Everett 1983). Perhaps even more important than learning the immensely complex verb structure of Pirahã, is the learning of the evidentiality suffixes that are found at the rightmost end of the verb. Pirahã stories are about immediate experience (D. Everett 2005a, 2008, 2012a, 2012b, etc.), and the function of these suffixes is to communicate that the stated or even reported is based on the evidence of hearsay, deduction, or direct observation (D. Everett 1983). Because of their unusual constraints on storytelling and verb structure, as well as the importance of evidence, the Pirahã restrict their sentences to largely single verb frames (D. Everett 2012a, 2012b). That is, they lack recursive sentence structures (D. Everett 2005a, 2012a, 2012b; Futrell et al., forthcoming). The unusual features of discourse topics, absence of sentential recursion, the prosodic complexity of the language, unique sounds—not to mention that Pirahã is a language isolate and that the people are still monolingual (though this is changing)—mean that their language sets the Pirahãs apart from other populations.

Language is the ultimate tool of attachment and group identification for all Pirahãs. If you speak Pirahã natively, you are a Pirahã. If you don't, you aren't. But what counts as "speaking" is not merely grammatical structure. Mastery of grammar is a necessary for being a Pirahã, in their terms. But ability to use that grammar to tell appropriate stories is the truly crucial skill, blending both language and culture, as discussed in Everett (2005a, 2008, 2012a). Thus, in this view, attachment is a process of defining the self, one's place in society, and the separateness of one's group and culture from others. Theories of attachment, like theories of language and culture, can only benefit from careful descriptive field studies by attachment specialists. Since I am not that kind of specialist, this description of Pirahã is intended only as a first step, perhaps an indication of empirical riches to be uncovered by such fieldwork.[11]

Pirahã children begin life with an attachment to their mother, developing stronger and closer ties with their community as they age. Mothers are affectionate with their children, from infancy on. Their tenderness with infants superficially contrasts with the hard-edge stage of weaning and the toddler age. But their affection never wavers. They are always ready to come to the aid of their child if it is genuinely and seriously threatened. At the same time, the entire community recognizes the necessity of physical and survival characteristics that do not come easy. In an environment with no doctors, no dentists, no police—no one but yourself, your family, and your fellow Pirahãs to depend on—the imposition of toughness on children is not from machismo but from necessity. There are no attachment-related discourses that I am aware of, developing or explicating the growing relationship between child and mother, child and family, child and village, and so on. There might be, but I haven't observed any. Rather, we see that the growing responsibility of the child forces upon it the need to form wider friendships and support within the village and the larger community in order to survive. As with so much of cultural learning, imitation and relationship building are essential and without any special language.

I have often thought that the Pirahãs are among the few people anywhere on earth where just about any member of the society could walk naked into the center of the jungle and emerge well fed, healthy, clothed (after a fashion), and armed. When I am there, I depend on them and am always grateful for their knowledge, their willingness to instruct me, and their ability to teach me (as I ask questions—i.e., at my initiative). Like many traditional peoples around the world, they represent a richness of life and possess a set of solutions to life's problems that can never be recovered if this people or their culture is lost to the world. Unfortunately, they are now under greater threat than ever before.

Cultural transmission, like genetic transmission, is always corrupted in some way, leading to "mutations" (cf. Newson, Richerson, and Boyd 2007; Schönpflug 2008). For example, among the Sateré people of the Amazon, there is a famous wooden club, the *poranting*, that has writing/marking on both sides. The people say that one side tells them the good they should do and the other has the bad things they are to avoid. The problem is that they have forgotten which side is which. Assuming that it was ever clear what each side was for, this is a case of serious transmission breakdown (or a useful legend that was never intended to have a resolution). But cases of smaller magnitude abound. Consider something as concrete as the making of a blowgun. I have witnessed the transmission of this skill in Arawan societies of the Amazon from father to son. Sons observe, imitate, and work alongside their fathers. Surprisingly little linguistic instruction takes place in this skill transmission (at least, relative to anything my dad ever taught me how to do). The wood for the blowgun comes from a narrow range of wood species. The vine used to tie the blowgun and render it airtight is limited to a couple of types. The needle used for the darts likewise requires highly specific knowledge of local flora. The kind of large jungle vine used to extract the poison (strychnine) and the other ingredients of the poison that help it enter the bloodstream more effectively: all of these steps and bits of knowledge, even without language, can be transmitted faithfully or inaccurately. For example, someone might accidentally use different type of wood. Or a different way of tying the blowgun. A different binding agent for the poison. Error or innovation may occur at any step of the transmission process in one father-son pair, leading to a divergence from the cultural norm. From the perspective of the culture, it doesn't matter whether the deviation was intentional or not; there is a deviation, a potential for mutation—a different type of blowgun or an inferior or superior weapon. Clearly such deviations have occurred, because in closely related Arawan languages, blowguns differ (as do the languages themselves!) in nontrivial ways. The technology varied and the language varied due to imperfect imitation and innovation. Such examples show that not all (however important) cultural knowledge is propositional.

One of the more interesting monograph-length studies of social and linguistic interactions to emerge in recent years is Enfield's (2013b) *Relationship Thinking*. As he says:

> Social interaction, typically in its real-time, face-to-face format, is a privileged domain for studying social relations; it is our most experience-near locus of sociality, and is where we may directly observe processes of learning, production, comprehension, diffusion, change, convergence, and diversification in language and other aspects of culture. (xvi)

Enfield refers to the creation of relations in real-time interactions as *enchrony*. His work reveals a great deal of significance for language and society—showing, for example, how meanings are created and negotiated and how their development is constrained by the cultures in which particular relationships take place. Part of attachment's contribution to dark matter is the communication patterns the infant uses and acquires to establish its ever-widening circles of relationships. Thus the study of attachment and relationships is vital to understanding and perceiving the emergence of linguistic forms and meanings, apart from how cultural values and arrangements of knowledge and responsibilities might emerge from these interactions. Enfield's work takes us closer to an understanding of Sapir's (1934, 5) statement that "culture is then not something given but something to be gradually and gropingly discovered."

Summary

This chapter offered a case study of the acquisition of dark matter, by means of a detailed description from my own field research on cultural learning among the Pirahãs. From this we learned that the entire individual—not merely a disembodied mind—learns and acquires dark matter, reinforcing the discussion of the previous chapter on the inability of computers are to ever be able to think or talk. The main reason is the conglomeration of factors that make human dark matter what it is: consciousness, emotions, apperceptions, cognitive plasticity, culture, society, and physiology. Without these, there is no thinking. The chapter also reviewed the necessary requirements to develop dark matter: a body, "culturing," a flexible brain, an emotional brain, semantics, and human intelligence.

4

Dark Matter as Hermeneutics

The perceiver (*aisthêtikon*) is potentially what the perceptible object (*aisthêton*) actually is already, as we have said. When it is being affected, then, it is unlike the object; but when it has been affected it has been made like the object and has acquired its quality.

ARISTOTLE, *Metaphysics*

To this point I have developed the idea of dark matter as a nuanced set of knowledges that are acquired via action and reflection, through lived experiences, attachment, personal apperceptional experiences (interpretation of memories), and by living culturally (participating in a community via social roles, shared and ranked values, conventions, symbols, beliefs, tools, etc.— the material, symbolic, and psychological integration of an individual into a community). Now we are prepared to apply the concept of dark matter to our perambulations through a given society in conjunction with the culture of that society.

The case I make in the following discussion of dark matter is that this unspoken knowledge is our primary tool for interpreting all our experiences. In building my argument, much of what I have to say has been somewhat anticipated in classic works such as Hall ([1959] 1973, 1976, 1990), Bakhtin (1984), Vygotsky (1978), Geertz (1973), Ryle ([1949] 2002), Pelli (1999), and others. The discussion here is novel, however, in the way it attempts to unify mind and culture.

Let's begin with an example from American racial attitudes. In American society, as among most other industrial societies, our value judgments of subgroups of the society—as well as the society as a whole—are formed tacitly through media, myths, and the universal "talk like who you talk with" principle mentioned at several points, as well as the related idea that you "prefer people like yourself." Human infants early on develop preferences for people who look like, talk like, eat like, smell like, and otherwise seem like the people they most often see and engage in these activities with.

Along with sociocultural attachment, our perceptions of others is formed by educational institutions, movies (the portrayal of different members of distinct groups, e.g., women, African-Americans, gays), books (where we learn initially of heroes, villains, settings, supporting characters and value-laden plots, etc.), magazines (erotic, food, race, ethnicity, etc.), among many other sources of enculturation and socialization. These perceptions lead us to adopt and rank values relative to other people and to place them in perceived cultural roles and statuses (valuations of roles). In one study (Eberhardt et al. 2004), it was demonstrated that stereotypes of "Black Americans as violent and criminal" are implicated when subjects are presented with certain concepts such as "basketball" and "crime." Gazing at black faces can decrease the perceptual threshold for recognizing crime-relevant objects. That is, there can exist bidirectional cognitive pathways between stereotyped groups and concepts such that either may trigger the other.

Extending this, as we see directly in more detail, we interpret the world around us based on our unspoken knowns (Gregory 1970). For more than a century, for example, smoking tobacco was interpreted as a part of the *trajé* of the cosmopolitan intellectual, an expression of cool, of joie de vivre. A pipe and leather-patched tweed coat were part of the semeiotic of the "thinker," from Sherlock Holmes to the professor of literature. Today, smoking is perceived differently in the Western world, largely as a filthy, self-destructive habit of nicotine addicts, with nothing cool about it at all. Showing the same images of beatniks or French philosophers from the 1950s and 1960s to people younger than thirty today would evoke quite different interpretations and value judgments for the simple reason that our values have changed dramatically—something about our culturing has led to different dark matters in younger members of the same societies that originally valued smoking.

Thick Descriptions of Cultural Experience

In fact, dark matter determines not only how we interpret images, but whether we can interpret them at all. The cross-cultural ability to interpret photographs is directly relevant to the idea that culture might provide a hermeneutics for interpreting the world. A bit of reflection suggests that differential perceptual ability in this regard might not be unexpected. After all, in the natural world, there are few if any two-dimensional visual experiences, aside perhaps from reflections in water. In cultures without two-dimensional visual arts—exposure to photography, or literacy, for example—the interpretation of photography could provide us with information on the sources of some visual representation, whether they are learned culturally (at least in part) or innate.

I had noticed, after taking Polaroid photos of the Pirahãs or bringing in developed photos of them from town, that they would stare at the photos and then ask me what or who a given picture was about, even when the photo was a clear portrait of the beholder or a loved one. I commented on this later to a few colleagues, expressing my belief that this was "because they haven't had much experience with pictures." Some of my psychologist colleagues thought the observations were worth following up on. But before going into a discussion of the Pirahãs' interpretation of two-dimensional objects, it is worth noting that their difficulty in this regard is no different than Westerners' effort to understand art, from modern to impressionistic to realist, it all must find a place in the observer's cultural matrix to be interpreted. Susan Sontag (2013, 1) insightfully observes that "in teaching us a new visual code, photographs alter and enlarge our notions of what is worth looking at and what we have a right to observe. They are a grammar and, even more importantly, an ethics of seeing." And also: "Finally, the most grandiose result of the photographic enterprise is to give us the sense that we can hold the whole world in our heads—as an anthology of images."

Anthropologist Glifford Geertz (1973) borrowed a very important idea from philosopher Gilbert Ryle (1968) and used it to dramatically alter the notion of anthropology: the idea of "thick description." In a thick description, the scientist provides sufficient information for an outside observer to *grasp the meaning* of the events being described. Another way of putting this is that the description should provide an emic understanding of events, not merely contenting itself with the etics of what is being described.

An example from linguistics can illustrate this. A "thin" description might say, for example, that English has aspirated occlusive consonants, going so far as to list the words in which these consonants occur, or at least a large sample. A thick description would include the fact that aspirated stops occur only in syllable-initial position and that native speakers hear the unaspirated and aspirated variants as *the same sound* whenever they adhere to the above-given expected distribution of syllable-initial and non-syllable-initial (that is, it would be at least a phonological analysis).

For something like, say, a Pirahã dance, we can describe this etically simply as the Pirahãs walking in a circle singing. An etic description would note things like how many dance at once, who dances, when they dance, where they dance, how long they dance, under what circumstances, and so on—all those aspects of their dance that are physically measurable. The etic description would include the fact that some of the men dance naked or clothed in palm-leaf skirts with headbands woven from palm leaves or the bark of the buruti palm. A thick description would include all of this and more. For example, it would describe/explain the role of the naked or palm-attired men, the expressions on the faces of the dancers, the people's explanation of why they walk in a circle, and so on. That is, the thick description of Pirahã dance enables us to understand the emics of the dance, the purpose and interpretation of the dance by its participants—what do the songs, circular walking, different dress or undress of some male participants *mean* to the dance and the dance's role in the overall life of the Pirahãs?

Pirahãs "dance" by walking in a tight circle. If there are a couple of villages dancing together, they may walk in concentric circles. This can last for a short while or as long as seventy-two hours (in my experience) without stopping (though individuals come and go, stop and eat, etc., the circle continues unbroken). Walking in a circle produces a slight dizziness, especially as dancers accelerate from time to time. The circle can be left or (re)joined during the duration of the event. The "songs" the people "sing" as they "dance" (i.e., walk in circles) are in fact prose retellings of remembered events, with the "melody" coming from the tones of the words themselves. The rhythm of normal speech is altered and the frequencies of tones are intensified (impressionistically high tones have higher frequencies than normal speech, and low tones have lower frequency).

Again, the purpose of thick descriptions is to provide information sufficient to research the meaning—the emic analysis of the event. For Geertz, the purpose of anthropology is to interpret the meanings of cultures, not to offer up a "scientific theory" of them. But even if we (reasonably) were to hope for a bit more than Geertz's anthropological hermeneutics as the output of anthropological studies, it is clear that in their individual development, members of a community will—through their practices, words, and individual experiences—come to acquire and imbue their bodies (including their brains) with dark matter.

All around us, the absence of thick descriptions affects our understanding of others, even those central to our own culture. For example, the lack of a thick description or any inkling of an emic perspective of biblical culture by the average, contemporary US church member produces job security for ministers, preachers, and priests, since most of their public role is explaining what their sects' scriptures mean.

Consider, for instance, the Apostle Paul's admonition that "women should keep silent in the churches." This comes from the first letter of Paul to the church at Corinth, chapter 14, verse 34, and is often quoted in isolation as "Women should remain silent in the churches. They are not allowed to speak, but must be in submission, as the law says." Taken thus out of textual, historical, and cultural context, these words seem to be a straightforward prohibition by the apostle against any leadership role for women in church life. But this would be mistaken. Consider, for example, the larger textual context:

> I Corinthians 14:33–35:
>
> For God is not a God of confusion but of peace, as in all the churches of the saints. The women are to keep silent in the churches; for they are not permitted to speak, but are to subject themselves, just as the Law also says. If they desire to learn anything, let them ask their own husbands at home; for it is improper for a woman to speak in church.

The passage talks about "confusion" and "learning at home." The women are not to keep silent merely because they should "shut up in church" but for reasons having to do with less confusion and a more appropriate learning environment. What could this mean? Well, this textual context has raised the issue of the local cultural context at the time of Paul's writing. Many commentaries observe that the woman's place in first-century Greek and Hebrew societies was low, more like property than personhood.[1] This treatment of women produced additional, compounded effects on the social inequality of women.

The lingua franca of the first-century world in which the Apostle Paul was writing was Koiné (common) Greek. Itinerant Christian preachers were becoming more and more frequent. Koiné was favored in the churches because there were so many local languages and dialects throughout the Mediterranean, just as there are today; a traveling preacher couldn't be expected to learn them all. However, women did not usually speak Koiné because their social roles kept them in the house and with very little opportunity to learn any language other than their local dialect. In addition, women and men often sat on separate sides of the meeting space. Thus as a trade-language sermon was being preached, some women would shout across the divide to ask their husbands for a translation. The husbands might answer or yell out the translation. The effect of this back-and-forth on church decorum and organization was not salutary, in Paul's opinion. Hence he admonished the women to be silent—to wait until they returned home to ask for a translation if they needed one. Ecclesiastical roles for women were not in focus here per se. Without this knowledge of the culture, there is no proper interpretation of the verses. A thick description—or emic perspective, in my interpretation of this Ryleian phrase—of the culture of the source and the target cultures is essential for translation to take place accurately or effectively (see chap. 8).

But culture in the popular sense, as we saw in chapter 2, is not really an entity in the real world. In this case, when we refer to culture, we are really generalizing over dark matters. Culture is an abstract network of ideas manifested through individual cultural members. It is not to be found outside those members. Thus what is usually referred to as "culture" is nothing more than individual dark matters (including roles, values, structures). So then, how does this dark matter come to function as a hermeneutic?

As in the development of attachment to one's caregivers and larger society (chap. 3), dark matter arises from the construction of concentric circles of relationships where the members of ever-widening groups are engaged in activities that the individual learns to interpret and to repeat, very much like they do in acquiring their first language. Culture is the dark matter that is acquired from these activities of culturing, social acting, languaging. Culturing, languaging—the various games, utterances, speech acts, discourses, symbols, icons, and so on—are of fundamental importance to the acquisition of dark matter, to interpret the world around us. They are our hermeneutics. To see this, let's return to consider the interesting issue of perception among the Pirahãs.

Culture and Perception

The discussion that follows is drawn partially from my own field research. But the technical argumentation, much of the reasoning, and some of the wording are taken largely from Yoon, Whitthoft, et al. (2014). The question asked here, again, is "Does our dark matter—derived from culture and psychology—help or impede our ability to perceive the world around us?" The short answer is that it does both. But to see this more clearly, I will first examine my own difficulties in seeing what Amazonian peoples see. Next I will look at their difficulties in seeing what I see.

In the rainy season, jungle paths flood. Snakes exit their holes. Caimans come further inland. Sting rays, electric eels, and all manner of creatures can then be found on what in the dry season are wide, dry paths. It is hard to walk down these paths in daylight during the rainy season, covered as they are by knee-deep, even chest-high water (though I have had to walk for hours in such conditions). At night, these paths become intimidating to some of us. As I walk with the Pirahãs, I am usually wearing shoes, whereas they go barefoot. Two memories stand out here. The first was me almost stepping on a small (three feet long) caiman. The second was me almost stepping on a bushmaster (there are many other memories as dangerous). In both cases, my life or at least a limb was saved by Pirahãs who, shocked that I did not or could not see these obvious dangers, pulled me back at the last moment, exhorting me to pay more attention to where I stepped. Such examples were frequent in my decades with Amazonian and Mesoamerican peoples. And each time, they were astonished at my apparent blindness.

I discovered, however, that "blindness" affected the Pirahãs as well. There were things that, like me, they looked at but couldn't "see." These were objects from my culture, such as pictures. (Interestingly, the Pirahã expression for being good at some skill is to "see well.")

Other researchers became interested in Pirahã perception after hearing that the Pirahãs seemed to be unable to recognize even pictures of themselves. Thus in a collaborative effort, Mike Frank, Ted Gibson, and I conducted a number of experiments among the Pirahãs in 2007 (in cooperation with the coauthors of Yoon, Winawer, et al. 2007). We reached several important conclusions, summarizing our findings thus:

> A core principle of vision science is that perception is not simply a passive reflection of the external world, but a process of constructive interpretation of inherently ambiguous input. Consider a shadow projected onto a wall. The same silhouette can be created by different objects of different sizes at different distances from the viewer. Images projected onto the retina have the same inherent ambiguity, and a wide range of perceptual judgments ranging from lightness (Adelson 1993), to color, to depth, to shape and identity, are the result of "unconscious inferences" by the visual system (Helmholtz 1878). Such inferences are often presumed to be automatic and culturally universal (Gregory 2005; Kohler 1929; Spelke 1990). (Yoon, Whitthoft, et al. 2014, 1)

As we interpret the world around us, the problem is not seeing the details, but putting together what we are seeing into coherent percept, or gestalt. This "putting together" occurs effortlessly and without awareness. What we see are the etics of vision. The gestalt, the interpretation, the connecting the dots of what we perceive is emic vision. Properly emicized, we see the whole "better"—seeing things that are not there and not seeing things that are. For example, consider the two-tone ocelot in figure 4.1 (right column, second row from the top).[2] People often fail to recognize the two-tone image; when shown the corresponding photograph, however, they find the two-tone often transforms suddenly into a coherent percept. Are they using emic knowledge to interpret etic images, or are they simply getting better information, unconnected to outsider or insider knowledge? Observers viewing the ocelot in the two-tone will often make figure-ground errors, incorrectly assigning some background regions to the figure, some figure regions to the background. Reconfiguring figure-ground assignments after viewing the photograph is to "reorganize" one's initial grouping to achieve a different perceptual state

FIGURE 4.1

(Kovacs and Eisenberg 2004). If the viewer ultimately recognizes the previously unrecognized image, perception reorganization is said to have been successful.[3]

It is not the case that all images we see—even degraded two-tone ones—require access to full cultural knowledge or require perceptual reorganization. For example, advertisers work to develop symbols, "logos," that the average person of most cultures (especially Western cultures) can recognize early, even when they are two-toned (e.g., the World Wildlife Fund's panda logo). Similarly, some simple black-and-white line drawings are immediately recognizable. At the same time, it is possible to construct two-tone images that are difficult to perceive from the emic perspective. Interestingly, however, when such images are presented with the corresponding full photo cue, they readily trigger perceptual reorganization (Dolan et al. 1997; Hsieh, Vul, and Kanwisher 2010; Ludmer, Dudai, and Rubin 2011).

The question that motivated this research project was whether principles underlying perceptual organization are universal. We know that there is evidence that very young infants and remote cultures show certain principles of perceptual (re)organization (Pica et al. 2011; Spelke 1990). And we also find a body of evidence that there is variable susceptibility to phenomena across different populations. This kind of variability, infant studies not withstanding, suggest an important role for culturally variable factors in perception. For example, there is evidence that experience with photographs (Segall, Campbell, and Herskovits 1966), digital clocks (Whitaker and McGraw 2000), culture-specific processing biases (de Fockert et al. 2007), and exposure to urban vs. rural vistas (Leibowitz et al. 1969; for a review of older work, see Jones and Hagan 1980) all produce different abilities of perceptual organization in the relevant domains.

All researchers agree that culturally invariant mechanisms of development such as the physiological maturation of the visual system ought to produce perceptual differences between children and adults. But children may also become more strongly enculturated into the practices of perceptual inference and interpretation accepted in their particular community over time, similarly predicting differences in how children and adults perceive the world (Vygotsky 1978). For example, a particularly striking phenomenon in perceptual development is the deficient recognition of two-tone images in young children. This is in spite of the fact that adults recognize them easily (Kovacs and Eisenberg 2004; Yoon, Winauer, et al. 2007). When faced with images like figure 4.1 (even ones containing familiar creatures), children—like adults—often struggle to recognize the animal. Significantly, however, children have much greater difficulty recognizing the animal even when the two-tone image is placed side by side with the original picture.

An important question that arises in the present discussion, then, is whether the perceptual reorganization reported by adults results from their perceptual maturation, or whether it is the result of dark matter acquired in specific cultural contexts (and particular individual histories). This was the motivation for my and my coauthors' interest in experiments concerning the Pirahãs. Like young children in a modern industrial culture, Pirahã adults have little experience or knowledge of the visual transformation that links a photo with a two-tone image. On the other hand, Pirahã adults do possess both physiologically mature visual systems and a lifetime of experience with complex visual tasks, such as hunting and fishing.

In our experiments, we tested Pirahã adults and English-speaking controls on their ability to recognize two-tone images given the corresponding photographs as cues (fig 4.1). My own prediction was that, like children and US adults, the Pirahãs would have difficulty recognizing two-tone images. In fact, we all believed that if expertise in interpreting symbolic visual materials is a key factor in photo-cued two-tone reorganization, then the Pirahã—like children but unlike US adults—would have trouble recognizing the cued image even in the presence of the photo.[4] For the experiments, my coauthors used Photoshop to create ten two-tone images by blurring and posterizing (reducing the number of distinct grayscale values, in this case to two: black and white) grayscale photographs of animals and individuals found in the Pirahã participants' everyday environment (fig. 4.1).[5] We additionally tested two other image pairs that did not include two-tones and for which the correspondence was easier to see (fig. 4.2).[6] We used these to get Pirahãs engaged and warmed up and to make sure that they understood what we were asking them to do.

Each trial proceeded in three stages. In stage 1, participants were shown a two-tone image and asked to indicate their recognition by pointing to the location of the eye or Pirahã person in the picture (fig. 4.2). Responses were marked by placing a sticker at the indicated locations. Trials in which the target was not initially identified were considered "candidate reorganization trials." These trials were of particular interest, as they provided a test of whether an initially unrecognized two-tone image could be successfully reinterpreted after seeing the corresponding photo. These trials proceeded to stages 2 and 3. In stage 2, participants were shown the corresponding photograph alone and asked to point to the location of the eye or Pirahã person. In stage 3, the two-tone image and photograph were shown side by side. The experimenter

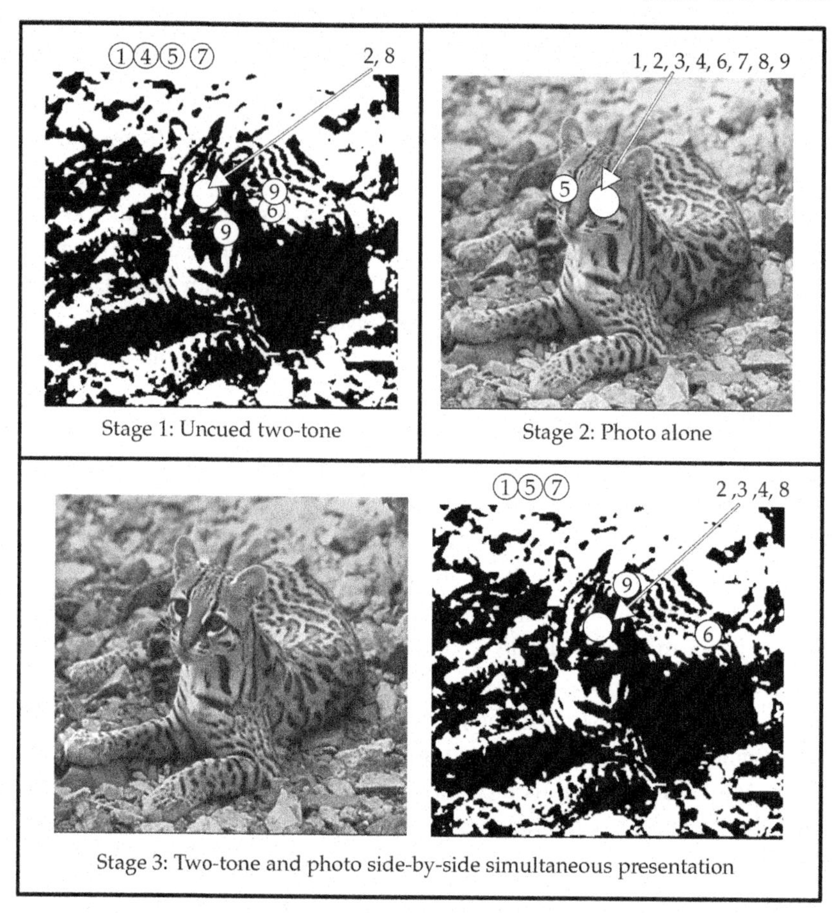

Stage 1: Uncued two-tone

Stage 2: Photo alone

Stage 3: Two-tone and photo side-by-side simultaneous presentation

FIGURE 4.2

then pointed back and forth between the two images using the Pirahã word for "same" to convey the correspondence between photo and two-tone. After this instruction, the subject was again asked to point to the location of the eyes or the person in the two-tone image. I held most of the pictures about a foot and a half to three feet from each subject. This slight variation is unlikely to have had any significant effect. In a separate control study to test for the possibility that close viewing interfered with perceptual reorganization, US adults viewed two-tones from a much closer viewing distance than seen in any participants (nine inches) and performed at ceiling on candidate reorganization trials (100 percent). In addition, US preschoolers, a similarly low reorganization population, viewed two-tones from distances of two and four feet with no difference in performance (Yoon, 2012).

My colleagues also tested Stanford students on an alignment manipulation task. This task controlled for the possibility that US participants' performance on the task was not due to recognizing the two-tone images, but merely to locating the point on the two-tone card in the same location as the corresponding point in the photograph. This study was identical to the main study, except that the images were cropped by 10 percent on two adjacent sides (e.g., top and left), chosen at random, with the constraint that the corresponding two-tone and photo were not cropped on the same two sides. (An example is shown below in fig. 4.4). Thus the eye or head was in a different location on the printed card in the photo and in the two-tone. If US participants were solving the task by pointing to the same location on the cards rather than by identifying the image features in the two-tone image, they would not have successfully located the eye in the two-tone image in this experiment.

As stated in Yoon, Whitthoft, et al. (2014):

> Pirahã participants and U.S. control participants on the same task successfully indicated the target locations (either eye or person) on the non-two-tone practice images without the corresponding photo cue (controls 100%, Pirahã 88.9%), showing participants understood the task (Figure 3, white bars [fig. 4.3 in this volume[7]]). Controls located the targets successfully in uncued two-tone images on 72.5% of trials. Initial recognition in Pirahã participants was less frequent (22.5% of trials). Controls identified the targets in the corresponding, untransformed photos 100% of the time and the Pirahã 90.3% of the time (Figure 3, black bars). All Pirahã participants correctly indicated the target on at least 7 of the 10 photos. Data from trials where the Pirahã did not correctly recognize the photo were excluded from subsequent analysis.[8]

What we were most interested in is termed *candidate reorganization trials*. These were cases in which the participants in our study failed to initially locate the target in the two-tone (incorrect stage 1), but did locate it when presented with the photo (correct stage 2). We next calculated the percentage of two-tones recognized after viewing the photo and then dividing by the total number of these trials. Interestingly, our US controls consistently demonstrated the ability to reorganize what they were perceiving, by accurately recognizing the eye or the Pirahã person on previously unrecognized two-tones. Pirahã participants, on the other, hand succeeded on such reorganization trials only 31.6 percent of the time. In fact, two Pirahã participants never were able to perceive the content of the two-tone images. The best recognition performance that any Pirahã achieved was 60 percent. Thus, what we did in these experiments was to test whether Pirahãs were able to perceptually reorganize

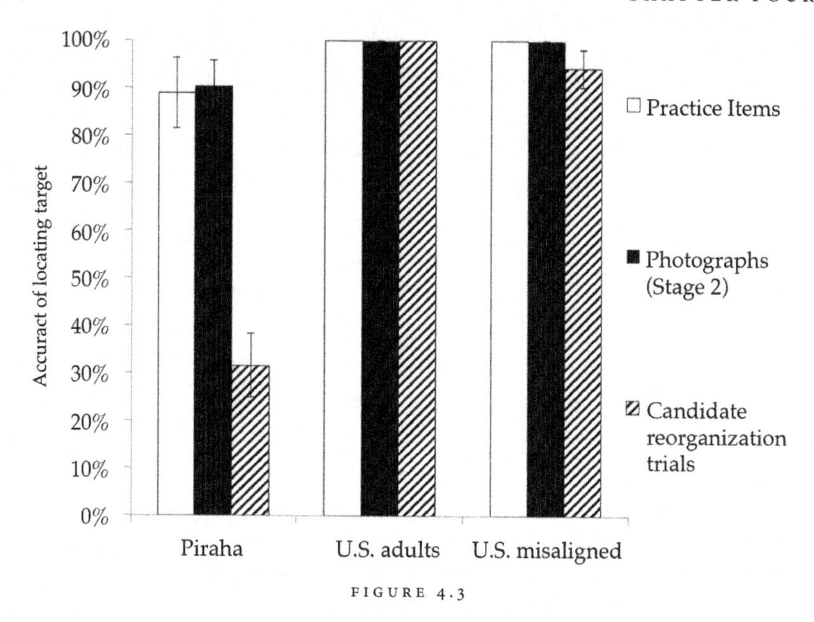

FIGURE 4.3

two-tone images when they were viewing the latter along with the original (unphotoshopped) photos. US participants performed nearly perfectly. The Pirahãs, on the other hand, struggled. The contrast was striking.

The question that we get to then is why this recognition and perceptual reorganization task was so much harder for the Pirahãs. There are a couple of potential explanations for our findings. These include how well the Pirahãs understood the task, their familiarity with the stimuli they were asked to judge, and the difficulty of the task. After deciding what it was that we were observing, our next step was to consider the range of possible differences in perception and discuss possible conceptual or experiential sources of differences in the groups' perceptual reorganization.

We determined that US adults are accurate at detecting the correspondence between photos and corresponding photoshopped two-tone images even when the images no longer share a predictable coordinate frame relative to one another (e.g., as in fig. 4.4).[9] This means that the US adults had to use emic understanding of the concept of two-dimensional representations—that is, perceptual reorganization—in order to identify the unpredictably displaced location in the two-tone image within the figure. We accounted for the US vs. Pirahã performance differences in terms of "perceptual literacy," attributing to Pirahã and US performance differences to cultural differences in training and education with visual symbolic materials.

Our data are incompatible with the idea that the Pirahãs did not under-

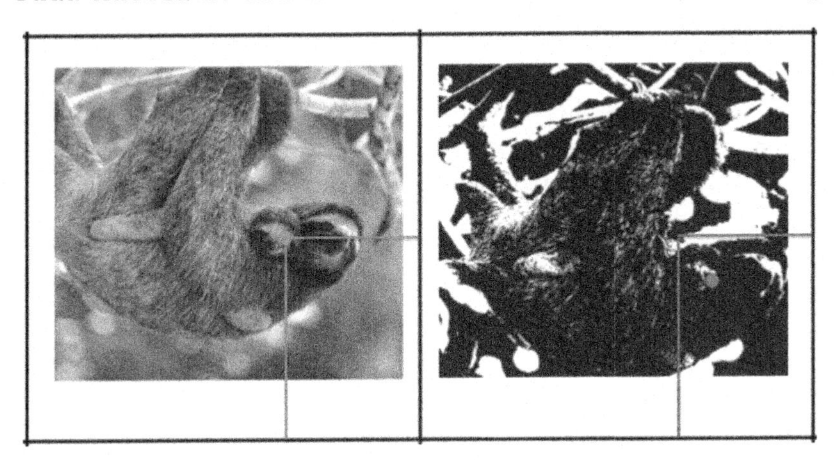

FIGURE 4.4

stand the task. The Pirahãs' excellent performance on the practice trials and on the photos themselves demonstrates that the Pirahã understood the general task instructions. As we ran the experiments, we were careful, both verbally and gesturally, to indicate that the photo and two-tone images were of "the same" subject, using the Pirahã phrase *ʔai sigíai*, "the same." On the other hand, suppose that Pirahã participants did not know how to interpret the experimenter's instructions. Nevertheless, they would have understood that the photo and two-tone images corresponded once they had correctly interpreted the two-tone image. Yet this is not what we found; success in one trial did not increase accuracy on subsequent trials. This indicates that even when recognizing a picture, there was no emic knowledge of two-dimensional space interpretation (these are my words, not those of my coauthors). Moreover, since the photographs we used were of people and animals the Pirahãs knew their performance, it is also unlikely that the result is due to a lack of familiarity with the pictured items. In fact, the Pirahãs knew the items, fauna, and people better than the US control subjects.

The kind of two-tone image recognition we experimented with could, of course, be missed for other reasons. Another possibility is that the Pirahãs are not able to do hard mental work. My experience with the Pirahãs' learning of math, emic understanding of their environment, ability to predict my behavior, and so on, contradicts that possibility. The Pirahãs inability on the two-dimensional tasks—like mine on seeing dangerous animals in the forest—simply shows that a mature visual system is insufficient to guarantee recognition of what one sees. The mature system "sees" only the etic until it has undergone emicization into a particular culture, with particular experiences, expectations, and so on.

US citizens need to acquire an emic recognition of two-dimensional representations because these are ubiquitous and crucial sources of information in their cultures, because they are schooled for years in such recognition, and because society places a high value on this type of seeing ability. In other words, there is such a thing as emic understanding of perceptual experiences that ranges beyond the merely physical and etic limitations of the ocular nerve.

As we noted in our study, "One prediction of this account is that participants' degree of cultural ubiquity and expertise with decoding visual symbolic materials should relate to their degree of ease and automaticity using visual cues to interpret ambiguous and impoverished images." And this is what we find, at least for the phenomena investigated.

Culture is essential for even seeing the world around us. Sights begin as etic experiences and are eventually turned into components of emic understanding. We are born like the blind man who told Jesus, "I see men as trees walking." There are therefore implications of the idea that dark matter is our hermeneutic, even for basic perception. We receive input far beyond our capacity to attend to it. Noise is noise, after all, because we *decide* it is not part of a message, or not part of what we are attending to at a given moment. Illusions, optical and aural, occur because we attend to incorrect cues, believing falsely that they are relevant to the inpretation or perception of the entity, thought, sound, and so in in focus (Rock, Hall, and Davis 1994). Individual experience and culture (what is relevant, what is known, etc.) are easily the most important components of our perceptual cognition. And they guide us because we have moved from an etic perspective to an emic one, embedded the perceptual categories and strategies into ourselves as unspoken material of reasoning.

Beyond Counterexamples and Exceptions: Dark Matter and Science

As a final example of the lessons of this chapter, I want to examine the notions of "counterexample" and "exception" in scientific research, in order to demonstrate the reach of culture and dark matter into this hallowed domain, beginning with a familiar form of argument, the syllogism. So consider the following statements:

All swans are white.
Johnny is a swan.
Johnny is white.

But what if Johnny is not white? What if Johnny is a black swan? This is a case that "All swans are white" cannot account for. Do we reject "All swans are white," or do we have to reject the idea that Johnny's blackness is relevant for the statement that "All swans are white?" What if Johnny had just rolled in black crude oil? Surely that would be superficially a violation of "All swans are white" but not worth abandoning the syllogism. Somehow we have the intuition that if the swan is black because it rolled in oil, then there is no violation of "All swans are white," though if the swan is black because it is born that way, then there is a violation. But even this is simplistic. What if the swan that was born black had a genetic anomaly of some sort?

Ultimately a black swan is evaluated relative to a cultural system. For some, a black swan will be an innocuous exception. For others, it will be a devastating counterexample. This cultural interpretation of exceptions vs. counterexamples is important to the central thesis of this book, because it gets at the very heart of the scientific enterprise that is the centerpiece of many modern societies. Science, we might say, ought to be exempt from dark matter. Yet that is much harder to claim than to demonstrate.

Throughout the course of this book, we have considered different approaches to the science of Homo sapiens, what binds us together, our marvelously group-individual psyches, and the ways in which business, culture, psychology, textual understanding, translation, the formation of linguistic systems, and so on, are functions of dark matter and the role of emicization in forming individual psyches to an understanding of aspects of the sociology of science. Science is among the highest attainments of some civilizations. Not all societies value it, though every society feels its effects, from the isolated, uncontacted Amazonian tribe being filmed from the air by drone to the religious fanatics decrying science while driving pickups. And scientists—like other intellectuals, businesspeople, extreme athletes, or in fact the members of any community—are subject to the formation of dark matter that can affect their work qua scientists without them being fully aware of it. Moreover, this subtle, invisible hand in scholarship has lessons that epitomize the thesis of this book and the applications of this thesis to life.

This chapter illustrates the work of the previous chapters by considering the concept of scientific progress in light of a theory of dark matter. To take a concrete example of a science, we focus on linguistics, because this discipline straddles the borders between the sciences, humanities, and social sciences. The basic idea to be explored is this: because counterexamples and exceptions are culturally determined in linguistics, as in all sciences, scientific progress is the output of cultural values. These values differ even within the same discipline (e.g., linguistics), however, and can lead to different notions of progress in science. To mitigate this problem, therefore, to return to linguistics research as our primary example, our inquiry should be informed by multiple theories, with a focus on languageS rather than Language. To generalize, this would mean a focus on the particular rather than the general in many cases. Such a focus (in spite of the contrast between this and many scientists' view that generalizations are the goal of science) develops a robust empirical basis while helping to distinguish local theoretical culture from broader, transculturally agreed-upon desiderata of science—an issue that theories of language, in a way arguably more extreme than in other disciplines, struggle to tease apart.

The reason that a discussion of science and dark matter is important here is to probe the significance and meaning of dark matter, culture, and psychology in the more comfortable, familiar territory of the reader, to understand that what we are contemplating here is not limited to cultures unlike our own, but affects every person, every endeavor of Homo sapiens, even the hallowed enterprise of science. This is not to say that science is merely a cultural illusion. This chapter has nothing to do with postmodernist epistemological relativity. But it does aim to show that science is not "pure rationality," autonomous from its cultural matrix.

In 1996, the late Peter Ladefoged—at the time of his death perhaps the world's leading phonetician—and I published an article in the main journal of linguistics, *Language*, entitled "The Status of Phonetic Rarities" (Ladefoged and Everett 1996). In this paper we discussed unique sounds I had brought to the attention of linguists, which had been found in the Amazonian languages Pirahã, Wari', and Oro Win, but in no other language of the world. The sounds we discussed were [t͡ʙ̥] and [l̺̆]. Both sounds are unique to Wari' and Oro Win on the one hand, and Pirahã on the other. [t͡ʙ̥] is a voiceless alveolar-bilabial trill and [l̺̆] is a voiced apico-labial double flap.

To some, these sounds would be little more than outliers in the range of human sounds, merely curiosities. But to phoneticians, they are more than this. Phoneticians and phonologists have theories of how sounds come to be incorporated into human languages and how such sounds fit into phonetic or phonological theories more broadly. The majority of phonologists believe that all sounds in human languages are decomposable into "distinctive features." Thus a sound like [t] is a [−voiced, + coronal, −continuant].[10] What Ladefoged and I showed, however, was that no extant combination of distinctive features could describe these rare sounds. Thus we were faced with a choice: we could refer to these sounds as exceptions or counterexamples to modern phonological theory, or we could modify the otherwise universally accepted list of distinctive features in order to accommodate these sounds. The former possibility would in effect be claiming that we should throw out all distinctive features on the basis of two sounds. The latter would be to so widen the theory that it could accommodate any sounds, or so we argued, but of course a theory that can describe anything explains nothing. We argued then that these sounds were exceptions, not counterexamples, but exceptions that could not be incorporated into the theory in any interesting way. What they show us, we argued, is that theories cannot always account for everything. Theories leak.

But this conclusion illustrates exactly the point of the black swan—namely, that counterexamples and exceptions are etically the same, but emically very different. For example, if we expand a universal phonology to account for all phonetic rarities, we weaken it because in so doing, it will predict things never found. At the same time, these rarities show us that our theory can never account for everything. We must resign ourselves to having exceptions at all times that in principle ought not to be incorporated into or analyzed by the theory. To see what I mean, let's examine these phonetic rarities in Pirahã and Wari' in more detail.

When I was first planning to visit the Pirahãs, I read what little information there was about them. I learned that among other things they have a sound [l], which apparently is not found in any other language in the world. This had first been written up by an American missionary, Arlo Heinrichs, and subsequently observed by Heinrichs's successor, Steve Sheldon. When I arrived among the Pirahãs in 1977, therefore, I was eager to hear this sound. Since Sheldon and Heinrichs had left lists of words in which this sound appeared, I asked the Pirahãs for exactly those words that featured [l]. After observing the word myself, I realized that no one had ever published an article on the sound, so I wrote up a small paper and submitted to the Journal of the International Phonetics Association, JIPA, which published my four-page description of the sound, as a voiced apico-labial double flap. It did not occur to me that this sound could have theoretical significance; it was merely a discovery that I wanted to share. And there was little reaction to the sound until I began to work with Ladefoged.

Likewise for the [tp] trill of Wari' and its closely related language Oro Win. The phoneme had been known for several decades by missionary-linguists who worked on these languages. Yet no one had realized that both [l] and [tp] were not predicted by anyone's phonological or phonetic theory. This failure to predict the sounds means that no phonological theory had any "slot" or "matrix" in which these sounds could fit—they were anomalies. But were they counterexamples or exceptions? If a counterexample is an analysis of facts that a particular theory cannot explain, what would an exception be? Exactly the same: an exception is an analysis of facts that a particular theory cannot explain. The difference between a counterexample and an exception, of course, is that the former is taken to be a serious problem that a theory must address, while the latter is a hiccup that the theory might have to account for at some point, but for now can safely ignore.

(The subtle distinction I am attempting to draw here does not include pseudo-exceptions. For example, if I claim that I have discovered an arithmetic system in some isolated community in which 2 + 2 = 5, I have in fact made a mistake. By definition any such claims is a *misunderstanding* and thus is neither a counterexample nor an exception. Many purported "counterexamples" are simply misunderstandings. For example, claims like "*The word* haggis *is cultural, therefore grammar is influenced by culture*" would be the linguistic equivalent of an arithmetic error. Another example of an actual counterexample or exception is found in the general statement and specific statement pair: "All birds fly. Ostriches do not fly.")

Whether we classify an anomaly as counterexample or exception depends on our dark matter—our personal history plus cultural values, roles, and knowledge structures. And the consequences of our classification are also determined by culture and dark matter. Thus, by social consensus, exceptions fall outside the scope of the statements of a theory or are explicitly acknowledged by the theory to be "problems" or "mysteries." They are not immediate problems for the theory. Counterexamples, on the other hand, by social consensus render a statement false. They are immediately acknowledged as (at least potential) problems for any theory. Once again, counterexamples and exceptions are the same etically, though they are nearly polar opposites emically. Each is defined relative to a specific theoretical tradition, a specific set of values, knowledge structures, and roles—that is, a particular culture.

One bias that operates in theories, the confirmation bias, is the cultural value that a theory is true and therefore that experiments are going to strengthen it, confirm it, but not falsify it. Anomalies appearing in experiments conducted by adherents of a particular theory are much more likely to be interpreted as exceptions that might require some adjustments of the instruments, but nothing serious in terms of the foundational assumptions of the theory. On the other hand, when anomalies turn up in experiments by opponents of a theory, there will be a natural bias to interpret these as counterexamples that should lead to the abandonment of the theory. Other values that can come into play for the cultural/theoretical classification of an anomaly as a counterexample or an exception include "tolerance for cognitive dissonance," a value of the theory that says "maintain that the theory is right and, at least temporarily, set aside problematic facts," assuming that they will find a solution after the passage of a bit of time. Some theoreticians call this tolerance "Galilean science"—the willingness to set aside all problematic data because a theory seems right. Fair enough. But when, why, and for how long a theory seems right in the face of counterexamples is a cultural decision, not one that is based on facts alone. We have seen that the facts of a counterexample and an exception can be exactly the same. Part of the issue of course is that data, like their interpretations, are subject to emicization. We decide to see data with a meaning, ignoring the particular variations that some other theory might seize on as crucial. In linguistics, for example, if a theory (e.g., Chomskyan theory) says that all relevant grammatical facts stop at the boundary of the sentence, then related facts at the level of paragraphs, stories, and so on, are overlooked.

The cultural and dark matter forces determining the interpretation of anomalies in the data that lead one to abandon a theory and another to maintain it themselves create new social situations that confound the intellect and the sense of morality that often is associated with the practice of a particular theory. William James (1907, 198) summed up some of the reactions to his own work, as evidence of these reactions to the larger field of intellectual endeavors: "I fully expect to see the pragmatist view of truth run through the classic stages of a theory's career. First, you know, a new theory is attacked as absurd; then it is admitted to be true, but obvious and insignificant; finally it is seen to be so important that its adversaries claim that they themselves discovered it."

In recent years, due to my research and claims regarding the grammar of the Amazonian Pirahã—that this language lacks recursion—I have been called a charlatan and a dull wit who has misunderstood. It has been (somewhat inconsistently) further claimed that my results are predicted (Chomsky 2010, 2014); it has been claimed that an alternative notion of recursion, Merge, was what the authors had in mind is saying that recursion is the foundation of human languages; and so on. And my results have been claimed to be irrelevant.

Yet in spite of such characterizations, the discussion of whether a language has recursion or not is vital to psychology and linguistics (Futrell et al., forthcoming). And if, as I claim, the manifestation of recursion or other properties of grammar are constrained by culture and dark matter, it is also vital to anthropology. One of the oldest and most important empirical programs in cognitive science and linguistics aims to characterize the range of possible human languages. Linguistic universals—if any exist—would point to deep properties of the cognitive mechanisms supporting language; at the same time, the search for possible universals and violations of universals creates rich data for linguistic theory.

One aspect of this controversy is whether a purported universal (recursion, in this case) actually needs to be observed in every language. Some linguists think so, depending on the nature of the universal, while others claim that linguistic universality is a claim of abstract cognitive abilities and not about the formal inventory of any specific language. The question is not whether this or that language actually has recursion, for example, but whether the speakers of that language are capable in principle of speaking a recursive language. Thus is a language without recursion a counterexample, an exception, or irrelevant to current syntactic theory?

To try to see how this is answered culturally, we must begin with two culturally distinct understandings of universals of language, Greenbergian or Chomskyan universals. One of the most common objections raised by critics of the idea that Pirahã falsifies the suggestion that recursion is universal is that the absence of recursion superficially (in what people actually say) does not mean that the language could not be derived from a recursive process mentally. And this is correct. (See chap. 5 for technical discussion.) Some then proceed from this banal observation to conclude that the claim that Pirahã lacks recursion is either deliberately or ignorantly failing to understand the difference between Greenbergian and Chomskyan universals. This is an old accusation, one that I and others have rebutted in numerous publications.

The late Joseph Greenberg was a professor at Stanford University and was the first researcher to make serious proposals on linguistic universals: forms or implications between forms *actually observed* in all or most of the world's studied languages. Thus Greenbergian universals refer to things that can actually be observed and thus easily falsified. Chomskyan universals are quite different. Chomsky's concept of universals includes the notion of what he refers to as "formal universals." Formal universals are grammatical principles, processes, or constraints common to all languages—that is, supposedly following from UG—*at some level of abstraction from the observable data*. Thus these refer to things that cannot be seen except by the appropriate theoretician. Unfortunately, this makes formal universals difficult to falsify because they can always be rescued by abstract, unseen principles or entities; for example, so-called empty categories (which frankly I find reminiscent of Kepler's "epicycles"—invisible to all but the initiate).

Take recursion. The Chomskyan claim would be that all languages are formed by a recursive process, even though the superficial manifestation of that process may not look recursive to the untrained eye. So long as we can say that a sentence is the output of Merge, limited in some way, then it was produced recursively, even though superficially non-recursive. The Greenbergian way, on the other hand, would be to say that either you see recursion or it is not there.

Both positions are completely rational and sensible. But the Chomskyan view renders the specific claim that all languages are formed by Merge/recursion untestable. In Chomsky's earlier writings, he claimed that if two grammars produce the same surface strings (weak generative capacity), we can still test them by examining the predictions of the structures they predict for the strings (strong generative capacity). Since most of my work on Pirahã recursion has been to show that the predictions Merge makes are all falsified (see chap. 5), I have dealt exclusively with strong generative capacity. Of course, clever lads and lasses can always add epicycles to the accounts to save Merge but, again, with two effects: (i) it loses all predictive power, and (ii) it provides a longer, hence less parsimonious, account of the same structures.

The Chomsky-Greenberg split is only apparent in this case. Pirahã falsifies the Chomskyan formal universals predictions/account (sans epicycles, i.e., the bare claim of Hauser, Chomsky, and Fitch [2002]) and is irrelevant to a Greenbergian account, exactly the opposite of the normal dialogue occurring among my critics.

This cultural divide between those whose value ranking is THEORY >> DATA vs. DATA >> THEORY surfaces again in theoreticians' criticisms of N. Evans and Levinson (2009). Again, a feature could be abstractly present, even though it is superficially absent. This is correct. But the critics then make the mistake of moving from this banal observation to conclude that they/we are either deliberately or ignorantly failing to understand the difference between Greenbergian and Chomskyan universals.

In the value system of the theory of dark matter here, there are the following values:

1. Understanding particulars is vital and is the first step in developing an etically valid basis for emic science.
2. There is no atheoretical research, so be informed.
3. Using insights from multiple theories can mitigate the counterexample vs. exception quandary.
4. Never be too sure.
5. The same structure can be a counterexample in one language but a pseudo-exception in another, depending on the "field/matrix" view (or in some cases, the "dynamic" or "wave" view).

Summary

This chapter applied the concept of dark matter as the primary tool for interpreting all of our experiences, focusing on visual perception among various societies, with a special focus on Pirahãs' perception of two-dimensional representations. The chapter demonstrated that Pirahãs' cultural background makes it difficult for them to recognize photographs that are degraded in rather minor ways, contrasting strongly with the perception of other populations. This was argued to be a result of the different apperceptional background of Pirahãs compared to, say, North Americans. At the same time, the chapter discussed ways in which North Americans (me in particular) are unable to perceive things that the Pirahãs are in fact able to perceive quite easily.

PART TWO

Dark Matter and Language

5

The Presupposed Dark Matter of Texts

> I believe that for each scholar and each writer, the particular way he or she thinks and writes opens a new outlook on mankind. And the fact that I personally have this idiosyncrasy perhaps entitles me to point to something which is valid, while the way in which my colleagues think opens different outlooks, all of which are equally valid.
>
> CLAUDE LEVI-STRAUSS, *Myth and Meaning: Cracking the Code of Culture*

Part 1 of this book laid the foundation for parts 2 and 3. It developed a pedigree of the notion of dark matter, and it offered a new theory of culture as knowledge structures, social roles, and ranked values (among other things). In this second part of the book, we take on the role of dark matter in language: in texts, translation, grammar, and gestures. This first chapter examines the content of texts of different sorts in English and Pirahã, showing how the implicit, unspoken material of texts contains some of the most important dark matter of culture—knowledge and values in particular—and the individual.

Implicit Values in Texts

As we saw in the previous chapter, in spite of their lack of expertise with visual symbols, young US children have arguably even more experience with two-dimensional representations than either adult or nonadult Pirahãs, acquiring emic understanding of two-dimensional representations early on.

There are other forms of emically interpreted experiences. Emicization and apperception are the two "power tools" for systematizing our knowledge about the world around us. We have just seen in the previous chapter that dark matter profoundly affects our ability to perceive the world. Yet it seems to me that dark matter exerts itself most powerfully in our stories. And our stories are fundamental to building our concepts of self and culture. Therefore, in what follows I want to examine texts at differing levels of detail from American and Pirahã culture to bring out the role of dark matter in their storytelling, writing, and interpretation.

To begin slowly, I offer some straightforward comments on an editorial from the *Wall Street Journal* in 1969 on the Woodstock Festival that marked the high-water point of the 1960s hippie movement. My cultural interpretations are embedded in the text in italic, enclosed in brackets. A story on the same event from the *New York Times* immediately follows this one. I make no comments on that one because I believe the contrasts between the two texts are clear—striking, in fact—and that the reader familiar with US culture can fill in the dark matter of the second text as I did partially for the first. Of course, an editorial is a type of text whose purpose is to express opinion, whereas a report—the case of the *NYT* text—is supposed to be more objective. For readers, though, dark matter is created in both cases via implicit values communicated. These texts present two very different sets of value rankings, which we can represent simply for discussion as:

INNOVATION IN VALUES >> STATUS QUO
STATUS QUO >> INNOVATION IN VALUES

Following these two stories, we examine in detail two Pirahã texts: one procedural (how to make an arrow) and the other narrative (about the exploitation of the Pirahãs by the Brazilian river traders that used to ply the Maici River). These were two of the first texts I ever collected from the Pirahãs, in 1978.

Let's consider, then, the *Wall Street Journal* report on Woodstock, from August 1969. Most of the values embedded in the narrative are easy to discover, though I want to point them out for discussion anyway because we often fail to register such implicit information. In comments on other texts in this chapter, I focus on dark matter *knowledge*. Here I focus on *values* in texts. Imagine the potential effect of reading such stories as part of a newspaper one reads on a regular basis over time—reading them passively, perhaps, while sipping coffee at the breakfast table or on the train or bus to work. Their effect could be largely subliminal, reinforced with every story that assumes similar values, which is the case of particular newspapers, the *WSJ* being a voice of conservatism and the *Times* speaking for liberalism. We begin with an examination of the dark matter in the WSJ report on Woodstock.

By Squalor Possessed

Wall Street Journal, August 28, 1969, 6

The so-called [*The author denies by this phrase the reality of the claim.*] generation gap is not really so much a matter of age as it is a gap between more civilized and less civilized tastes. [*"Civilized" is a strong value judgment. It is vague, unanchored, and unspecified, but equates the Woodstock participants with barbarians.*] As such, it may be more serious, both culturally and politically, than it first appeared. [*The original coinage of the term "generation gap" referred to a lack of communication, a change in values, a shift in culture. Talk of civilization making the phrase more "serious" indicates that the author sees this as a decline in the quality of life, a weakening of civilization.*]

Starting with the relatively small hippie movement several years ago, the drug-sex-rock-squalor "culture" [*The author manages to denigrate even the word "culture" in this line, placing it in scare quotes. He then equates this culture with drugs, sex, rock, and squalor. He doesn't ask whether the squalor—and there was some at Woodstock—was the result of the hygiene of the participants or the poor planning and failure of the organizers of Woodstock to anticipate the huge crowds. If the latter, then there is no value of squalor shared by all attendees. Everyone is sorry for it and would rather avoid it. Are soldiers living in squalor in combat to be equally denounced? Is such forbearance even admirable or courageous? Notice that no evidence for any assertion has yet been given.*] now permeates colleges and high schools. When 300,000 or 400,000 young people, most apparently from middle-class homes, [*Apparently coming from middle-class homes indicates something negative, such as that they "should have known better" or, more likely, that these hippies are "spoiled" or pampered. But in fact, the author offers no evidence whatsoever for the claim that the participants come from any particular social class.*] can gather at a single rock festival in New York State, it is plainly a phenomenon of considerable size and significance.

We would not want to exaggerate. Probably a goodly number will grow out of it, in the old-fashioned phrase. On campus, the anti-radicals [*Why think the hippies are radicals? Their politics? Their dress? This is not explained, merely used as a scary word. And aren't the opposite of radicals usually taken to be "reactionaries?" Why not call the anti-radicals reactionaries?*] seem to be gaining strength, and it may well be that these more conservative youngsters will be the people who will be moving America in the future. [*"More conservative" apparently is intended to be a positive value judgment.*]

But that prospect is by no means certain enough to encourage complacency. [*By "complacency" the author seems to indicate that there is clearly something we should be worried about and taking action on, not merely reading the news about. These Woodstock people represent a problem.*] For various reasons it is being suggested that many rebels will not abandon their "life-styles" (the clichés in this field!) [*This rejection of a new term seems to imply that variation from the norm is not a "style" but a movement of anti-civilization, as per above. It is most certainly a negative judgement in content. And orthographically it is also sarcastic, with the use of the scare quotes around "life-styles."*] and that there are enough of them to assume some of the levers of power [*something we should be concerned about, apparently*] in the future American society. It would be a curious America if the unwashed, [*a double-entendre—meaning both unworthy and dirty*] more or less permanently stoned [*What observation could justify this statement? What percentage of Woodstock participants were high—and how high—for what percentage of the festival? And why would we think that they would replicate their festival behavior in the workplace? This is again a value judgement based on the author's personal taste, not research.*] on pot or LSD, were running very many things. [*Where does the evidence emerge that people were permanently stoned? The fact that they may have been stoned off and on during the festival is translated into "permanently" via dark matter values. Of course, now the "Woodstock Generation"* is *running everything.*] Even if the trend merely continues among young people in the years ahead, it will be at best a culturally poorer America and maybe a politically degenerated America. [*What values bad for the culture or country do these hippies hold? Again, this is unanchored value judgment with no evidence, no quantification at all.*]

Now taste is that amorphous quality about which one is not supposed to dispute, [*But in merely saying this, once again the author expresses a negative judgment.*] so we won't argue whether rock is a debased form of music; we don't like it, but never mind. [*By selecting this form of the statement, the author's opinion is expressed but there is no need to defend it.*] Without pursuing that argument, it is possible, we think, to say a couple of things quite categorically [*The author believes that knowledge is a matter of certainty.*] about rock and related manifestations. [*The author, like most of us, sees values as part of a network. With this I agree.*]

One is that a preference for a particular kind of music is not necessarily a matter of age. [*Here the author denies that ages can have their own cultures. Simultaneously he claims that there is no generation gap but, rather, a "taste gap"—he again offers a value judgment.*] In times past many young people were drawn to classical music and retained that taste as they grew older. Today the young's addiction to rock is at the same time a rejection of classical and the more subdued types of popular music, and considering the way rock is presented it must be counted a step down on culture's ladder. [*There is no evidence provided for the assertion that rock is worse than classical music, which is what this assertion is intended to convey. One could have argued, for example, that classical music is more complex than rock. But if that makes it better, jazz might even be superior to classical; Miles Davis could arguably be compared in talent to Mozart. Thus there is no attempt to argue—merely, again, to assert a value judgement.*]

That is our second point: The orgiastic [*The author first assumes that orgies are bad. Why? Second, the author asserts something similar to the claim that the movements, body postures, screaming, etc., are not music or dance so much as lack of proper sexual inhibitions.*] presentation on the part of some of the best-known groups. It is not prudish, we take it, [*Meaning: "Wouldn't it be silly to criticize me for what I am about to say?"*] to suggest that a certain amount of restraint [*What kind of restraint? Are you sure that conductors of symphonies are not orgiastic in their movements at times? This is simply another value judg-*

ment.] is appropriate in these matters. But then, the whole "life-style" of many of the performers is incredible—disgusting or pitiful or both, but certainly hoggish. [*more unanchored value judgments*]

[*From this point in the text, I make no more comments. It should be clear that both the remainder of this text and the next are (like all texts) chock-a-block with value judgments.*]

The same applies to public sex in the audience, also in evidence at the mammoth Woodstock festival. It is not necessary to be a Puritan to say that such displays are regressive from the point of view of civilization. As for the ubiquitous drugs—well, we guess on that score we feel more sorry for the kids than anything else.

What perhaps gets us most is the infatuation with squalor, the slovenly clothes and the dirt; at Woodstock they were literally wallowing in mud. How anybody of any age can want that passes our understanding. Again, though, it's not a question of age. A person doesn't have to be young to be a hobo. He does, however, have to have certain tastes and values (or non-tastes and non-values) which are not generally regarded as being of a civilizing nature.

Now we are aware of all the cant about how these young people are rejecting traditional tastes and values because society has bitterly disappointed them, and we would be the last to deny the faults in contemporary society. It is nonetheless true that their anarchic approach holds no hope at all.

They won't listen, but if they, and some of the unduly sympathetic adults around, would listen, here are some words worth hearing. They occur in a speech by Professor Lawrence Lee to a social fraternity at the University of Pittsburgh, quoted in National Review:

"You have been told, and you have come to believe, that you are the brightest of generations . . . You are, rather, one of the most self-centered, self-pitying, confused generations . . .

"The generation gap is one of the delusions of your generation—and to some men of my generation . . . The only generation gap is that we have lived longer, we know more than you do from having lived, and we are so far ahead of you that it will take you a lifetime to have the same relative knowledge and wisdom. You had better learn from us while you can . . .

"It is not mawkish to love one's country. The country, with all of its agony and all of its faults, is still the most generous and the most open society on earth . . . All generations need the help of all others. Ours is asking yours to be men rather than children, before some frightened tyrant with the aid of other frightened and ignorant men seeks to make all of us slaves in reaction to your irresponsibility."

In any event, opting for physical, intellectual and cultural squalor seems an odd way to advance civilization.

Well, we know now not only about this writer's (professed) values, but also about the values he assumes will resonate with the larger readership of the WSJ, a culture in its own right. The *New York Times* text, as a report rather than an editorial, presents more numbers and actual facts than the *Journal*. Still, that reporter chose not to comment on some of the more prurient aspects of Woodstock that so exorcised the *Wall Street Journal* (e.g. "public sex," etc.). This means that dark matter also guided his focus, as it guided the *WSJ* editorial.

In other words, values are also shown in the foci of the separate texts, as well as in their words. That is, dark matter values are also seen in the answers to questions like "What did they choose to write about what they did?" It is like a cameraman who sees a naked woman in one corner of a room and a man giving a lecture in another. Which will he turn his camera onto if he has to choose? And what values does the choice reveal? I reproduce here only a small portion of the original report, as the lesson emerges quickly.

200,000 Thronging to Rock Festival Jams Roads Upstate

By Barnard L. Collier, special to the *New York Times*, August 16, 1969

Bethel, N.Y., Aug 15—A crowd estimated at more than 200,000 poured into this Catskill Mountain hamlet today for a three-day rock and folk music festival, creating massive traffic jams and a potentially serious security problem.

The police reported that an increasing number of cars were being abandoned by motorists on the shoulders of highways as drivers and passengers decided to walk to Bethel. Estimates of the total number of people both at the festival and in the surrounding area were as high as 400,000.

Wes Pomeroy, director of security for the Woodstock Music and Art Fair, who issued the 200,000 figure, warned late afternoon that Bethel should be avoided.

"Great Big Parking Lot"

"Anybody who tries to come here is crazy," he said. "Sullivan County is a great big parking lot."

At about midnight, the festival promoters announced that, with their full cooperation, the state police and local authorities would start turning back all cars heading for the fairgrounds. This would primarily affect vehicles attempting to reach Route 17B from the Quickway (Route 17).

A state police official said, "We're just going to re-route everybody. Sullivan County is filled up." . . .

. . . The police and the festival's promoters both expressed amazement that despite the size of the crowd—the largest gathering of its kind ever held—there had been neither violence nor any serious incident.

As a state police lieutenant put it, "There hasn't been anybody yelling pig at the cops and when they asked directions they are polite and none of them has really given us any trouble yet." . . .

. . . The security force has been augmented by 100 members of the Santa Fe, N.M. Hog Farm Commune, whose members wear colorful clothing, beads and beards, with orange armbands depicting winged pigs perched atop guitar frets.

Sullivan County Sheriff Louis Ratner said today "We don't want any confrontations," indicating that the police were not seeking to make mass arrests.

So far there have been about 50 arrests, most of them for possession of such drugs as LSD, barbiturates and amphetamines. The number of arrests could only be estimated, Sheriff Ratner said, because they have been made at various places through the county, and those arrested were arraigned before various judges. . . .

. . . "As far as I know the narcotics guys are not arresting anyone for grass. If we did there isn't enough space in Sullivan or the next three counties to put them in." . . .

Scheduled for the first performance, which will run into tomorrow's early hours, are Joan Baez, Ravi Shankar and Sly and the Family Stone. Tickets for the three-day program cost $18 each. . . .

. . . The directors of the fair, who invested $500,000 in the promotion and organization of the event, reported late today that a fleet of trucks carrying packaged food was en route to Bethel. Meanwhile, free rice kitchens were available to festival visitors. Restaurants in the area are being hard pressed to provide food and service.

Six wells have been dug in the field and water runs continuously from spigots driven down to the water supply. In addition, water tanks have been placed in the parking lots surrounding the farm. Six hundred portable toilets have been brought to the fair site.

As the audience waited for the afternoon performance to begin—it was delayed as workmen struggled to finish outfitting the 80-foot wide stage—it was diverted by strolling musicians, improvised group performances, and debates among long-haired girls and their bearded boy friends. The debate topics included Vietnam, campus disorders and the merits of various music groups. . . .

It seems clear from these two texts that the perception of the world around us (as we saw earlier in the studies of visual perception among the Pirahãs and Americans)—whether the ability to see snakes under leaves or caimans in water on dark nights, or two-dimensional visual representations, or the simple activities of a rock concert—is not the result of a maturation of the physical perceptual system alone, but also of the content and formation of dark matter, via apperceptions and culturing.

There is a rich anthropological tradition that looks to stories to find culture. The two stories above, from two major US newspapers, underscore the value of textual study (see Longacre (1976), Grimes (1975), Ochs and Capps (2002), Silverstein (2003), Tedlock and Mannheim (1995), Sherzer (1991), Urban (2000), and many others). In every story we tell or hear, there are built-in, assumed or presupposed values, roles, knowledges, expectations, conventions, indexicals, and so on. We exploit form, content, and culture in every communicational exchange.

YOU TALK LIKE WHO YOU TALK WITH

Getting back to the two newspaper reports of Woodstock, they both reflect the oft-repeated principle "You talk like who you talk with." The philosophers Thomas Kuhn (1996) and Paul Feyerabend (2010) independently (and with different foci) argued that scientific advances are not cumulative. Rather, scientists form groups with their own dialects, based on their satisfaction or not with another group's way of talking about the world. Journalism offers one of many examples of this phenomenon. Science another. Linguistics and anthropology others. One could take just about any period in linguistic history over the last 150 or so, for example, to illustrate the same points.

For example, prior to the work of the Swiss linguist Ferdinand de Saussure, most linguists thought of the study of language as the study of diachronic (historical) variation and change—languages are always in a state of change. Therefore, we must understand previous states of any language under study if we are to understand the present state. Saussure claimed that the structure of a language could be meaningfully and successfully studied at any particular point in time, largely independent of the language's historical development, and that understanding of this particular *synchronic* slice of language is independent of that language's diachronic evolution. Not all linguists accept(ed) Saussure's synchronic linguistics way of talking. But because many did, he was part of the formation of a new group, the *structuralists*—those concerned with talking about where language is now, not where it has been or where it is going.

Later, Chomsky became dissatisfied with the linguistics of his student days, which he later characterized as overly preoccupied with linguistic description, ignoring the important work of theorization. His writings introduced yet another new way of talking, a fresh dialect, forms, and dialogue on language. Some linguists did not enter his dialogue or dialect and thus did not talk with him or like him about language. But for years, Chomsky's was arguably the dialect of the majority in the field, the standard dialect, what linguistics call the *superstrate dialect*. Nowadays, however, this former superstrate is becoming the substrate—more and more linguists are talking in a different dialect, with different foci and different values and different forms than Chomsky's, and that earlier powerful dialect is now becoming less prestigious—certainly less pervasive. More significant is that over the years, several of the dialects of linguistics have grown so far apart that in Kuhn's terminology they are now incommensurable.

Another perspective on the difficulty of cross-cultural discussions comes from the article "What Is Philosophy?" by Deleuze and Guattari:

> Philosophers have very little time for discussion. Every philosopher runs away when he or she hears someone say, "Let's discuss this." Discussions are fine for roundtable talks, but philosophy throws its numbered dice on another table. The best one can say about discussions is that they take things no farther, since the participants never talk about the same thing. Of what concern is it to philosophy that someone has such a view, and thinks this or that, if the problems at stake are not stated? And when they are stated, it is no longer a matter of discussing but rather one of creating concepts for the undiscussable problem posed. Communication always comes too early or too late, and when it comes to creating, conversation is always superfluous. To criticize is only to establish that a concept vanishes when it is thrust into a new milieu, losing some of its components, or acquiring others that transform it. But those who criticize without creating, those who are content to defend the vanished concept without being able to give it the forces it needs to return to life, are the plague of philosophy. All of these debaters and communicators are inspired by *ressentiment*. They speak only of themselves when they set empty generalizations against one another. Philosophy has a horror of discussions. (1996, 28–29; emphasis in the original)

The authors here manage to come very close to the perspective on dark matter–imbued discourse I am advocating here. Another way of putting this is found in McDowell (2013, 465), where he cites Friedman: "Are we not faced, in particular, with the threat that there is not one space of reasons but many different ones—each adapted to its own cultural tradition and each constituency in its own 'world'?"

The crucial question that emerges from all of this is how everyday discourse (journalism, scientific discourse, quotidian discourse, or otherwise) is shaped by the dark matter of the background that particular cultures and individuals develop. Dark matter is formed in individuals via participation in speech communities and cultural groups, and it therefore comes to affect what we talk about, what we look at, how we look, how we talk and so on. These are in turn formed, again, by the ubiquitous phrase "You talk like who you talk with."

Knowledge is not dark matter per se—not even knowledge that is accessible but not actively shared by all members of community. For example, the Pirahãs have an encyclopedic knowledge of nature, while my own personal knowledge of nature, especially of Amazonian flora and fauna, is very limited. Yet I have Wikipedia and the Pirahãs do not. This knowledge—that I can consult sources of information stored outside my body—is part of my dark matter, though the information so stored is not. But the Pirahãs do not share this dark matter with me. All knowledge to them is inside their or other Pirahãs' heads—not impersonal sources. Overt knowledge can be found inside and outside our brains, but dark matter is found only within the individual. This holds true of Americans, Pirahãs, French, and all people.

Business Culture

Another interesting example of culture, the individual, and emicized dark matter—perhaps closer to everyday life for some, though equally exotic—is the (idea of) cultures of business in the United States. As noted earlier, US corporations often expend energy talking and writing to employees about the "culture" of their particular company. As an example, consider the following from the "big four" accounting firm Ernst and Young:[1]

> We are already proud of our people culture, and we are committed to doing even more. Our people tell us that our culture of global teaming and our focus on building a better working world make EY a great place to build their careers. . . .
>
> . . . We are investing in three key elements of our culture that enhance what is important to our clients and our people:
>
> Inclusiveness—Recruiting outstanding people is just the start. Inclusiveness means making sure all our people's voices are heard and valued. This not only helps attract and retain the best people, but also it helps get better answers for our clients and our organization.
>
> Development—Our approach to development involves offering the learning, experiences and coaching all our people need to enrich their careers and deliver the best results for clients, as well as offering additional programs for current and future leaders of our organization.
>
> Engagement—We want all our people to feel enthused by their work and their colleagues and to be comfortable in an organization that gives

> them the flexibility to achieve their professional and personal aspirations. We engage our people in countless ways, from selecting the right people to lead major change, to taking an interest in our people as individuals, to being sure to say thank you for a job well done.

The phrase *people culture* as presented here is a plausible culture. To be a real culture, however, the values stated here would need to be folded into the company's defining knowledge, structures, and various roles. But is a statement of culture enough to conclude scientifically that a corporation actually has the culture so described?

Think of other parts of corporate daily life that could also be included in trying to understand the culture, regardless of what they declare in brochures and websites. What are the roles of employees? Who is hired? How are they hired? What tasks and roles are most rewarded (with salaries, bonuses, commissions, stock options, etc.)? What are the relative roles of shareholders vs. stakeholders? What are the company stories in the boardroom, the washroom, parties, and the lunchroom? We cannot understand culture through questionnaires and public pronouncements alone. We must engage in intense participant observation, as in Karen Ho (2009), or in careful analysis of intended results, as in LiPuma and Lee (2004). Businesses use "culture" as an advertising or esprit de corps–building tool, but this doesn't mean that their advertising is necessarily false nor that the culture is as the companies present it. Other considerations include how people dress; how they talk to one another; what their behavioral norms are; their conventions; how employees are trained/emicized; whether they are international or national only, and so on.

Business is a fascinating area for intellectual inquiry of all kinds. From the mathematics behind marketing, accountancy, and finance to the cultural theory and sociology of management, product creations, company values, and so on, it is a rich source of social understanding. But business leaders often do not understand the term *culture* that they so like to use. To see what I mean, consider one example in the news as of late—the attempt to remove hierarchy from the roles in a business, supposedly sharing power throughout the organization, in a "culture-changing" model known as *holacracy*.[2] In this model, managers are eliminated for the most part; the employees are structured around specific jobs, with final say, supposedly, for their part of the job.

This "culture change" is claimed to be based on the following sharp contrasts between "normal" business and a holacratic business:

> Roles are defined around the work, not people, and are updated regularly.
> People fill several roles (serially or simultaneously).

Authority is held by teams and roles specific to a particular job, not by a permanent management corps. This means that decisions can be made more quickly and locally, by the team doing a particular job.
The organizational structure of the organization is regularly revised as each local team self-organizes around a particular job.
The same rules bind all, including the CEO. Everyone knows and can easily find the rules.

Does this model actually produce a new culture? Recall that the discovery of values, knowledge structures, and social roles—the core components of culture, in my sense—is at minimum a two-stage process. First, we interview members of a society to get their understanding of these three components. Second, we observe them, as participant-observers, to see in what ways their stated understanding agrees with their practice. In other words, what culture is proclaimed and what is actually implemented?

In July 17, 2015, the *New York Times* discussed the implementation of a holacratic model by the company Zappos. But neither that article nor the proponents of holacracy know how to evaluate whether it is indeed a new culture or merely tweaking operational methods or just painting over the same wall. Holacratic change affects roles (superficially, at least, but it does not change the core of a company's culture, namely, its core values, nor its knowledge structures. Let us say that the primary value of a company is profit. And perhaps another value is hierarchical role structures—everyone should "know their place." Another value might be fluidity, for some jobs roles will shift. Thus the value ranking for a common company might be:

PROFIT >> HIERARCHY >> FLUIDITY

The ranking for a company applying holacracy would be:

PROFIT >> FLUIDITY >> HIERARCHY[3]

Looked at in this way, the "cultural shift" here is relatively minor structurally (though it might have a major impact operationally, of course). So long as the highest-ranked value is PROFIT, companies will look more alike than different. And their shareholders and stakeholders will ultimately evaluate them in very similar ways (how much profit is coming in, what the future of the company is—i.e. will I have a job—based on current profit, etc.). Moreover, the role structures are affected superficially, nowhere as profoundly as claimed. To see this, consider the holacratic model versus the non-holacratic model (figs. 5.1 and 5.2). While the non-holacratic model includes buffers—managers—between the employees and the CEO, the CEO (regardless of what the company brochure might say) is still the CEO and can still change

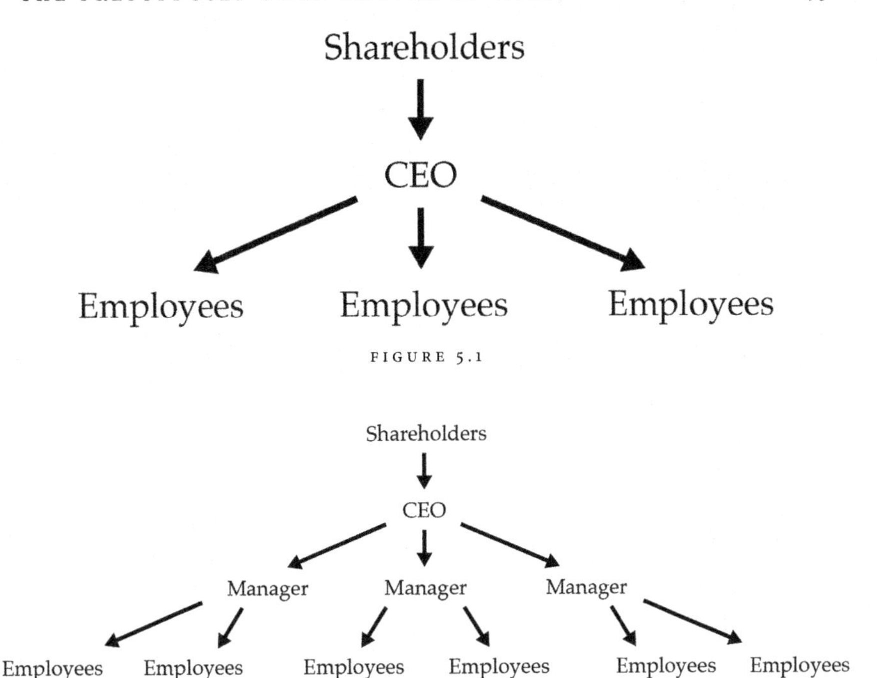

FIGURE 5.1

FIGURE 5.2

management style or overrule employees. There is definitely a change. And it could have profound benefits for the company. But from the perspective of the theory of culture and dark matter being developed here, the difference is very slight, with the value FLUIDITY allowing in fact relatively little variation.

This gets us to a more radical pair of alternative rankings that could profoundly affect a company's culture. For quite a few years one of the central topics of business (especially as folks in business use the word *culture*) is whether businesses should be focused on serving shareholders or stakeholders. The shareholders are the owners of the equity of a company. The stakeholders are those who engage the company in some way—customers, managers, employees, shareholders, subcontractors, stores dependent on purchases by company employees, the environment, and so on. All shareholders are stakeholders, but not all stakeholders are shareholders.

Let's say that you take ten dollars from me and that you tell me that you are going to use this to purchase some candy bars, set up a roadside stand from which to sell them, and then make a profit of 15 percent per candy bar. You and I agree to be partners—I putting up the capital, you the labor. We agree to divide risk and profits fifty-fifty. While you are selling the candy, a young boy offers to clean off your display table and help find customers if you agree to pay him one dollar per day. You agree. Then, at the end of the day, you are so impressed with his work that you give him a free candy bar and an extra fifty cents from *our* profits. He is happy. You are happy. But am *I* happy? No way. Technically, you gave away *my* money without my permission. You *stole* my money.

In this little fictional story, the boy who helped is a stakeholder and I am a shareholder. You might respond to my anger with something like "Chill out! Giving away some inventory and being generous is good for business." I reply, "That is not for you to decide. It is for us to decide together. And what is your evidence that giving away my money is good for business? I think I am going to have to sue you."

What has happened? Part of my value ranking is: SHAREHOLDER/PROFIT >> STAKEHOLDER, whereas the corresponding part of your value-ranking would appear to be STAKEHOLDER >> SHAREHOLDER/PROFIT.[4] Rankings can produce slight differences or major differences, depending on how high up the hierarchy the particular values sit. In any case, the idea that businesses have cultures is likely correct, based on the understanding of culture developed here, but it is on a much smaller scale than the companies claim for themselves. Given the importance of business and businesses in economically modern societies, to understand their claims and practices regarding values, value rankings, knowledge structures, and social roles is a worthwhile and important undertaking.

Implicit Knowledge in Texts

In the texts of the *WSJ* and *NYT*, we focused on values implicit in texts accessible via dark matter. Here I want to consider knowledge, as another form of dark matter accessible through texts. Hermeneutics is a key itself, emerging from emicized knowledge. For example, consider a sample English text such as the following, from the *New York Review of Books*:

> Federal and local law enforcement agencies have revealed fourteen plots that have either failed or been foiled since September 11, 2001. It would be impossible to quantify the role the NYPD has played in this record. For example, the would-be Times Square bomber of May 2010, who reportedly had ties with the Taliban, was thwarted by a hot dog vendor who spotted smoke from a lit fuse in the parked SUV that held the terrorist's bomb, and immediately reported it to nearby police. (M. Greenberg 2012)

Looking very superficially at these lines, merely for illustration here, we see that they are rich with shared dark matter of cultural knowledge. Such knowledge cannot be googled for the most part, but must be learned by life in a particular culture. For example:

1. Societies can name pre-measured periods of times and record these (calendars).
2. Calendars are important forms of shared knowledge, allowing members of a society to temporally situate non-current events with relative precision.
3. Some cultures quantify (keep numerical records of) social practices by various subgroups.
4. There are people who earn income.
5. Income can be earned by selling food to people who do not always make their own.
6. There are nonnatural, processed foods.
7. There are multiethnic communities.
8. There are social subgroups that are not family, yet are smaller than the whole society.
9. There is such a thing as a political organization.
10. Smoke can indicate imminent death as well as fire for cooking and warmth.
11. Words should be placed in the order subject-verb-object unless intentionally signaling something beyond the literal meaning of the words.
12. One can put one sentence inside another to more effectively communicate old information (relative clauses).
13. Consonants are aspirated at the beginning of certain syllables.

None of these items—just barely scratching the surface of implicit information in the sentences—are universal. "Plots" are not universal. "September 11" is a particular historical event known through Western culture. Times Square, the Taliban, fuse, SUV, terrorist, and so on, are all cultural and nonuniversal—concepts and categories that are neither explained in the text nor universally known. They draw upon dark matter, unstated cultural experience.

Another example of dark matter knowledge can be found in Mark chapter 1, verse 4:

> And so John the Baptist appeared in the wilderness, preaching a baptism of repentance for the forgiveness of sins. The whole Judean countryside and all the people of Jerusalem went out to him. Confessing their sins, they were baptized by him in the Jordan River.

Here we see at least the following contents of American dark matter, based upon American interpretation of this ancient religious text:

1. We do bad things that we must be forgiven for.
2. We can be forgiven by telling others about our bad things.

3. It is possible for one person to speak to God.
4. There is an entity known as God. God is male. God is judgmental. God is to be feared.
5. A story about another culture that has not existed for more than two thousand years can be crucially relevant to an extremely different modern industrialized culture.
6. There are deserts.

INDEXICALS IN DARK MATTER

Alongside dark matter knowledge that is triggered by texts, we use additional kinds of overt references to dark matter outside of formal texts (e.g., conversations, art, and other manifestations of culture and Culture). For example, we often use placeholders for cultural values, often in the guise of knowledge, which serves primarily to underscore the values. Important work on these referents refers to them as *indexicals*, and they are also important for our understanding of how dark matter structures our interpretation of the world, being as they are a subset of our unspoken, occasionally ineffable, knowledge.

The term *indexical* initially was developed to single out the behavior of words—especially pronouns—with systematically shifting references. For example, the first-person singular pronoun, *I*, refers to whoever is speaking, shifting back and forth in a conversation to refer to the different speaker. It can also be used by someone speaking when quoting another person speaking, as in

> "John said, '*I* [i.e., John, not the person telling us what John said] can't make it.'"

Or this indexical can refer to the speaker, as in:

> "John said I can't make it because he needs to use the car."

Likewise, the indexical *now* refers to the moment of utterance. *There* refers to a place away from where the speaker is. In other words, the literal meaning of the word is not enough to know what it is referring to: *which* first-person singular is it singling out? Which speaker? To interpret indexicals, one needs knowledge about the current context of utterance. Indexicals often straddle the boundary between language and culture. Ochs and Capps (2002) and others have shown, for example, that body posture during conversation can index gender or respect for authority in particular cultures.

The idea of indexicals emerges from the work of C. S. Peirce (1977) and, to a lesser degree, from Saussure ([1916] 2012). All signs are combinations of form and meaning. Peirce identifies three types of sign: the *icon*, the *in-*

dex, and the *symbol*. Indexes are connected physically to what they signal or "mean." For example, smoke is usually physically connected to/caused by fire, and so smoke is an index of fire. A footprint is an index of the person who made it. An icon is not physically connected to its referent or meaning, but it somehow resembles it, however idealistically. A picture, for example, is an icon for a person by physically (reflection of light waves, etc.) resembling that person. Finally, though, there are the signs that make human language possible: symbols. The forms of symbols are connected *conventionally* to their meanings. The form-meaning connection is thus a result of dark matter induced by cultural practices—culturing, more generally. I utter the phonemic sequence *snow* when I wish to describe crystalized frozen precipitate; another language would use another phonemic sequence. These are all well-known facts, of course. I review them here to better understand the more elaborate notion of indexicals that has emerged from the work of Silverstein (2003), Eckert (2008), and others.

In one of Silverstein's (2003, 193–229) most important discussions, he argues that indexicals are ordered hierarchically. He makes this point via an analysis of how some people talk about wine. As he puts it (193): "'Indexical order' is the concept necessary to showing us how to relate the micro-social to the macro-social frames of analysis of any sociolinguistic phenomenon." In developing an intricate theory of hierarchical and recursive indexing functions linking hierarchically structured social contexts, Silverstein illustrates the knowledge-and-value intertwining of dark matter, showing how a confluence of the cultural and the individual shapes the way we think, talk, and present ourselves in particular social contexts.

Like the populations Silverstein has studied (indeed, any population), the Pirahãs also index sociocultural components in their speech and behavior. For example, consider what I have elsewhere (D. Everett 1979, 1983, 1986, 2005a) referred to as "phonetic posture" and phonological inventory, used to distinguish male vs. female speech. Pirahã women usually speak with more constriction at the back of their throat, producing the impression of a more "guttural" speech than men. Moreover, the points of articulation of most Pirahã women (at least, at the villages I focused on regarding this) tend to be retracted, or farther back in the mouth, relative to men's speech. Moreover, women speak usually with one less phoneme than men, as discussed in more detail in the next chapter. Thus phonetics and phonology are employed as gender indexicals in Pirahã. Moreover, the use of evidentials and the principal of "immediacy of experience" (D. Everett 2012b) are both further indexicals at a higher level, indexing Pirahã culture as distinct from outsider culture.

DARK MATTER OF PROCEDURES AND COMPLAINTS

Returning to texts, however, perhaps the richest and most accessible source of indexicals is texts. We saw rich evidence of dark matter earlier looking at texts on Woodstock, and the Bible. Now I want to examine textual cues of dark matter in Pirahã discourses. The text that follows is an expository discourse from Pirahã, collected in a small office in Belém, Brazil, in March 1979. The speaker, Kaabogí, had traveled with me from his village, in order to accompany his niece, the approximately twelve-year old Paáxai, who was to receive surgery and physical therapy on her legs because she had contracted poliomyelitis while a baby in the village (from an unknown source).

Since her surgery was scheduled several weeks later in Brasilía, the FUNAI (National Indian Foundation of Brazil) had asked me to allow the Pirahãs to stay with me until surgery and then to transport them (at my expense) from Porto Velho to Brasília. I took advantage of the time to learn about the Pirahãs from Kaabogí and to help him learn more about Brazilian culture. We stayed in the large city of Belém.

I had brought some Pirahã bows and arrows from the village as souvenirs, and they were in the office. So I asked Kaabogí to tell me how to make an arrow. As I switched on the recorder, he went right into instruction on how to make the serrated tip arrow they use to kill monkeys. Fluently, though impromptu, he explained how to make such an arrow and said a bit about its purpose. The Pirahãs often hesitate when talking about such things because they assume that everyone knows how to make arrows, hunt, and so on. But Kaabogí was an experienced teacher and knew that when I asked for information, I was after any information I could get.

What is explained is the *visible* portion of the arrows. What is unspoken is their unseen functions. The functions of the different arrow types are never explained. It seems obvious to a Pirahã, for example, what the different tips are for—that is, their functions are derivable from their forms (to one with an emic perspective). For example, the serrated tips are useful for monkeys. This is because some monkeys, especially large spider monkeys and woolly monkeys, will pull arrows out of their bodies when shot, fleeing to another part of the jungle. They cannot pull out the notched or serrated-tipped arrows, however, without inflicting much more pain and suffering on themselves. So the five-foot-long arrow remains in them, preventing them from escaping into the thick branches. They either fall to the ground or die in place, where they can be easily recovered. This is all background information that would not normally be made explicit, as indeed it is not in the following text.

Arrow-Making Text

Kaabogí Pirahã, March 1979
Recorded by Daniel L. Everett

1. *Sahaí itababi hi aagá. Hoí hi gái.*
   He did [asked about] tying an arrow. A couple he said.
2. *Sitababi hi aagá hói. Ti soioágahái ʔogáogába gaí. Ti ʔigáísai.*
   He asked about tying an arrow. I need thread to put there. I spoke.
3. *ʔi ʔi soioágai ʔiga. ʔáítatíi ʔogabógaáti.*
   Thing, thing, thread. [Or] cotton spun line you need.
4. *Tíigíi poioaagá gai. Ti tipóita gái. Gáobáháá gaííhí.*
   Hard tip you need there. I tip there. Put right there.
5. *Ti aiíʔí bóíóoeooe hiaóbáhá gaií. Ti ʔbaáʔai ʔií.*
   I do bamboo. You put there. I make it good.
6. *ʔíáaogió ʔipoíooe hiaó bí sogi. TipóíbogaiʔipÍso.*
   There you need to put [it]. I therefore put hardwood on [the] arrow tip.
7. *Kabáahagáí. Ti ʔíi poiʔaoágai. Poi báakoi.*
   Nothing else. I put hardwood. Good hardwood.
8. *ʔogáísai. Hi gáísai ʔoogiái. Poi báasí káipaá. Sitaí.*
   Dan. [I] speak [to] him, Dan. Put good hardwood on it. Feathers too.
9. *Píaií. Hoí toíʔí táogaá gaii hoítoí. Hoítoíʔí táooágai.*
   Also. Currasow feathers tie on. Currasow. Tie on currasow feathers.
10. *ʔígai ʔígai ti gáabáá ʔígai. Kahaiʔíooí, kahaiʔíooí.*
    I tie it next. The arrow shaft, the arrow shaft.
11. *ʔaáágaii. Kahaiʔiooí ʔiʔáoihoi. ʔiʔáóíʔo ʔíooíhii.*
    I place it. The arrow shaft. Species of grass [from which is made the arrow shaft].
12. *Tigáobá gaiii. Tii ʔíi poiaáagaii tipói tagáigáobai.*
    I finish tying it. I put on the hard tip. I tie on the hard tip.
13. *ʔoogiái hi, káhiógisai. Kahai booʔabísa póoii.*
    Dan wants an arrow. Put on the bamboo tip.
14. *TagáigáobáháʔaI. Higáísai ʔoogiái koapóí tagaigábógááti.*
    Finish tying it. [I] told Dan. Thusly tie it on.
15. *Tagai gábógááti. ʔibóihoi píaii. Póii píaii.*
    Tie it on well. The bamboo. The hardwood too.
16. *Póii ʔaáati poiii. Hoíbogaai paháxai.*
    Put on the hardwood. For when you shoot the bow.
17. *ʔípói bogaai páʔai. Ti ísitaí gáoʔoa.*
    The hardwood on the bamboo. I tied feathers next.
18. *Sitaíta pitaí hi abaíʔai. Pita ógííá.*
    Eagle feathers to make it pretty. Get eagle feathers.
19. *Kogaʔí kabahákogá. Hoítoí ʔí táogá hoítoí.*
    Nevertheless [eagle feathers] ran out. [Just] tie it with currasow feathers. Curassow.

FIGURE 5.3. Thread spun by Pirahã women holds the tip of the arrow together, binding feathers to the arrow.

20. *Sitaíxi ʔáoi hoiʔipóooi poi. ʔipóii kahaiʔioíi ʔáaá kahaiʔíooí.*
    Tie the feathers on to the end. Tie the feathers onto the end of the arrow.'
21. *Kahaiʔíooí póii pói bogaiipahái. Gaii pói koʔoí ʔabáaáísai.*
    Put a hard tip on the end of the arrow shaft. It is good for killing spider monkeys.
22. *Pói koʔoí. ʔabáaísai póibogai. Hi gáisai. Poibogaáti poi.*
    The serrated tip is good for spider monkeys. I speak to him. [Use] a serrated arrow tip.
23. *Sabáaáíhai saopíkoí. ʔai ʔigíaiʔa.*
    It enters securely. OK, that's all.

What is the dark matter of this text? Well, it turns out that the cotton thread that holds the arrow tip and feathers to the arrow is spun by women. A man

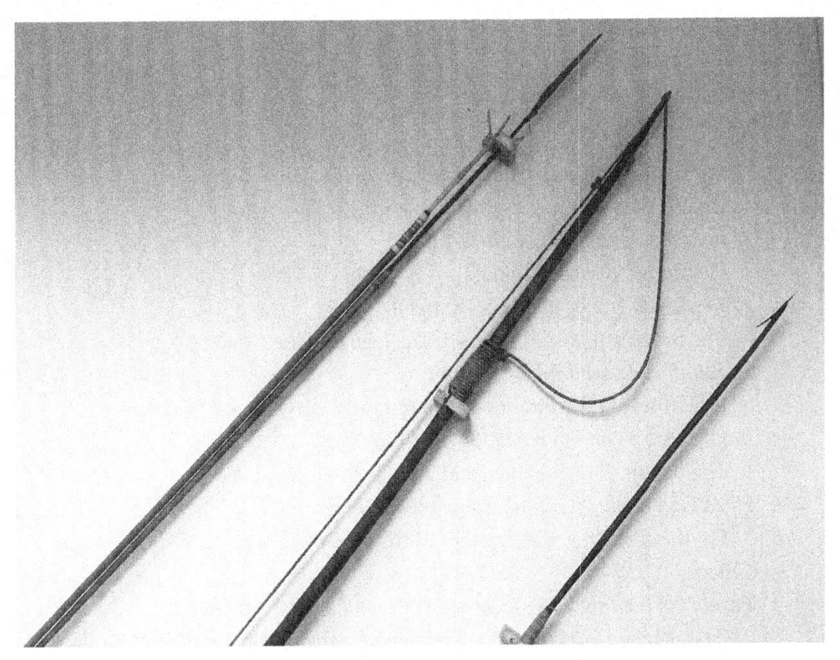

FIGURE 5.4. Set of Pirahã arrows: bird, big game, and fishing arrows.

cannot have an arrow without a woman in his life—mother, sister, or wife. Arrows need a hard primary tip to hold on the additional specialized tip for the type of game (see figs. 5.3 and 5.4). This tip is crucial. Another fact is that there are various types of tip material available in the jungle, but some of it warps and then the tip is less than optimal. Thus the spoken text is merely superficial information about arrow making. What was not spoken is assumed. But most of it is certainly able to be discussed, as we just did here. Deeper dark matter would be the need to kill in the first place.

Another Pirahã story reveals a more complex set of values, built around the Pirahãs' ambivalence toward river traders. On the one hand, they want the trade items they bring, but on the other, they do not like the traders taking things from what the Pirahãs perceive as their own land, the flood plain of the Maici. The trading has been halted since around 2002, as FUNAI takes the Pirahãs to town by road to trade and purchase goods. But at the time of this story, a group of men whose families had for generations traded with the Pirahãs—all based nearby at a settlement called Nossa Senhora Auxiliadora came regularly, several times per month during the Brazil nut season (December to March) to trade.

Stolen Brazil Nut Groves

Informant: Kaabogí
Linguist: Daniel Everett
Date: ca. March 1979

1. *Hi ʔooagaií ʔooagaií hi hiabaaí hiabisóaa.*
   He did not pay Xoágaii [Kaabogí's brother].
2. *ʔo hi hiabaaí hiabisóai hiaitíihiʔí.*
   He did not pay the Pirahã[s].
3. *ʔaiia saagábagaá kagáíiai hoaʔáí sigíai.*
   The Brazil nut grove is named. The jaguar place.
4. *Hiaitíihíí ʔaiia sabá ʃííko (Chico).*
   [The thief of] the Pirahãs' Brazil nut grove [is] Chico [Alecrim].
5. *Báí ʔao ʔao ʔo híi hia baí hiabahá hi ʔao ʃííko.*
   The parent, foreigner, he, paid not. He foreigner, Chico.
6. *Hi ʔao hói hiába ʔo ʔáíiasi hiabaí hiabáí ʃííko.*
   He the foreigner did not do a little. It Brazil nut grove did not pay Chico.
7. *Pasabí ʔaí hiaitíihí ʔaíia sagábagáá. Kagáíia hoaʔáí sigiʔáíao.*
   Passar Bem [foreigners] [with respect/to the detriment of] the Pirahãs call the Brazil nut grove.
8. *ʔa ʔao ʔáíiasii hiabaí, hiabahá. ʔao ʔao ʔáíia soáo báóoa. ʔáóiitá.*
   It, foreigner Brazil nut grove did not pay. Foreigner Brazil nut grove João last name. Didn't pay.
9. *ʔooá hi ʔogaaá ʔí ʔai. ʔa ʔa ogoó.*
   He wants the place. That one. Foreigner wants.
10. *Bigái ʔi ʔaaí ʔi hi Boisíi ʔa híai hi.*
    Bernardo, then Moisés also want.
11. *ʔa Boisíi ʔa ʔao ʔáíasi hiabaaíhiabá.*
    Foreigner Moisés foreigner Brazil nut grove did not pay. [Moisés did not pay for the Brazil nut grove.]
12. *Sogohóíga ʔai píaii.*
    [The Brazil nut grove] Sogohóíga also [he did not pay for]. ["Sogohóíga" is a Portuguese loanword, but I am not sure what it is.]
13. *Hiaitíihí ʔí ʔaiia saagabagaá ʔai. Si kaá sogohóíga ʔáí.*
    Pirahãs call the Brazil nut grove that. Its name is Sogohóíga.
14. *Hiatíihí ʔí ʔaiia sabá sogohóíga.*
    Pirahãs' Brazil nut grove is called Sogohóíga.
15. *ʔáí ipoógii hii gíai. ʔoogií hoiigíaí ʔao ʔáíiasi hiabaaíhiaba.*
    Then Xipoógi hi [said to] you. In the deep jungle, in that area, he did not pay for the Brazil nut grove.
16. *Hi ao ʔao hi I hoaihiabahá. ʔaohii hoaí hiabahá.*
    He, foreigner, foreigner, he didn't talk. The foreigner did not talk.

17. *ʔo hii hoaihiabahá ʔa ʔo ʔa ʔa ʔo ʔoga ʔaó a ʔo.ʔáíí gaaába.*
   [About the Brazil nut grove] he did not speak. He wanted the foreigner the Brazil nut grove. Thus [not speaking] he remained.
18. *Batíóoi ʔaabaábái.ʔaiiisai gíai gaá Batíóoi.*
   Martinho almost remained [not speaking]. Did not speak. Martinho.
19. *ʔogáísai híiga hi obaihiabihái ʔoogiái hi ʔáiia hiahoáti.*
   [About] [the Brazil nut grove] he [did not] go. He did not see. Dan speak [to him about our] Brazil nut grove.
20. *ʔoogiái. Batíóoa ʔíga ʔaisahá. Tibaáóbahá.*
   Dan. Martinho spoke. I shoot you.
21. *Hi Batíóoi ʔo gáísaiiá. ʔoogái hi obaihibihaí.*
   He, Martinho, spoke. He will not see the field.
22. *Higíhi hiaópikáha. ʔáapagi ʔi tiihíi.*
   The man is angry. There are lots of Brazil nuts.
23. *ʔisáí hiagohá ʔ Batío. ʔa ʔaí hiogísaʔái.*
   It [is too] hot to go. Martinho [said]. Doesn't he want to go?
24. *Tíaihaí ʔip ʔipógí. Hiogísa. ʔaitíihi ʔaí ʔipógii.*
   I will go Xipoógí (said). He wants to. The Pirahã, Xipógií.
25. *Hiogísa. ʔaitíihi ʔaí ʔipógii. Potagíipa ʔai ʔí.*
   He wants to. The Pirahã Xipógií. [He wants to go to] Ponta Limpa.
26. *ʔaapago bií tihí ʔai ʔi saóoi ʔahoagía.*
   He paid well for the Brazil nut [pagou bem—Portuguese]. Xiaóoi spoke.
27. *ʔai ʔaitáobaaa giso. Bagisai. ʔiáágaí ʔisoáo.*
   He laid down much [of] this. Giving. He gave much, João.
28. *Baagáha ʔisoáo. ʔi ʔipógií hia ʔipógi hiáo.*
   He gave much, João. He [paid] much to Xipógií.
29. *Hi oagaoa kaisa. Hóʔai ʔaoáto bohóí hóihii.*
   He gave [many] box[es]. The guy. The [guy in the] small boat.
30. *ʔahía tobooá hóááobáá ʔagaíísi ʔao.*
   He paid many sacks [of] manioc meal, the foreigner.
31. *ʔo ʔo ʔáái hiabáaí hiaba. Hi ʔáihi ʔaogosa.*
   Brazil nut grove he paid not. He Brazil nut grove [did not pay] the foreigner.
32. *ʔagá Paohósa gaísai ʔaooíi hiabaaihiabisaíʔáaga.*
   Thus pão rosa [rosewood] speaking. The foreigner remained without paying.
33. *ʔaoooiíi hiabaaí hiabisaiʔága. Paohósa ʔigaísaí.*
   The foreigner remained not paying. Talking [about] rosewood.
34. *ʔaooiíii hi píai ʔógaabá. ʔáooiíi hi píai.*
   Foreigner he also wanted it. Foreigner too [wanted it].
35. *ʔii gaabáhá paohósa. Gai hiahóá ʔao hoʔo.*
   He took the rosewood. Then he took it.

36. *Hi aagáaoákaisi. ʔóogísi ʔaagaoa kaihiabaaí.*
    He [promised to] make a canoe. [We] wanted a canoe. He did not make one.
37. *ʔisáooi ʔaooá. ʔaooí hihiabaaí. Hiabísóai.*
    Xisáooi, foreigner. [That] foreigner really paid not. He just paid not.
38. *ʔí soʔáo ʔai kagáíahoaʔáí sigíai.*
    João [paid for] it, [the Brazil nut grove] Jaguar place.
39. *Hiaitíihi ʔiáiiakoíʔaahá. ʔioága.*
    The Pirahãs hunger. [They are] hungry.
40. *ʔoógihoigíaihi. ʔáíia. ʔáíia baagabaikoí.*
    In the jungle. The Brazil nut grove. The Brazil nut grove. He gave it.
41. *ʔaboitóhoi. Hiboitóhoi. Gísóógábáí.*
    Foreigner's boat. His boat. [I] want this. [Says the foreigner.]
42. *ʔihiʔaa ʔáíiaasi hiábaaíhiaba. ʔigaahiá pasabíi.*
    He did not pay the Pirahã's Brazil nut grove. The one called Passar Bem.
43. *ʔiai hi sogohóíiga. ʔai goatáa ʔai.*
    The Brazil nut grove Sogohóíiga. [Nor] the Brazil nut grove Guatá.
44. *ʔiʔi ʔopísi ʔao ʔáíia soabógabái ʔaháa.*
    The foreigner wanted to take Xopísi's Brazil nut grove.
45. *ʔopísihi ʔáíia sabáisiiʔi Bogáʔaiiabai.*
    Xopísi's Brazil nut grove has a name. Bogáʔaiiabai.
46. *ʔaoi tóhoi gaogabaii. ʔopísi hi gáisáí. ʔáíiaabagi.*
    The foreigner wanted to take his boat [to the Brazil nut grove to get Brazil nuts]. Xopísi spoke. About the Brazil nut grove.
47. *Toisáaoaí hiaitíihi ʔáiapi. ʔabáógooíhi.*
    Tuchawa is the Pirahãs' Brazil nut grove too. The foreigner really wants it.
48. *ʔáóoi itopáhátaío. ʔabásaiʔaagáti ʔaooíi.*
    Foreigners take from it. The foreigners remain.
49. *Topápá. ʔabá boitoisáaoaʔáí. Hoiíí píaii híʔai.*
    They have taken. They probably took [Brazil nuts] in their boat. Óther products too.
50. *Hi ʔopísi ʔaga ʔáíias abóaiisihaá. Higáísai.*
    Xopísi did. He took the Brazil nut grove. He spoke.
51. *Hibáógoohigí ʔáooáapísaohaʔáí. Síinti (Vicente).*
    He really wants it. Will he take it? Vicente?
52. *ʔo ʔáooaáti gíʔai. ʔaiia kabísiʔaiípí. Hisíʔáí.*
    Speak to [the foreigners about our] Brazil nut grove. The Brazil nut grove will be gone. The sun will be hot.
53. *Nósi (Não sei)*
    I don't know.

This story was told to me on my second visit to the Pirahã community in January to March of 1979, after a life-threatening trip, recounted in Everett (2008).

Again, the storyteller was Kaabogí, a man perhaps two or three years older than me, one of the two most experienced language teachers of the community, the other being my main teacher, Kohoibiíihiai (Kohoi), and both having worked extensively with my predecessor among the Pirahãs, SIL missionary Steven N. Sheldon.

The attitudes of Kohoi and Kaabogí were very different. Kohoi was willing to help me learn his language, but to him it often seemed like just a job, one of the ways he provided for his family. Kaabogí was always joking during our sessions but trying to enlist my help for the community as a whole (he had no children at the time). He was particularly concerned for the relationships between the Pirahãs and the various river traders that came up the Maici to purchase jungle products from the Pirahãs or to hire them as laborers. He wanted a richer source of manufactured goods, an issue that appealed to some Pirahãs more than others (for example, Kohoi wanted mainly that outsiders leave the Pirahãs alone).

For his first story to me, therefore, Kaabogí chose to talk about various river traders and how they were exploiting the Pirahãs by taking Brazil nuts, rosewood, and other products out of the Maici without either paying the Pirahãs or paying them insufficiently for this access. I did not understand this very well at the time, both because I spoke very little Pirahã and because I had no knowledge or understanding of the complex relationships between a dozen or so individual Pirahãs and regular river traders.

As a general introduction to the text—material that I would have liked to have known at the time in order to interpret it—we need to understand some background. First, there are many *caboclos*, Brazilians who live along the banks of the Madeira, the Amazon, and other rivers of Northern Brazil. These people make their living from the land—cultivating, fishing, and hunting, but also by owning small boats powered by diesel engines that they use for traveling in smaller tributaries to collect jungle products, such as Brazil nuts, sorva, rubber, rosewood, mahogany, fish (that they salt for later sale), and palm oil. At one time they also brought back exotic birds, live jungle cats, otter skins, alligator skins, and so on, but these have been prohibited, dramatically affecting some caboclos' income. Some caboclos became river traders. Many of these have become rich, though their visible standard of living is not dramatically affected. I have met simple-living caboclos with incomes of tens of thousands of dollars per week, made from owning several boats (even some very large boats, perhaps two hundred feet long and fifty feet wide), who purchase products from other traders and sell them in Manaus to the main dealers in jungle products for most of Brazil. However, most river traders are poor, barely able to support their families, working hard in their fields and hunting in the jungle when they are not out on their boats. It is an extremely tough life.

Rich caboclos live like poorer caboclos because of deeply held values shared in caboclo culture regarding wealth. Wealth is a sign of sloth and dishonesty among caboclos. To be wealthy is to have taken money away from workers for your own use, without earning it. This is because all respected forms of earning money to a caboclo must include physical labor—making fields, fishing, hunting, building a house, clearing land, trading, and so forth. Thus to ostensibly demonstrate their honesty and work ethic, even well-off caboclos live in (occasionally) slightly larger homes, dress like other caboclos (in old clothes and flip-flops), and use their machetes daily to clear land. Their wealth is shown in the cities, away from other caboclos, where many of these wealthier caboclos have large homes, drive cars, send their children to private schools, and so on. Wealthy caboclos thus, in the contrast between city and country lifestyles, illustrate a difference the field worker and theorist must be aware of: the difference between *professed values* and *practical values*.

All the traders who work among the Pirahãs live in small settlements, like the Auxiliadora—a Salesian-founded community on the banks of the Madeira River, about thirty miles north of the mouth of the Maici, where the Pirahãs live. These include the people named in the text above. These caboclos have interesting attitudes toward the Indians. They believe that the government and missionaries pamper the Indians by offering them services and support that the caboclos themselves have no access to. Indians such as the Pirahãs seem lazy and purposeless to the caboclos because they have only relatively tiny fields, spend a great deal of their day (to the outsider's eyes at least) lying around chatting, find physical labor "repugnant"—to quote a river trader—and have no desire to accumulate wealth. Since the caboclos do want and need outside goods, think that they work harder, think that they speak a "real" language, and are all around more deserving than the Pirahãs, they see nothing wrong with taking jungle products and not paying for them, or paying very little for them. The most common form of payment is cheap rum (*cachaça*) at about US$1 per liter. Six liters would at the time have paid an entire village for several days of product gathering labor.

To these traders, outsiders like me—who help the Pirahãs charge higher prices for their goods and services—are their enemies, taking from them, the deserving, to give to a group of lazy, barely human, worthless creatures. Tensions can be quite high, based on these misunderstandings. Such misunderstandings, which serve as the background to this text, result from different emicized values, knowledge structures, and roles. They may not be ineffable. But they are not usually talked about. They are simply obvious to both parties.

Other dark matter behind these texts is seen in the sophisticated way that Pirahãs (like other indigenous groups I have worked with) try to play outsiders off against one another. Neither the caboclos nor I are Pirahãs. We are all outsiders. Moreover, the Pirahãs understandably see themselves as the people that matter most. And they perceive me as wealthy and able to do even more for them than the river traders. Why else would they allow me to remain with them, other than for me to offer assistance to them? To many Pirahãs, I am as natural a friend as a Martian would be to the average American. They know I am not dangerous, so then, am I useful? Social scientists who fail to ask why they are being told this or that story relative to their own status in the community (and they always have one), is in fact not going to understand very well the context, purpose, or background behind what they are being told.

In my experience, a very common tactic in the encounter between outsiders and autochthonous populations (e.g., anthropologists or linguists and Amazonians) is the "You're the best friend we have ever had" tactic. The people will tell the field researcher how much better that researcher treats them than previous researchers. They make him or her want to even outdo him- or herself in giving more, buying more, doing more for the people. The researcher who wants others to respect "their" people but who does not recognize the subtle psychology of some of the stories he or she is told, their content, and their theme is being no less disrespectful. But it is hard to see this because it is the indigenes playing on the dark matter of the researcher—the Western-culture syndrome. The researcher often believes or seems to believe that his or her motives are pure and untainted by desire for personal gain (too often forgetting or ignoring the fact that tenure decisions, promotions, book deals, salary increments, etc., are based on the researcher's success at fieldwork). As the people see the researcher's tendency to want to be their "ally" in this I sense, the stories include more accounts that call for help. There is nothing wrong with this. And the need for help is almost always real, acute, and essential. However, it is important that the researcher recognize all of this—his or her place in the cultural encounter—in order to better understand what is being done and thus better understand the intentionality and meaning behind it. But this requires meta-reflection on the dark matter of the text as well as the dark matter of the intricate dance between scientist and subject—roles that are more fluid than the "scientist" realizes (the Pirahãs, for example, were always faster to understand and predict my behavior than I theirs). Because it is such a rich source of insight into dark matter, I want to examine just over a third of this text line by line, bearing in mind that all we have just discussed is part of the background.

In line 1, we start off with a vague reference "He (a Brazilian trader) did not pay Xoágaii (Kaabogí's brother)." Who was this thief? What was he stealing? Although some clarification is provided in the course of the text, much is assumed. The Pirahãs' harvest their major jungle crop—Brazil nuts—between December and March, the height of the rainy season. The work is not easy. Brazil nut trees are among the tallest in the forest, up to and exceeding 160 feet, up to seven feet or more in diameter. Brazil nuts grow as wedges of an orange in very hard, thick-walled pods. The pods require several hard blows from a machete to be opened. The Brazil nuts must then be scooped out, placed in a basket, and later cleaned of the brown goop that surrounds them within the pod. This is labor intensive. When the trees are deep in the jungle, they can be located only by people with expert knowledge of that part of the jungle. Because the Pirahãs see the jungle along the entire length of the Maici, from its mouth up to and including the Trans-Amazonian Highway (Brazilian road BR-230), they expect to be compensated in some way for their knowledge in showing the trees. And they expect this (more dark matter) every single time the trees are accessed.

Local caboclos refer to the collection of Brazil nuts as *quebrando castanha* (literally, "breaking Brazil nuts"). This is because of the work required to break open the pods. Occasionally, traders want to get their nuts to the market first, so they ask men, such as the Pirahãs, to climb up the 160-foot tree to shake or cut the pods loose, ignoring the danger. There are no branches on the trunk of a Brazil nut tree to get a foot or handhold on. A climber will usually use a strap of bark tied around his ankles to shimmy up the tree with, holding on for dear life. One Pirahã man, Kaxaxái, fell from the top of the tree and bounced like a ball (in the vivid descriptions of Pirahã witnesses) off the jungle floor. A missionary, Steve Sheldon, took him out right away by plane. After surgery and months of recuperation, he returned, largely recovered, but never quite the same mentally or physically afterward. The Pirahãs took special care of him. Stories like this underscore the difficulty and occasional danger of Brazil nut gathering in the Amazon and help us understand why the people expect compensation for their dangerous efforts. Also, Brazil nuts are harvested in the rainy season—the absolutely most dangerous period of the year for snakes and other dangerous animals, since their holes, dens, and nests are often flooded by the rainwater and they must come out onto the forest floor, buried under dead leaves, or in puddles along the path (I once saw a baby bushmaster strike at a friend of mine from out of a brown puddle in the middle of the path, just missing its intended victim. What looked like an innocuous puddle harbored death.)

On the other hand, the location of Brazil nut groves along the Maici is now known to all of the regular river traders; these traders also know of the risks and are more than able to gather the nuts themselves, bypassing the Pirahãs entirely. These caboclos believe that they have paid the Pirahãs for permanent access to the Brazil nut groves, though they do offer the Pirahãs liquor and small things as they are able, but they are too poor to pay the amounts of trade goods that the Pirahãs want. Moreover, the price of Brazil nuts can fall dramatically in a year, hardly making it worth the cost of the fuel for the boat to come up the river after them. These economic realities are in the background and conscious knowledge of the traders but by their very nature fall outside the scope of Pirahã knowledge, involving numbers, economy, and so on. Thus some of the traders' activities are understandable to the Pirahãs, while others are not. On the other hand, given that the traders think the Pirahãs are irresponsible and lazy—or to some, even subhuman—most of them see no need to deprive themselves of full profits in order to pay people who do not even (in the traders' minds) understand payment. But the traders miss the fact that the Pirahãs *do* understand *volume* of payment, that they do want things from the outside world, and that the land is under their control.

But the traders do no better in understanding the Pirahãs. This story would make no sense to a caboclo trader without understanding the Pirahãs' dark matter perspective—namely, that permanent land possession is inconceivable. So the first question, again, is: *Why* is the speaker telling me [Dan] this story? Because he wants me to help him and the Pirahãs. It is fascinating that he introduces (as all Pirahãs do in all texts) proper names of people with no further description, even those (at that time, all the names) I have never met. Interestingly, the effect on me, tacitly, was that when I actually *did* meet these men, it was something like meeting a celebrity. I had come to think of them as celebrity pirates because of Pirahã stories (though, as I later learned, "pirate" fit some of them well and others not at all—I had my own life threatened seriously by a couple of them.)

Line 2 tells us that this person did not pay the Pirahãs. Notice that because Pirahã lacks plural nouns or number of any kind, this sentence is ambiguous between a reading in which there is only one Pirahã, Xoágaii, and a reading in which all Pirahãs are intended. Which reading is correct? If the latter, then this means that Pirahãs have collective control of all their land. If the former, then individual Pirahãs have control of specific Brazil nut groves. This is clearly an ambiguity that requires understanding of Pirahã culture. Yet this understanding would be very hard to get at merely by asking the Pirahãs (I tried), because they would likely either answer both or neither. This is because there simply is no place for ownership as we know it—the lines blur between Pirahãs but are stark between Pirahãs and outsiders.

Notice also that in lines 1 and 2 of this story, the speaker is setting up these broad topics. There will be several specific points, but the general thrust is that Brazilians are using Pirahã Brazil nut groves without payment.

Lines 3 and 4 begin to narrow things to specific incidents. We learn that the Brazilians—in this case Chico (Alecrim)—have their own names for the Brazil nut groves they use most.

Line 5 is very interesting in its use of the kinship term *baíʔi*, "parent" (*báí* is a shortened form of *baíʔi.*), to refer to Chico. The Pirahãs themselves translate this into Portuguese when directly addressing a trader, referring to the ones they know best as *papai*, "father." The traders think that by the use of this term, the Pirahãs intend to communicate that they are children and the Brazilian is a deeply beloved adult. But this is not what it means at all. The Pirahãs can refer to either parent or to anyone who has control over the speaker or an addressee the speaker wants something from. This subtle point leads to great misunderstandings, interpreting even anger as goodwill because of the improper translation of a term. Of course, the Pirahãs do not know that their usage of the Portuguese term does not match the Brazilian usage of the word, nor vice versa (more dark matter–based misunderstanding).

In line 6 we hear, "He did not do a little." This means that he took a lot of Brazil nuts from the grove (yet did not pay for the use of the grove). Notice here, again, the major, familiar misunderstanding of land payment. The Pirahãs do not see land as sellable, only available for paid use on a case-by-case basis, much as the aborigines at Medicine Lodge. Chico believes that he has purchased this Brazil nut grove for a payment in perpetuity, the Pirahãs think of him as a thief. (Ultimately the Brazilian government agreed with the Pirahãs, forcing him out of the land without indemnification after I complained.)

Now clearly, these two perspectives are not ineffable. I believe I have just made them more or less explicit. But to the Pirahãs, the transactions' assumptions would be difficult to state because the presupposition that land cannot be sold is so deeply embedded. This is an interesting point with regard to emicization and dark matter. Emicization often results in making the retrieval of knowledge more difficult. Emicized knowledge can be fairly obvious as such to an outsider, but nearly impossible for an insider to talk about because they have "absorbed" the knowledge—they have stored it in their (collective and individual) subconscious.

In line 7 the speaker gives the name of the Brazil nut grove in both Pirahã and Portuguese, to make it clear what place is being discussed.

In line 8 another foreigner is introduced, João Baowa. (The surname here is a Pirahã pronunciation of a Portuguese word, but I haven't been able to further identify this person or the actual name.)

Line 9 lets us know that the character introduced wants a particular grove. Again, though, the literal translation fails to communicate the emic understanding of place, as something that can be enjoyed but not owned. As Nida (1964, 161) put it, "Adherence to the letter may kill the spirit." Dark matter is the spirit of a text.

On my first hearing of line 10, my question was "Who are Bernardo and Moisés?" It turns out that they are two brothers who live at the Auxiliadora. They are named without introduction, but as I said earlier, it is assumed that I know them. Is this failure to introduce them due to the speaker's assumption that I must know them? I think so. But why would he think I know them? The answer is part of the speaker's dark matter, an assumption that we all live in a "society of intimates"—like Pirahã society—where everyone knows everyone, rather than a "society of strangers" where many or even most of the people we see in public places are unknown to us. The Pirahãs' conception of the world as an intimate world is natural. It reflects their own society and the emicization of their knowledge of the social. Of course, the Pirahãs know that Brazilians and Americans are separate groups, like the Pirahãs and their Amazonian neighbors the Tenharim and the Parintintin. The issue, though, is their knowledge of how large those groups might be—surely I know all Brazilians. And this view persists even though Pirahãs, especially, Kaabogí, have traveled with me to several Brazilian cities, including Porto Velho, Belém, Brasilía, and São Paulo. In Brasilía, Kaabogí watched a procession headed by the then-president of Brazil, João Batista Figueirdo. I referred to Figueirdo as the "Tuchawa" of the Brazilians. Since the Pirahãs used this Tupi loanword for "chief" I mistakenly thought that they would understand this use of it for head Brazilian. But no, the Pirahãs attach no significance to this term other than as how they understand land use–temporary, not permanent. In other words, in Pirahã someone can be a leader in a particular situation, as on a hunt or talking to a river boat trader, or a parade or procession, but there is no concept of a permanent leader, or royalty, or a chief. Kaabogí just asked if we could leave the sun and go into the shade to get a soft drink. No one in

Pirahã culture has a concept of "famous" or "powerful" or "rich." Therefore all the cultural trappings of a head of state were meaningless to Kaabogí.

This view of an egalitarian, intimate humanity where everyone knows everyone else is a form of tacit knowledge, rather than a tacit value per se (though it likely is that, too). It is a filter on the Pirahãs' understanding and interaction with the world that simply cannot be fully explicated by those affected by the filter because of its deep emicization. Not only being raised in a society of intimates but also learning to interpret all interactions in the assumption that no one is superior to any other socially has a profound effect on what Pirahãs see, hear, and understand. For example, proper names in texts are never introduced with identifying information; but notice that there are not even any relative clauses in this or any other Pirahã story. This is interesting because, as I have argued elsewhere (D. Everett 2005a, 2008, 2012, 2014b), it shows the architectonic effects of Pirahã culture on Pirahã grammar—*core grammar*, in the Chomskyan sense.

The function (not the only one, but arguably the main one) of relative clauses is to help the hearer identify the speaker's intended referent by expressing information that the hearer shares with the speaker about the referent. For example, consider these English examples:

> The man is in the room *vs.* The man who is tall is in the room.
> John, who said hi to you yesterday, punched John, the guy of the same name who works in the pizza parlor and insulted the first John's girlfriend.

In these examples, the restrictive relative clause (*who is tall*) and the appositive relative clauses (*who said hi to you yesterday* and *the guy of the same name . . .*) have a variety of functions in English discourse. But their main function here is to let the hearer know who the speaker is talking about and which part of her speech is about which referent—that is, they *restrict reference.*

The Pirahãs do not avail themselves of such devices. Though they can use parenthetical clauses to help with identification, such parentheticals are never part of the syntax of the clauses or clause constituents that they modify.[5] The Pirahãs' assumption that all people live in societies of intimates is emically derived dark matter that they would find difficult to explain to an outsider (in fact, my own wording suggests that even for someone who has experienced a different emicization of the understanding of societal relations, it is not an obvious concept).

Line 11 continues the discussion of the illicit use of Pirahã Brazil nut groves.

Line 12 identifies another Brazil nut grove, Sogohóiga.[6] Notice the repetition in lines 13 and 14. Repetition in Pirahã discourse usually has two functions. First, it emphasizes an important line of the story, such as the transition here from Brazil nut grove, *Passar Bem*, to Sogohóiga. Second, the repetition functions meta-communicationally to overcome environmental noise. There is, by European languages' standards, enormous redundancy in Pirahã discourse. But it plays an important role in a noisy, aliterate environment.

Line 15 raises the question of why Kaabogí bothers to tell us that the grove is located in the deep jungle. There are several Brazil nut groves not far from the banks of the Maici. But theft is paradoxically more common from the deep jungle groves. This is because back trails that wind for miles away from the shore can be used as a way to hide one's actions, to execute more furtive theft. That one would walk miles in the jungle, one way, to steal (from the Pirahãs' perspective), say, two hundred pounds of Brazil nuts may seem extravagant. But when such products are the only source of cash and extreme poverty is the norm for river traders and their employees, it makes sense.

Lines 16 through 18 underscore the need for river traders to secure Pirahã permission before taking out Brazil nuts. Forget the fact that the Pirahãs are almost totally monolingual and do not speak Portuguese, aside from a small vocabulary of trade words overlain on a Pirahã grammar (Sakel 2012a, 2012b). They know enough to recognize *pagar*, "to pay," and *castanha*, "Brazil nut," and to understand when someone is acknowledging Pirahã control of the land.

In lines 19 and 20, Kaabogí requests that I speak to the Brazilian, Martinho, even though he has threatened to shoot me. We are learning that a Pirahã text, like a *Wall Street Journal* article, is rife with dark matter.

DARK MATTER CONSTRUCTED BY POPULAR CULTURE

For members of certain cultures, perhaps more obvious source of dark matter and cultural values and knowledge comes from a special kind of text: popular song lyrics. Songs in various cultures crucially rely on shared dark matters to trigger emotions and values, and to develop attachment to the song. So consider the lyrics to the classic 1962 Marty Robbins song "Devil Woman":

Devil Woman
Marty Robbins, 1962

I told Mary about us.
I told her about our great sin.
Mary cried and forgave me,
Then Mary took me back again,
Said if I wanted my freedom

I could be free ever more.
But I don't want to be,
And I don't want to see
Mary cry anymore.

Oh, Devil Woman,
Devil Woman, let go of me.
Devil Woman, let me be,
And leave me alone.
I want to go home.

Mary is waitin' and weepin'
Down in our shack by the sea.
Even after I've hurt her,
Mary's still in love with me.
Devil Woman, it's over,
Trapped no more by your charms,
'Cause I don't want to stay.
I want to get away.
Woman, let go of my arm.

Oh, Devil Woman,
Devil Woman, let go of me.
Devil Woman, let me be,
And leave me alone.
I want to go home.

Devil Woman, you're evil,
Like the dark coral reef.
Like the winds that bring high tides,
You bring sorrow and grief.
You made me ashamed to face Mary.
Barely had the strength to tell.
Skies are not so black.
Mary took me back.
Mary has broken your spell.

Oh, Devil Woman,
Devil Woman, let go of me.
Devil Woman, let me be,
And leave me alone.
I want to go home.

Runnin' along by the seashore,
Runnin' as fast as I can.

Even the seagulls are happy,
Glad I'm comin' home again.
Never again will I ever
Cause another tear to fall.
Down the beach I see
What belongs to me,
The one I want most of all.

Oh, Devil Woman,
Devil Woman, don't follow me.
Devil Woman, let me be,
And leave me alone.
I'm goin' back home.

Popular songs like this reveal—perhaps more effectively than other texts would—common dark matter, including easily recognizable values. For example, in this text we are expected to agree with the values that wives should be told about sexual infidelity (Pirahãs would not agree), and that sexual infidelity is evil (Pirahãs would agree that it can bother a spouse). This would fail to resonate in many cultures. Other cultures might have values in which women should be faithful but not men, or vice versa. The lyrics further indicate that forgiveness for sex outside of wedlock is rare, but even when it occurs, sadness results. The singer also assumes that forgiveness by one party entails an obligation on the part of the other party.

We see numerous other values represented—for example, the idea that two people hold authority over each other's right to end or begin sexual relationships with others. The song also supposes that women who sleep with men married to other women bear more of the blame for the relationship than the man who slept with her. Thus in the line "Let go of me," the man blames the "other woman" for his attraction rather than himself—not completely, perhaps, but it is an old trope originating from a similar value. We also assent, if we share the values of the text, that a wife must have exclusive sexual access to her husband. The woman who is not the man's spouse is evil. The man is now tired of her or his attraction has worn off (he claims). He states this almost like a performative speech act—"Mary has broken your spell"—stating something to make it so. It seems a rather shabby treatment of the other woman, but its expression here demonstrates an expectation of resonance from hearers. The other woman is bad. She is the temptress. The man is but her victim. And this message gets driven home each time this song or any song with a similar message is heard.[7]

Asking more questions of the lyrics, we are expected to know that the answer to the question of why the other woman is evil is that she tempted the man—in other words, because he wants to bear little or no responsibility for their mutual relationship. It is taboo for a nonspouse to expect the same level of attention as a spouse, regardless of emotional attachment. His wife, Mary, however, came to his rescue by breaking the "spell" and once again placing him under obligation by taking him back.

The idea that the seashore and its trappings are symbols of freedom and escape is also assumed as shared dark matter, as is the singer's transcendent joy at the moral victory of leaving one woman for the original woman. And we see lines near the end, like Western marriage vows ("till death do us part"), that are grandiose promises of future commitment ("Down the beach I see / What belongs to me").

Now the husband and the wife, we are supposed to agree, are both free—the wife from his sin, he from sin and the evil other woman. The phrase "Don't follow me" we are to recognize as a sign of his understood weakness and continued attraction to the other woman despite best intentions, as well as the evil power of the other woman. It's almost humorous, in fact, but we are all supposed to share many of these values or at least share the knowledge that they are common, or the song would make no sense.

We are infused by dark matter from all the communication we attend to and much that we do not—through indexicals, presupposed values, conventional expressions, discourse assumptions, popular tropes, and so on. And as we use these, we reinforce our own dark matter and strengthen and expand the scope of this psychoculturalunspoken knowledge among our hearers and in ourselves.

Summary

In this chapter we saw that texts can be interpreted only against a background of structured knowledge, social roles, and ranked values. We saw this in the different roles implicated in the various texts explored: speakers, writers, audience, reporters of texts (me for the texts of this chapter), singers, composers, and others.

We also considered the role of texts and ideas about business culture—an important new issue in modern capitalism—arguing that though the use of the term *culture* by modern businesses is not completely misguided, the concept of culture upon which business ideas are based is an anemic one, usually focusing strictly on roles (as in holacracy), values (shareholders vs. stakeholders), or knowledge structures (marketing research would be one example). No company has thought this issue or concept through with sufficient care to realize that (i) a culture is built on not merely one, but all three of these three components; and (ii) what a company (or anyone else) professes its culture to be must be compared with what the culture is like in practice.

6

The Dark Matter of Grammar

> If we adopt this point of view, language seems to be one of the most instructive fields of inquiry in an investigation of the formation of the fundamental ethnic ideas. The great advantage that linguistics offer in this respect in this respect is the fact that, on the whole, the categories which are formed always remain unconscious, and that for this reason the processes which lead to their formation can be followed without the misleading and disturbing factors of secondary explanations, which are so common in ethnology, so much so that they generally obscure the real history of the development of the ideas entirely.
>
> FRANZ BOAS, *Introduction to Handbook of American Indian Languages*

In this chapter, we examine the architectonic effect of culture on language—in syntax, phonology, lexicon, morphology, semantics, and so on. This, I argue, was the original and laudable position of North American linguistics, as founded by Franz Boas and Edward Sapir, but due to a not entirely salutary reification of linguistics research that began with Noam Chomsky in the 1950s, the connection between culture and language was forgotten. In effect, linguistics decided to pursue an alternative path, one that I argue here was and is severely misguided.

The chapter's empirical core is the phonology and syntax of the Pirahã language of Amazonas, Brazil, arguing that their syntax and phonology, for example, must be reconceived as ethnosyntax and ethnophonology.

Symbols and Signs

Silverstein (1979, 2003, 2004) has argued convincingly that every component of grammar must be analyzed at multiple planes simultaneously—the social and the grammatical at an absolute minimum. This work complements my own insistence that grammar and culture work with the social (to the degree that that might be separate from culture) synergistically to produce language. He has argued that among the functions of language, we must acknowledge a semantic function (function$_1$) and an indexical function (function$_2$) of language. Function$_1$ includes such ideas as, what do words *really* mean? That is, it is concerned with the situationally invariant (though this is a fiction that Wittgenstein, among others, attacked) meanings of words, from their diachronic development, etymologies, and so forth. Function$_2$, on the other hand, looks at the use of language in dynamic social and ideological uses.

Silverstein (1979, 193) brings into his discussion of the two major functions of language, the structural principles that Whorf introduced decades earlier, the "cryptotype" and the "phenotype." However, I think that Pike developed these ideas, though independently, much better in the etic and the emic, so I will ignore this component of Silverstein's theory, though it is compatible with how we have been discussing language to this point. Another researcher already mentioned with compatible ideas on language is Enfield, especially in his work on relationship thinking (Enfield 2014).

Communication is a diffuse area of investigation. At its simplest, communication is just the transfer of information (Shannon 1949). If this is all that communication is, however, then even a thermostat could be described as "communicating" with its environment, taking in information about the temperature and producing an appropriate response. It "knows" what is relevant to it. It uses concepts. It responds to a stimulus. Yet we resist this characterization because there is no intentionality involved in the thermostat's response to its environment. Without this function of intent, the judgment that thermostats communicate does not hold (Searle 1983). Intentionality seems a reasonable threshold for labeling an act "communicative." In this sense, many animals communicate, including dogs, cats, nonhuman primates, birds, frogs, and so on. Communication does not entail language, though language is the apex of communication systems. But what is it that makes the special subset of communication we call "language"?

The first crucial invention that is required for language is the *symbol*. The importance of symbols for language has been recognized at least since the philosophical work of Peirce (1977) and the linguistic development of the sign by Saussure ([1916] 2012). In recent years, Terrence Deacon's (1998) work has offered more insights into the development of symbols for human language. But perhaps the most significant advances in understanding the symbol's centrality to human language come from the theoretical framework known as construction grammar (for the founding work, see Lakoff 1977 and Fillmore 1988, among others; Croft 2001 and Goldberg 2006 are two of the most important developments of the theory). Construction grammar rejects the strict dichotomy between signs and grammar that has played such an important role in formalist theoretical linguistics, especially of the Chomskyan variety.

One type of symbol is what Saussure ([1916] 2012) refers to as "signs" (these are the linguistically crucial symbols without which no human languages would exist). Signs have two components: their form (physical instantiation) and the meaning (sense and reference; Frege 1980). A symbol is a form combined with a meaning in a culturally specific (i.e., non-logically required) way. So *dog* is a form associated with the meaning "canine," though a priori there is no special logical or other connection between the three phonemes of *dog* and the meaning "canine."

However crucial symbols may be, the development of symbols is perhaps not as difficult a breakthrough for larger hominin minds as one might think. In fact, these two components are crucial to every entity that has relevance to hominins. A special path, a loved one, a drawing on a cave wall, marks on bone—each must have both a form and meaning. The form may or not be invented. It could be, for example, a tree, a stream, a rock. But the meaning must be created, largely through apperceptions and social awareness as individuals in a group learn that the stream is the "place of good hunting" or the rock is "the place of the eagle."

Other forms are created by creatures and given meaning by the same creatures or others. A vervet monkey makes a sound when a bird of prey is overhead and alerts other vervets. If a human utters "ugh" and means by that "no," and another human is subsequently attacked for their failure to recognize the "ugh" symbol's meaning, that symbol is more likely to be recognized the next time it is encountered (by witnesses, at least). The difference between the vervet's and the human's symbols is that the former requires natural selection for its spread, whereas for the human culture, it is sufficient for the origin and spread of a symbol (and perhaps for other species—see the literature discussed in chap. 10). But although what animals produce may be symbols in a general, nontechnical sense, perhaps no other animal produces signs other than humans. The distinction, of course, from our earlier discussion, is that the relation of form to meaning in signs is not as straightforward. It can range from highly indirect (the classic example of different forms in different languages, such as *dog*, *perro*, *cão*, and other words for the animal {dog})—to the more direct (such as sound-symbolic words like *snap*, *crackle*, *pop*).

Unlike the case of most other animal species, the sign in human languages takes many forms. Here are some of them:

Indexicals. We discussed these earlier as various forms of symbolizing particular cultural and individual values, from body language to word selection.

Sound symbols. These include things such as onomatopoeic words, like *zoom-zoom* for the sound of a motorcycle or *clippety-clop* for the sound of a horse's hooves as it walks. Such symbols also include the sound /h/ in related words such as *hut*, *house*, and *hovel*, or the /gl/ sequence in *glow*, *glimmer*, and *glisten*.

Icons. Drawings, pictures, statues, and so on, are special kinds of signs (in Peirce's sense) that are intended to convey a likeness of the meaning (referent).

Indexes. Direct physical evidence of the referent, such as tracks (for animals), smoke (for fire), cookie crumbs (for cookie eating), and so on.

Signs. Here the crucial notion is not merely that there is a form linked in the culture to a meaning, but that the form-meaning connection is almost exclusively cultural—that is, a decision made by the society, a type of social contract (Searle 2006)—and thus for the Martian linguist or the newborn baby, the form-meaning connection is arbitrary. Why is *fiddle* an instrument for bluegrass music, the *violin* an instrument for classical music, but *guitar* an instrument for either? Only culture, if anything at all can be found, can explain such differences, via history in this case (*fiddle* may originate from Old English *fithele*, deriving from the Latin *vitulari*, "to celebrate," and at one time came to refer to any instrument used in a party or celebration. Or it may be an English corruption of Old French *viele*—a predecessor of the violin).

To the question of how signs originate, part of the answer is that this is like asking who the first person was who told a particular joke. Neither the origination of a sign or a joke is the act of a single individual. All individuals think original thoughts and make up what are (at least, until they make them up and others use them) nonsense words, expressions, jokes, and such, that in potential could spread but do not. Chomsky's "Colorless green ideas sleep furiously" could itself be the origin of a new symbol in the language that could eventually become a sign. But for an odd form-meaning pair to become a symbol in a language, it must be adopted by the society. "Grammar" is a symbol to all linguists, though its meaning varies significantly according to linguistic subculture. For some it means that form and meaning are unrelated; in particular, that a phrase can in principle be both meaningless and grammatically well formed (as in Chomsky's example, though most linguists see this largely as a theoretically internal fact, based on unique concepts of meaning and form). For others, grammar is the way that the language conveys meanings and is formed principally by the meanings it conveys. For one, extreme grammar is a vast domain. For another it is a relatively small residue of history, cultural constructions, and so on.

What constitutes a sign, however, can vary tremendously from theory to theory. For some, grammatical constructions are not signs, but epiphenomena that result from grammatical rules taking individual words and putting them together in some arrangement. According to the Chomskyan conceptualization of syntax, once syntactic phrases are built up from words to larger units, they can be interpreted by the hearer top-down, using inverse compositional or decompositional semantics (Van Valin 2006).

However, as indicated in the title of Lakoff's famous 1977 paper, "Linguistic Gestalts," at least some (perhaps all) phrases and sentences have meanings that are properties of the constructions themselves, and *not* predictable from compositional semantics. An example might be the sentence *Keep tabs on John while he is out on parole.* The isolated word *tabs* has no literal meaning in this construction, but is crucial to the form of this idiom where *keep tabs on* means "supervise" or "watch." A related construction might be *Keep an eye on Mary while she is shopping,* where *keep an eye on* is not meant literally. Once constructions are recognized in the theory of grammars, the distinction becomes blurred between the "grammar proper" (the rules that generate grammatical forms without regard for meaning) and the "lexicon" (the speaker's brain-stored dictionary, based on their culturing experiences; the lexicon in this sense is one of the largest portions of our dark matter). Since signs are constructed culturally, this entails a greater role for culture in what has been traditionally known as grammar. In fact, as we see directly, culture's hand can be found even in the grammar itself—the way that phrases are formed, to the degree that they are, apart from constructions or meaning.

Grammar

Following intention and symbols, there must also be grammar for language to appear (D. Everett [2012a] discusses the various "platforms" that must be in place before language is possible.) Some linguists—most notably Chomsky—have a purposely narrow understanding of what constitutes grammar. And if we take grammar to be a strictly formal, computational system that at most filters meanings without being driven by meanings, then our understanding of language will be quite different from the views of those who see grammatical form as primarily a function of meaning. These varied views represent the central controversy of linguistics, arguably one unresolved at least since Sapir (1921) and Bloomfield (1933) presented very different pictures of language as sociocultural interactions with the mind, meaning, and structures vs. meaning-free and culture-free structure, respectively.

This "meaning movement" began in the modern era in the US with "generative semantics" in the 1970s. This was a theory developed by several of Chomsky's former champions and, ironically, was based on Chomsky's notion of "deep structure," the level of syntax in which the lexical and basic phrasal meanings were once thought to reside. Many linguists find it humorous that non-linguists frequently confuse deep structure with universal grammar. However, this is unfair; the mistake is in fact quite a natural one to make. Part of the reason is that in the literature on UG, the definition is usually unclear. Perhaps a main reason, though, for confusing UG and deep structure stems from Chomsky's competitor theory of the 1970s, generative semantics' *universal base hypothesis* (UBH). According to the UBH, all languages share the same semantic (deep) structure that serves as the foundation for their sentences—clearly a notion similar to UG.

Other meaning-based theories that emerged from generative semantics included *functionalism* (in its myriad forms), construction grammar (already mentioned), and role and reference grammar. Interestingly, however, even these meaning-based theories that rejected Chomskyan linguistics for ignoring the central role of meaning in grammatical structure failed to reach deeper for the source of meaning, appealing in Chomskyan fashion to universals of human meaning as if those are preordained in some way.

If I am correct, however, the meaning that shapes human grammars emerges at least partially from culture. This seems to be what Sapir (1929, 2) is getting at when he says: "Speech is a non-instinctive, acquired, 'cultural' function." Therefore, I want to look more closely at the idea that language is significantly shaped by culture. One way to interpret Sapir's statement is to say that grammars arise within particular systems of cultural values and that the values and patterns of conversation, telling stories, uttering phrases, and so on, themselves constrain the grammatical structure of a language.

Before entering in to a detailed discussion of dark matter, culture, and grammar (phonology, morphology, discourse, and syntax), however, we need to review some linguistic history in order to better understand the issues. It is also important to understand how linguistics has been led away from its traditional concern with language as partially constructed by culture. Principally, this shift occurred due to a reification of the field that began with Chomsky's earliest work, continuing through to the present state of formal linguistics more broadly.

From there, I argue that the child learns its culture and other forms of dark matter at least as early as—perhaps even before—it begins to learn its language. This non-linguistic learning affects the child in many ways, including its conception of how language is used for communication, a conception that can in turn affect her grammar.

Next, I want to look at Pirahã's segmental phonology. I argue that, contrary to most linguists' expectations, even the sound system of Pirahã is architectonically affected by culture. I conclude the discussion of Pirahã with a suggested methodology for establishing culture-grammar connections.

We then examine two additional Pirahã texts to exemplify how culture and grammar interact in discourse. From this we turn to a consideration of how Pirahã culture affects its syntax, via the overarching cultural value IMMEDIACY OF EXPERIENCE (IEP) (D. Everett 2005a, 2008, 2012b), and the reflex of this value in the evidentiality system of the grammar, through the POTENTIAL EVIDENTIALITY DOMAIN. This potential evidentiality domain (PED) is a culturally motivated principle of Pirahã grammar. A side effect of the PED is to bar recursion from the morphosyntax entirely.

The discussion leads us, I argue, to the formula

COGNITION, CULTURE, AND COMMUNICATION→GRAMMAR

arguing that Pirahã is not a unique case and that all languages will show culture-language connections if we look. But of course, it is hard to find such connections if we do not look for them.

REIFICATION OF LINGUISTICS

Until the 1950s, the common professional classifications and departmental homes for linguists were anthropology and foreign language departments. The idea that language was itself a component of culture, society, folklore, and so on, was shared on both sides of the Atlantic. Both Sapir (1921) and, later, Roman Jakobson (1990) wrote widely about language's various manifestations in discourse, poetry, conversation, sound systems, and so on, underscoring the symbiotic relationship between language and culture.

But Chomskyan linguistics turned away from interest in human culture in the late 1950s, leading to a marked reification, ignoring various intersections of culture and grammar (e.g., discourse structure, idioms, sound symbolism, and field research) in favor of what it intended to be a deeper understanding of the forms of language, as manifested in phonology, morphology, and syntax.

Chomsky famously designated the sentence as the "start symbol" (Σ) of grammar, ignoring the possibility that sentences might be constituents of discourses and conversations. Chomsky (1956, 1959) argues that a grammar will generate "all and only the sentences" of a language L. Thus not only does he make the (now apparent) error of making the sentence the foundation of any grammar, but then, as a necessary consequence of the initial assumption, all that grammars can generate are sentences—completely omitting discourse and conversations, arguably the most interesting units of any language. This in effect says that there is nothing of interest for the grammarian above the sentences.

And discourses and conversations are not simply arrangements of sentences, as a myriad of work shows.[1] The failure to look at discourse (and culture) in studying sentences is on par with the now-outdated position of earlier linguists who avoided incorporating morphological phenomena into the analysis of phonemic structures (Pike 1952). As many linguists have shown, beginning in particular with Givón (1983)—one cannot understand sentence structures well without understanding the discourses they are embedded in. Though generativists might insist on looking only at sentence-level phenomena, the rest of the (psycho)linguistic enterprise has long moved on from this artificial, self-imposed limitation.

Looking beyond the work of Sapir, Jakobson, and Pike, inter alia, recent research by Levinson (2006), Enfield (2002), Silverstein (2003) and many others has advanced our understanding of sentences as units of interaction. However, formalists respond that such research is orthogonal to the enterprise of generative linguistics, because it fails to explicitly focus on the basic grammatical architecture of languages. To address this objection, therefore, I discuss in what follows precisely the effects of culture on this so-called core grammar.

Mainstream generative studies of core grammar have focused on the forms of sentences, phrases, and words—thus offering simply a continuation of Bloomfieldian structuralism with the ideas of the mind and nativism added. Starting from the assumption of UG, the analyst proposes a deductive set of categories. Subsequent analyses in the same tradition then apply and tweak these categories or processes, with the aim of showing that they fit all languages in some way. Generative studies are said to differ from Bloomfieldian structuralism by paying greater attention to mental representations. But this claim can be dismissed at once because the mental is never causally implicated in any analysis of the theory, and UG is orthogonal to the analyses (D. Everett 2012b).

In this new Chomskyan structuralism, several assumptions have come to dominate thinking about syntax in theoretical linguistics: (i) all grammars are hierarchically organized by recursive procedures; (ii) all grammars involve derivations; (iii) all syntactic structures are formed by combining two units at a time to produce endocentric, binary branching (and hierarchical) structures; (iv) all grammars derive from a genetic endowment common to humans, called universal grammar; and (v) the domain of grammar is the sentence. Many linguists, however, would argue that all of these points have been falsified. Jackendoff and Wittenberg (forthcoming) have argued that Riau and Pirahã have nonrecursive syntax (see also Futrell et al., forthcoming). Robert Van Valin (2005) and others have argued that derivations are never necessary in any grammar. S. Frank, Bod, and Christiansen (2012) have even argued that hierarchy and recursion are unnecessary for the proper analysis of *any* natural language. Lieberman (2013) has developed a formidable case to the effect that there is no neurological support for the idea that grammars derive from language-specific innate principles. Culicover and Jackendoff (2005) have argued convincingly against (i)–(iii). And I myself have offered analyses of various languages—especially Pirahã (D. Everett 2005a) and Wari' (D. Everett 2005b, 2009b)—that appear to falsify (i)–(iv). (For a slightly different though still compatible perspective, see Corballis 2007.)

In any case, if we focus on grammar at the sentence level only, we miss important principles of the formal organization of language above the sentence, which is arguably also grammar but whose principles are more diverse. These include the principles of sentence-grammars as a subset. By way of example, I will consider some coarse-grained features of a couple of small Pirahã texts.

CULTURAL LEARNING

First, however, I want to review evidence that cultural learning takes place without language. This is important because such evidence falsifies the claim that culture is necessarily subsequent to language.

In chapter 3, we discussed how dark matter is formed. In this formation, a significant portion of cultural knowledge and dark matter are learned independently of language. I want to come at this issue again with evidence from my own field research with the Banawás of the Brazilian Amazon (Buller, Buller, and Everett 1993; Ladefoged, Ladefoged, and Everett 1997).

Banawá males must all master the making of blowguns, from gathering raw materials to assembling the weapon. Consider in this regard the non-linguistic aspects of learning of how to make a blowgun, which I have done while among the Banawás.[2] Sons observe, imitate, and work alongside their fathers. Surprisingly little linguistic instruction takes place in the transmission of this set of skills. The wood for the blowgun comes from a narrow range of wood species. The vine used to tie the blowgun and render it airtight is a specific kind found in certain places in the jungle. The needle used for the darts likewise requires highly specific knowledge of local flora; so does the kind of large jungle vine used to extract the poison (strychnine) and the other ingredients of the poison that help it enter the bloodstream more effectively. All of these steps and bits of knowledge can be transmitted without much language, by the son traveling with the father and observing. While learning how to find and gather blowgun components, the son also learns about hiking in the jungle, fortitude, bravery, flora and fauna, and so on. Often lessons are learned without a single word being exchanged between father and son.

The cultural transmission of dark matter, like genetic transmission, is nearly always corrupted in some way, leading to "mutations" (Newson, Richerson, and Boyd, 2007; Schönpflug, 2008). For example, a Banawá may be forced to use a different type of wood, or tie the blowgun slightly differently, or use a novel binding agent for the poison for some reason. Moreover, at any step of the transmission process error, or innovation can occur by the father or the son, in turn leading to changes in traditional blowgun construction. It does not matter whether the deviation was intentional or not—only that there has been a deviation that provides a potential for a mutation, whether resulting in an inferior or superior blowgun. Clearly such deviations have occurred, because in closely related Arawan communities (e.g., Jarawara and Jamamadi), blowguns differ (as do the languages themselves) in nontrivial ways. The technology varies and the language varies due to imperfect imitation and innovation. Although language enriches and accelerates the process of culture learning, allowing for the construction of different sorts of cultural institutions (e.g., the family, school, the military, church), differences in dark matter from one individual to another can arise without linguistic guidance.

Ethnophonology

I want to begin my discussion of the effects of culture on grammar with an ethnophonological analysis of Pirahã phonemes and "channels of discourse." Though I have discussed these data elsewhere (D. Everett 1979; 1985; 2008; 2015), it is worth reviewing them here to round out our picture of the effects of culture on grammar more generally. As pointed out in D. Everett (1979, 1982, 1985), Pirahã phonology cannot be fully described or understood without knowledge of how it interacts with culture. To demonstrate the independence of these prosodic features, let me present two examples of "ethnophonology" in Pirahã (taken from Everett [2014b]).

To set the stage, imagine that a language could have various systems/modalities of sound structure, beyond its phonetics and phonology. And then consider the possibility that one modality could affect another, but not necessarily via standard devices of phonological theory proper (e.g., rules). If so, then to understand the sound system of language, L, at any level (e.g., "what happens" or "what native speakers know when they know the sound system of their language") we must look carefully at the modalities of expression made available via an *ethnography of communication* (Hymes 1974) and not merely at a supposed universal formal apparatus. Corollaries of this scenario might include, for example, the appearance of new roles for old constraints. If this were true, then *coherent fieldwork* (D. Everett 2004)—the idea that not only

TABLE 6.1. Pirahã Phonemes

Consonants () = missing from women's speech

p	t	k	ʔ
b		g	
	(s)		h

Vowels

i		u
	a	

our analyses but in fact our intellectual life more generally must cohere with our personal and professional values in life and science—would evolve from a curiosity to a desideratum to an imperative. Is there such a case? Indeed. There is, as we see by examining the following facts about Pirahã phonology. We begin with its phonemes (table 6.1; also shown in chap. 1).

Pirahã's segmental inventory is one of the smallest in the world (the only two other languages with inventories of similar size are Rotokas and Hawaiian—though unlike Pirahã, Rotokas lacks tones and has an even smaller inventory overall). As a reminder, the /s/ is in parentheses because it is not generally found in women's speech, though always in men's (women use /h/ where men use /s/ and /h/).

Yet even though Pirahã has one of the simplest segmental phonemic inventories in the world, Pirahã prosody contrasts dramatically in complexity. The Pirahã stress rule is a good place to begin, since it is well known. This rule, from D. Everett and K. Everett (1984), is considered one of the more complex and unusual stress rules in the literature, mainly for its theoretical consequences (rather than, say, any difficulty in stating or recognizing it):

> *Pirahã stress rule:* Stress the rightmost token of the heaviest syllable type in the last three syllables of the word.

The phonetic basis of the "heaviness" referred to in this rule is that voiceless consonants are always longer than voiced consonants, and there are five syllable weights based partially on this contrast: Pirahã's five syllable weights are:

CVV>GVV>VV>CV>GV

(where *C* = voiceless consonant, *G* = voiced consonant, and *V* = vowel)

Pirahã is a tonal language as well, and stress, tone, and syllable weight vary independently in the language. (For a phonetic study of these features, see K. Everett 1998. Data in what follows is from K. Everett 1998 as well as my own field research.) To see this, I review the simple set of examples below. In these examples, tone is independent of stress. ´ = high tone; no mark over vowel = low tone. The stressed syllable is marked by !. There are no secondary stresses.

a. *!tígí* "small parrot"
b. *!pígí* "swift"
c. *!sábí* "mean, wild"
d. *!ʔábí* "to stay"
e. *tíí!híí* "bamboo"
f. *ʔí!tì* "forehead"
g. *tí!ʔí* "honeybee"
h. *tí!hí* "tobacco"

Thus, alongside Pirahã's extremely simple segmental phonology, the language possesses a rich prosodic system. This leads us to ask whether the language exploits this differential complexity in any way. Indeed, as D. Everett (1985) describes it, Pirahã communication makes crucial use of *channels of discourse*, where Hymes (1974) defines a channel as a "sociolinguistically constrained physical medium used to carry the message from the source to the receiver." The four principal modalities or channels in Pirahã after "normal" speech are:

Channel	**Functions**
a. Hum speech	Disguise Privacy Intimacy Talk when mouth is full Caregiver-child communication
b. Yell speech	Long distance Rainy days Most frequent use: between huts and across river
c. Musical speech ("big jaw") *Women produce this in language-teacher sessions more naturally than men. Women's musical speech shows much greater separation of high and low tones, greater volume.*	New information Spiritual communication Dancing, flirtation

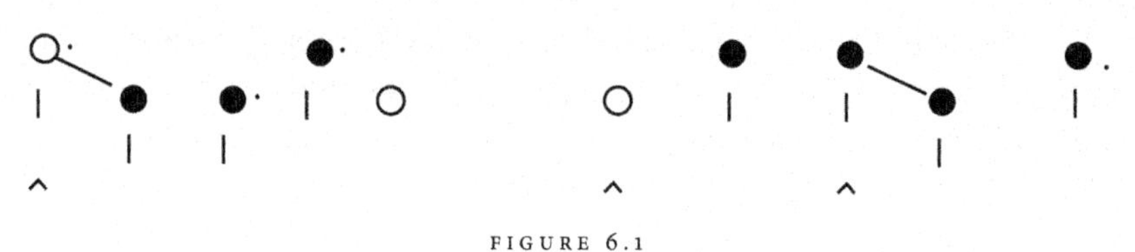

FIGURE 6.1

d. Whistle speech (sour or "pucker mouth"—same root as "to kiss" or shape of mouth after eating lemon)	Hunting Men only One unusual melody used for aggressive play

The example below illustrates how prosodic information in Pirahã is exploited to create these channels. The inventory above also partially shows how little the segments contribute to the total set of phonological information in a given Pirahã word. We see that the phrase *There is a paca there* has a quasi-musical tonal representation (where an acute accent over a vowel represents high tone and no mark over the vowel represents low tone), the basis for the channels just summarized (see fig. 6.1).

kaiʔihíʔao-ʔaaga gaihí
paca poss/exist-be there
There is a paca there.

All channels must include full prosodic information (stress, tone, length, intonation), though only the consonant and vowel channel needs to include the vowels and consonants. In the musical form, there is a falling tone followed by a short low, with a preceding break in the whistle (where the glottal stop, *ʔ*, would have been in *kaiʔihi*), followed by another short break (where the *h* would be) and a short high tone, and so on. Thus, the syllable boundaries are clearly present in whistle (and humming and yelling) channels, even though the segments themselves are missing. The syllable in this case indicates length and offers an abstract context for tone placement; the overall word is stressed according to syllable weight (D. Everett 1988). The syllable in these cases is vital to communication in differing channels, primarily in parsing the input.

One might now reasonably ask whether the discovery of such channels implies any causal interaction between culture and grammar. Or are these channels outside the grammar proper? Notice that the channels rely crucially on the syllable weights and stress rule above. So, if nothing else, they help account for what is otherwise an anomalous level of complexity in the stress rule, since this rule is possible only because of the larger complexity. Yet the facts cut deeper than this. Consider the following example of what D. Everett (1985) calls the "sloppy phoneme effect":

*tí píai ~ kí píai ~ kí kíai ~ pí píai ~ ʔí píai ~ ʔí ʔíai ~ tí píai, etc. (*tí tíai, * gí gíai, *bí bíai)* "me too"
*ʔapapaí ~ kapapaí ~ papapaí ~ ʔaʔaʔaí ~kakakaí (*tapapaí, *tatataí, *bababaí, *gagagaí)* "head"
ʔísiihoái ~ kísiihoái ~ písiihoái ~ píhiihoái ~kíhiihoái "liquid fuel"[3]

Pirahã allows for variation among voiceless occlusives, as shown, though not for consonants that are [+continuant] or [+voice]. This variation can be accounted for, but only if we refer to Pirahã's channels. (The ungrammatical examples above show that the features [continuant] and [voice] may never vary. Only place features may vary.) With no reference to channels, this variation is without explanation. But in light of the channels this follows because [+/–continuant] and [+/–voice] are necessary contrasts for stress placement (D. Everett 1988).

I am not claiming that the absence of variation for different values of [continuant] is predicted by the distinctive features alone. This case in fact demands that we further investigate the connection between [continuant] and [voice]. In other words, I am not proposing that ethnography should replace phonology (or syntax, morphology, etc.). But I *am* claiming that without the ethnographic study of channels, their role in Pirahã culture, and their place in Pirahãs' dark matter, then even an understanding of Pirahã's segmental phonology is impossible. These in turn must be preserved in every discourse channel, or the constraint below (D. Everett 1985) is violated:

Constraint on functional load and necessary contrast:
a. Greater dependence on the channel → Greater contrast required
b. Lesser dependence on the channel → Less contrast required

The lesson for the field researcher and theoretical linguist to be drawn from these examples is just this: first, language and culture should be studied together; second, as a modality-dependent channel, phonology may be subject to constraints that are (i) language specific and (ii) grounded not only in the physical properties of the instantiating modality (the phonetics) but also or alternatively on the culture-specific channels of discourse employed. This is a very important result because it shows that the 'interface conditions' of the *human computational system* (HC_L), in Chomsky's (1995) terms, may range beyond "phonological form" (PF) and "logical form" (LF), if we define an interface system as a system setting bounds on interpretability for HC_L. Such examples also show how coherent fieldwork can be useful for theory. Thus not only the fieldworker but also the phonologist must engage the grammar and language as forming a coherent whole with culture. And this in turn entails more culturally informed (ethno)linguistic fieldwork.

Of course, language is a crucial component of cultural transmission. As seen in many of the texts examined above, discourses from a given culture reveal how the culture talks about the world, what it talks about, and how this talk is organized (Silverstein 2003; Sherzer 1991; Quinn 2005, among many others). To reexamine and reinforce this view in the specific context of how grammar emerges, I want to examine two very brief Pirahã texts. Both of these examples were collected by Steven Sheldon, a missionary among the Pirahãs, in the mid-1970s. Sheldon, who speaks Pirahã fluently, did the initial transcriptions and most of the translations. (See Futrell et al. [forthcoming] and Piantadosi et al. [2012] for an analysis of the syntactic implications of these texts, as well as what follows).

Some things worth mentioning about the texts which follow include the fact that there is no special or formulaic language for beginning or ending texts (e.g., "Once upon a time," "Happily ever after"). This is, I believe, because Pirahã has no phatic language (D. Everett 2005a, 2008). This aspect of the discourses is thus consonant with the larger culture. Another important observation is that both of the texts show *thematic recursion*. For example, the first text includes three dreams (fat Brazilian woman, papayas, and bananas) as one larger text about dreaming. The second text places sentence-sized questions, answers, asides, and direct address into a single whole.

Another culturally shared assumption of the first text is that dreams and talk about them are worth doing—these are important experiences. The Pirahãs understand dreams as real experiences, though of a different kind from conscious thought. The first text is about a Pirahã man with the Brazilian name Casimiro, recounting to Steve Sheldon the contents of a dream he had.

Casimiro Dreams about Large Brazilian Woman

Told by Kaaboíbagí to Steven N. Sheldon, ca. 1970

1. *Ti aogií aipipaábahoagaí. Gíxai. Hai.*
   I Brazilian woman began to dream You. Hmm.
   I dreamed about Alfredo's wife [aside to Sheldon, "you probably know her"].
2. *Ti xaí Xaogií ai xaagá. Xapipaábahoagaí.*
   I thus. Brazilian woman there be began to dream
   I was thus. The Brazilian woman was there. I began to dream.

3. *Xao gáxaiaiao. Xapipaába. Xao hi igía abaáti.*
   she spoke dreamt Brazilian woman she with remain.
   [Casimiro] dreamt. The Brazilian woman spoke. "Stay with the Brazilian woman."
4. *Gíxa hi aoabikoí.*
   you him remain.
   You will stay with him!
5. *Ti xaigía. Xao ogígió ai hi ahápita.*
   I be:thus woman big well she went away
   I was thus. The big Brazilian woman disappeared.
6. *Xapipaá kagahaoogí. Poogíhiai.*
   dream papaya. bananas
   I dreamed about papaya. Bananas.

Background for this second text includes the fact that Bigixisitísi was a well-liked and well-respected man in the village. He was also one of the best of Sheldon's language teachers. Once when Steve and Linda Sheldon were gone from the tribe, Bigixisitísi became very ill and died. His death was caused by some unknown sickness, and the speaker felt that if Linda had been here perhaps Bigixisitísi could have been saved. Several of the details require cultural or implicit contextual knowledge; for example, the fact that "Xioitábi (Linda) was not there." As an American woman who had lived among the Pirahãs and treated their health for years, it would be known to all Pirahãs who this Linda was and why the fact that she was absent is significant to the story of this man's death.

Bigixisitísi Dies
Told by Itaíbigaí

1. *Bigixisitísi hi baábi. Kapío xiai.*
   Bigixisitísi he is sick. Other is
   Bigixisitísi has a different kind of sickness.
2. *Hi baábioxoi.*
   He sick interrogative
   What is his sickness?
3. *Hi aigía ko Xápaí. Xí kagi hi xaoabábai.*
   He thus hey Xapai her husband He nearly died
   He thus. Hey Xapai. Her husband nearly died.
4. *Hi ábahíoxioxoihí.*
   Unknown.sickness interrogative
   Did he have an unknown sickness?
5. *Hi aigía. Koaísiaihíai.*
   He thus became dead
   Well then. He was dead.

6. *Soxóa ti kabáo. Koaíso. Xai Bigixisitísi.*
   Already I finished dies (he) did. Bigixisitísi
   Bigixisitísi is already finished, affecting me. Bigixisitísi died. He did.
7. *Xabí Xioitábi*
   not there Linda [Xioitábi is her name in Pirahã]
   Linda was not there.
8. *Hi xabaí.*
   She not
   She was not here.
9. *Ti xaigía gáxai. Xai. Hi abikaáhaaga.*
   I thus speak. [I] do. He not be.
   I thus spoke. I did. He is no more.
10. *Hi oaíxi. Pixái.*
    He dead now
    He is dead now.

We see that these Pirahã texts—like texts in any language—reveal cultural values and knowledge(s), and require a culturally based hermeneutics. To understand these, we need to review some of the crucial issues in understanding the relationship of grammar to culture; in particular, how culture and grammar shape each other through their evolved symbiotic relationship.

When I say that culture and grammar are symbiotically related, I mean first that grammar (and language) is dependent on culture not only for its functions but also for the very forms it employs to carry out those functions. But I also mean that culture is codified, regulated, reinforced, and partially formed by grammar. Thus, though grammar and culture may be epistemologically and ontologically distinct, they are not independent in practice.

My notion of a grammar-culture symbiosis is not to be confused with the idea that either is supervenient on the other. Supervenience is a relationship such that "a set of properties A supervenes upon another set of properties B just in case no two things can differ with respect to A-properties without also differing with respect to their B-properties" (McLaughlin and Bennet 2011). Further, by symbiosis I mean that grammar and culture are each causally implicated in and dependent upon the other in their respective historical developments, even though there is no one-to-one mapping between them.

Linguists and anthropologists, as to be expected, hold different views on the connection between grammar and culture. There are those who argue that any interaction between the two is trivial, the total range of interactions not moving much farther than a few lexical choices and things such as polite vs. formal address forms. At the other end are those who think of grammar as little more than a cultural artifact. Neither extreme captures what I am after here.

CULTURAL CONSTRAINTS ON "CORE GRAMMAR"

For another example of culture affecting grammar, I want to revisit Pirahã's apparent lack of recursion (D. Everett 2005a, 2008, 2012b; Futrell et al., forthcoming). Most languages use recursive operations in the construction of their syntactic structures. This is so common cross-linguistically that in 2002 Marc Hauser, Noam Chomsky, and Tecumseh Fitch (HCF) made the startling claim that the single innate cognitive ability underlying language—possessed exclusively by Homo sapiens—was the ability to construct grammars recursively. Subsequently, this bold claim has since been falsified for being both too weak and too strong. The proposal is too weak because there is abundant data that humans are not the only species that use recursive cognitive or communicative operations (Corballis 2007; Golani 2012; Pepperberg 1992; Gentner et al. 2006; Rey et al. 2011). Second, the proposal is too strong because there are languages that lack recursion (D. Everett 2005a; Gil 1994; Jackendoff and Wittenberg 2012). To see what HCF mean by recursion, here is a statement from the original paper:

> FLN only includes recursion and is the only uniquely human component of the faculty of language. . . . In particular, animal communication systems lack the rich expressive and open-ended power of human language (based on humans' capacity for recursion). (Hauser, Chomsky, and Fitch 2002, 1569–70)

There are many potential senses of the term *recursion*, so it is vital to understand what HCF had in mind. Their paper leaves no doubt that they intended a process that applies to its own output without limit. This is clear in such statements as the following, where they claim that when a language has recursion, then "there is no longest sentence (any candidate sentence can be trumped by, for example, embedding it in 'Mary thinks that . . .'), and *there is no nonarbitrary upper bound to sentence length*" (reference 9, 1571; emphasis mine).

Although such quotes are straightforward, some Chomskyan syntacticians now claim for it a more esoteric meaning (as a reaction to my criticisms and empirical work [D. Everett 2005a, 2008, 2012a, inter alia]). According to this initiate exegesis, recursion means for HCF only a (singleton) subset of recursive operations internal to the program known as minimalism, what Chomsky (1995) calls "Merge."

This means that Merge is potentially falsified by any exocentric or nonbinary (ternary, quaternary, etc.) branching structure—for example, a structure with flat syntax. Culicover and Jackendoff (2005) have argued—to my mind, convincingly—that ternary structures exist in the syntax of some languages, and I (D. Everett 1988) have argued that non-derivable ternary structures exist in the metrical structure of Pirahã phonology. Further, I (D. Everett 2005b; 2009b) have elsewhere demonstrated that the syntax of the Wari' language of Brazil makes widespread use of non-endocentric constructions. Yet even though counterexamples exist, the authors and their followers continue to insist that Merge is what they meant by recursion.[4]

Nevertheless, the Merge interpretation has to strain to produce the "no longest sentence" clause of their earlier quotation, since that is a result of the more general notion of recursion. Even Chomsky (2010, 2014) allows that Merge itself may be blocked from repeating endlessly by language-specific stipulations. But such stipulations play no part in the mathematical notion of recursion.

I have argued elsewhere (D. Everett 2012b) that this post hoc, theory-internal reasoning is unhelpful. First, it excludes an important empirical space—namely, the class of languages that lack Merge but have other forms of recursion, such as languages with ternary branching but no longest sentence. Second, it ignores the possibility that some language may lack any form of syntactical recursion, such as Pirahã. Third, it overlooks what is to my mind the most important consideration in understanding the role of recursion in natural language—natural conversations, narratives, and other discourses.

Lobina and Garcia-Albea (2009) offer a helpful elucidation of various notions of recursion that have been employed in mathematics, computer science, linguistics, and cognitive sciences. They correctly observe that Merge itself need not be a recursive operation, since iteration does not properly fall within the standard mathematical or computational definitions of recursion. But if it is not recursive, there must be ancillary hypotheses to guarantee this, further weakening the hypotheses.

Languages with non-recursive grammars—such as Pirahã (D. Everett 2005a) and perhaps Riau (Gil 1994)—are far from irrelevant to the construction of syntactic theories. In HCF's sense, recursion is the *only* item in the "linguistic toolbox." Further, HCF claim that recursion is the only biological difference between humans and other animals that makes language possible. If they then concede that not all languages require recursion, their original claim loses any empirical force it might have had. If recursion is the fundamental component of universal grammar, how could it be lacking in any language? It would be a strange proposal indeed if the absence of the singular biological underpinning of language is treated as empirically irrelevant.[5] In fact, treating Riau, Pirahã, and other languages that lack recursion as exceptions would be like saying that finding a black swan or a penguin does not falsify the claims that all swans are white or that all birds fly.

Ironically, although I have repeatedly argued that Pirahã shows recursion in texts, texts lie outside the sentential syntax that has defined generative theory since its inception, where, again, the "start" symbol (Σ) for all syntactic operations has always been the sentence. (D. Everett [1994] offers a [Chomsky-inspired] discussion of the "sentential divide" in grammar and cognition.)

Linguists have long resisted the idea that culture is causally implicated in the formation of grammars, at least insofar as what Chomsky calls "core grammar"—the state of the language faculty after language-specific parameters have been set. Here I want to underscore arguments I have made elsewhere (D. Everett 2005a, 2008, 2012b). I will do this by examining the relationship between the morphosemantic notion of evidentiality, Pirahã culture, and Pirahã syntax.

Evidentiality—the semantic notion of evidence for an assertion—is found in all languages in one form or another. For example, if I say, "The man came in here," the default assumption in English is that I have direct evidence for this assertion. Evidentiality is arguably found in the pragmatics of every language because it helps the hearer distinguish speculation from evidence-based declarations, something that could save a lot of time in deciding where to hunt, build a village, and the like. However, for some cultures, evidentiality is not only a semantic or pragmatic fact but a morphosyntactic fact as well, encoded in some way on words (usually verbs) of the language. At some point in the development of such languages, speaker usage turned this near-universal semantic category into an overt symbol in their grammar. This is to me a cultural development, even if no speaker(s) consciously invented the evidentiality morphemes for their language (how cultural innovation or innovation in language arises is a complex issue, known as the "actuation" problem—how changes spread through a culture or language; Weinreich, Labov, and Herzog 1968). I believe that the relative importance of evidentiality, like any other category, can be calibrated for a given language from its effects on the morphosyntax and its role in the culture. The greater the effects, the more important it is; the fewer, the less important. These are determinations about dark matter based on observable behavior. They do not necessarily result from conscious manipulation of specific morphemes by speakers. This, then, is evidentiality. The next step for us is to understand how Pirahã evidentiality follows from Pirahã culture.

Everett (2005a) describes a range of unusual features of Pirahã culture and language, many of them never documented for other languages (though one would not be surprised if many other languages had similar features or lacked such features). These include: simplest kinship system known, lack of color words, lack of numbers and counting, no perfect tenses, no creation myths, no historical or fiction myths, being monolingual after more than three hundred years of regular contact with Brazilians, and *no recursion*. I proposed to account for all these facts by the immediacy of experience principal (IEP). This is a principle found in some degree of strength in many Amazonian languages. (See Gonçalves 2005 for a discussion of the pervasiveness of immediacy of experience as a cultural value throughout Amazonia.)

Dark matter's effects are far-reaching. In fact, the IEP affects Pirahã grammar profoundly. To see how, let's begin by restating this principle:

> *Immediacy of experience principle (IEP):* Declarative Pirahã utterances contain assertions only related directly to the moment of speech, either experienced (i.e., seen, overheard, deduced, etc., as per the range of Pirahã evidentials, as in D. Everett [1986, 289]) by the speaker or as witnessed by someone alive during the lifetime of the speaker.

D. Everett (2005a) offers a range of arguments for the IEP, based on the empirical points mentioned earlier, as well as (among other things) the culturally important notion of *xibipíío*, "experiential liminality," discussed in D. Everett (2008). This word is further evidence that liminality is an important cultural and individual concept in Pirahã. It is used to describe things that go in and out of vision or hearing, from the flickering of a match to the disappearance or appearance of a canoe around a bend in the river.

Moving from this initial cultural statement to the grammar, and later back to link them, the evidence that Pirahã lacks recursion (also discussed in Everett 2012a) is as follows (though see Perfors et al. [2010] for another type of approach to checking the grammars of languages):

First, the lack of recursion correctly predicts that factive and epistemic verbs will be absent. This follows because if Pirahã lacks recursion, then there is no way to express factive verbs as independent verbs, since these would require a complement clause. That would in turn require embedding and thus, ceteris paribus (in some analyses), a recursive rule in Pirahã syntax. Pirahã expresses such notions via verbal suffixes, consistent with the "no recursion" hypothesis, not with complement clauses.

Second, Pirahã has no marker of subordination. This is also predicted by my hypothesis, because if Pirahã lacks recursion, there is no subordination to mark.

Third, Pirahã has no coordinating disjunctive particles (e.g., *or*). The absence of explicit markers of disjunction is predicted by my hypothesis, since disjunction entails recursion.

Fourth, Pirahã has no coordinating conjunctive particle (e.g., *and*). There is only a more general particle, *píaii*, which may appear preverbal or sentence final and which means "is thus/simultaneous" (vague meaning), which never works like proper conjunction, but only supplies the information that these two things were simultaneous. Again, this is predicted by my analysis, since coordination also entails recursion.

Fifth, Pirahã has no syntactic complement clauses. If Pirahã has recursion, where is the unambiguous data? I have claimed that it lacks embedded clauses. Others claim, based on my own data and my own earlier analysis, that it has them (Nevins, Pesetsky, and Rodrigues 2009).[6] But although quotatives *could* be embedding, there are no multiple levels of embedding, which would be expected if Pirahã has recursion.

Sixth, Pirahã does not allow recursive possession. The point of Pirahã possessives that I have made is not simply that it lacks prenominal possessor recursion, but that it lacks recursion of possessors *anywhere* in the noun phrase. Nevins, Pesetsky, and Rodrigues (2009) might be correct to suggest that German, like Pirahã, lacks prenominal possessor recursion. But German *does* have postnominal possessor recursion, while Pirahã has *none*. The facts are therefore exactly as my analysis predicts them to be.

Seventh, Pirahã prohibits multiple modifications in the same phrase. As I have discussed above and in Everett (2008) and (2009a), there can at most be one modifier per word. You cannot say in Pirahã "many big dirty Brazil nuts." You'd need to say, "There are big Brazil nuts. There are many. They are dirty." This paratactic strategy is predicted by my analysis since multiple adjectives, as in English, would entail recursion. But the paratactic strategy does not.

Eighth, Pirahã semantics shows no scope from one clause into another (e.g., no "NEG-raising"). Pirahã lacks examples such as *John does not believe you left* (where *not* can negate *believe* or *left*, as in *It is not the case that John believes that you left* vs. *It is the case that John believes that you did not leave*). In this example, *not* can take scope over *believe* or *left*. That is not possible without recursion, so my analysis predicts the absence of such scope relations. This is also predicted, correctly, to be impossible in Pirahã under my account, since it would entail recursion.

Ninth, Pirahã shows no long-distance dependencies except between independent sentences—that is, discourse. The kinds of examples that are standardly adduced for long-distance dependencies include:

"Who do you think John believes ___ (that Bill saw ___)?"
"Ann, I think he told me he tried to like ___."

We have stated the IEP and rehearsed the evidence against syntactic recursion in Pirahã. It remains now to show how these fit together causally. It turns out that they engage like the teeth in cogs, via evidentiality. Pirahã, like many other languages (see, inter alia, Aikhenvald 2003 and Faller 2007), encodes evidential markers in its verbal morphology as affixes: *-híai*, "hearsay"; *-sibiga*, "deduction"; *-ha*, "complete certainty"; and *-o* (zero affix), "assumption of direct knowledge." The Pirahã IEP, in conjunction with its requirement that evidence be provided for all assertions, produces a narrow domain in which assertions and their constituents need to be warranted. Reminiscent of the potential focus domain developed by Van Valin (2005, 70ff), I label this domain in Pirahã (and presumably some version of this will exist in all languages—at least, those with evidentiality morphology) the *potential evidentiality domain* (PED); that is, the range of structures where the actual evidentiality domain could in principle fall. The actual domain of evidentiality in a given utterance will be as follows:

> *Evidentiality domain:* The syntactic domain in a sentence that expresses the evidentiality component of the pragmatically structured proposition.

The PED in Pirahã is limited to the lexical frame of the verb—the verb and its arguments (more technically, the phrasal nuclei of the predicate and its arguments in Van Valin's role and reference grammar [RRG] terminology)[7]. Let's assume that the IEP is one of the reasons that Pirahã has evidentiality markers and that it dramatically strengthens their effect by narrowing their scope to the PED just mentioned.

The PED then rules out syntactic recursion in Pirahã. As stated, the PED clearly depends on the main verb as the core of the speech act. The PED will include only nuclei (semantic-syntactic heads, not heads in the X-bar sense) directly licensed by the predicate (its semantic frame). No nuclei are allowed outside the PED of a containing sentence.

By the PED, there are no embedded possessors, no embedded predicates—only arguments licensed by the main predicate. For example, in a noun phrase like *John's house*, *house* is the nucleus—the semantic core, what this phrase is about. *John* is the possessor, a type of modifier of the nucleus house—the possessor tells us which house we are talking about. On the other hand, in a larger noun phrase such as *John's brother's house*, *house* and *brother* are each a nucleus of a separate containing phrase. *House* is the nucleus of the phrase

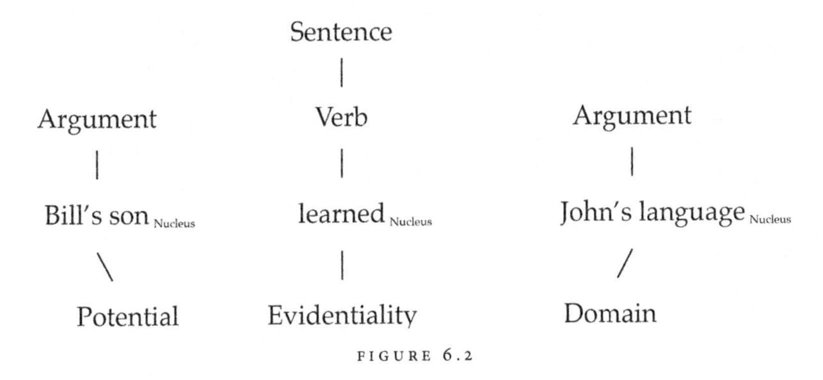

FIGURE 6.2

brother's house, and *brother* is the nucleus of the phrase *John's brother*. *John* is not a nucleus of any phrase. This means that *John*, not being the possessor of an argument of the main verb (it is a nucleus of *John's brother*, but *brother* is not a nucleus of the verb) is unwarranted in the PED and the sentence is disallowed. An embedded predicate would contain arguments not licensed by main predicate. Therefore, there can be no phrases within phrases and no sentences within sentences in Pirahã. There can also be no productive compounding in the morphology. Such apparent compounds as are found are in fact synchronic or diachronic phrases. This is exemplified in fig. 6.2, in a theory-neutral representation.

This example is allowed because each Nucleus is found in the semantic frame of the verb, represented along the lines of the following lexical representation: [BECOME *know* (son, language)]. This is a very strict evidentiality requirement. It predicts that the number of arguments in a sentence cannot exceed the number allowed by a standard (e.g., RRG) verbal frame. It rules out all embedding and all syntactic recursion.

The lexical representation of an "accomplishment verb"—for example, *learn* ([BECOME *know*] indicates the change of state of knowledge) projects three nuclei to the syntax: the verb *learn*, and the nominal nuclei/arguments *son* and *language*. Each of the nominal nuclei is possessed by a non-nuclear nominal. So the requirements of the PED are met. However, in the example below, there are two non-warranted nuclei—appearing in the PED without being found in the lexical representation shown in fig. 6.3.

This sentence would therefore be ungrammatical in Pirahã, though it is fine in English. My analysis claims that the existence of evidentials, their scope, and the consequent lack of recursion are all reflexes of the cultural value IEP in Pirahã grammar.

Although the PED (forced by the IEP) rules out recursion in Pirahã, my

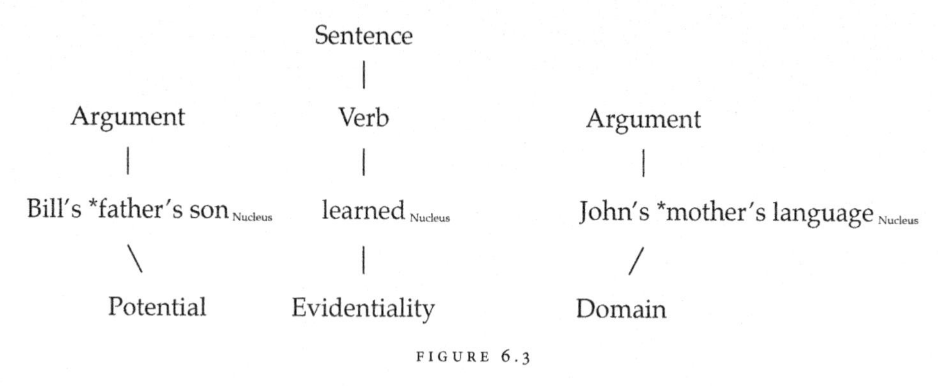

FIGURE 6.3

analysis does not require that any another language, e.g. Riau, necessarily derives the absence of recursion in the same way. Recursion serves several purposes (D. Everett 2012b) and thus there is more than a single reason why a language might use or not use recursion in its sentential syntax. For example, Riau might simply rank a value of slower information rate above a value favoring recursive sentences in its language. Many oral traditions use repetition and slower information rate as aids to communication in the noisy environments of human speech. So this is a cultural explanation of some very complex syntactic facts that affect the Pirahã language as a whole.

Methodology for Studying the Culture-Grammar Interface

Before ending this discussion, I would like to present a set of methodological suggestions for studying the connection between grammar and culture, building on suggestions by Saville-Troike (1982, 108ff). The beginning steps for the ethnography of communication are to: (i) identify recurrent events; (ii) analyze these events, examining their function, form, and relationships between different constituents; and (iii) examine the relationship between these events to other speech events and to the society and culture in which they occur.

As an example, let's consider the use of whistle speech as it is found on the Canary Islands. One variety, Silbo Gomero, is used in and around La Gomera. In relation to (i), each use of whistle speech is thus an event. Some questions that might be asked about these events, (ii), (iii), and so on, are: When is it used? Who uses it? What are the constraints on its intelligibility? (E.g., can two people understand Silbo under any circumstances, or does a topic of conversation need to be established first to provide context?) How many other channels of discourse are there among speakers who use Silbo? Are there contents or types of discourse in which the people prefer to use Silbo?

Are there contents or types of discourse in which the people prefer not to use Silbo? What are the phonetic details of Silbo, and how is whistle speech possible in this language (since the language it is based on is not tonal, does it use inherent segmental frequencies as a basis, intonation, etc.)? How does it relate to the consonant-and-vowel channel (i.e., normal speech)?

Beyond these suggestions, there are further methodological preliminaries for investigating the culture-language connection. These preliminaries include at least the following:

1. Are there irregularities that have no obvious structural explanation?
2. Are there examples of "free variation"—that is, where there are choices between two structures which are not constrained by the structures or the grammar, insofar as can be determined?
3. Are there unusual facts about the cultural events, values, or explanations that involve principles or phenomena that at any level look similar to principles operative in the grammar?

As to the methodology that follows from such questions, Enfield (2002, 14ff) offers some cogent and very important considerations and suggestions for the study of ethnogrammar. First, he recommends that the fieldworker "examine specific morphosyntactic structures and/or resources and make explicit hypotheses as to their meaning." Second, following development of this and related methodological considerations, he raises the crucial issue of "linkage"—namely, how can we establish a causal connection between facts of culture and facts of grammar? I turn to this directly. Before doing this, however, I want to point out what seems to be the biggest lacuna in the study of ethnogrammar, whether in the studies in Enfield (2002) or elsewhere. This is the effect of values, especially cultural taboos like the IEP above, in restricting both culture and grammar. That is, previous studies, like those in Enfield (2002), while reasonably focusing on meaning—which is, after all, a principal contribution of culture (i.e., guiding its members in imputing meaning to the world)—fail to consider cultural prohibitions or injunctions, however deeply or shallowly embedded in the community system of values. The Pirahã example of this section is evidence that such values should also be considered in ethnophonological as well as ethnosyntactic studies—hence "ethnogrammar." However, before we can draw any conclusions at all about ethnogrammar in a given language, we need to consider the vital issue that Enfield refers to as "linkage"—the establishment of a causal connection between culture and language. That is, how can we convince someone (or, at least, effectively argue) that property *p* of culture C causally determines feature *f* of grammar G? As

Enfield (2002, 18ff) summarizes, there are four prerequisites to establishing linkage between culture and language:

1. *Empirical grounding:* Are the phenomena clear and well established?
2. *Structure independence:* Are the cultural and grammatical structures or principles independently needed in the grammar?
3. *Theoretical coherence:* Does the analysis follow from a clear theory?
4. *Noncircularity:* Has the analysis used independently justified values to explain independently justified structures?

An example of a circular argument in ethnogrammatical studies might be to claim that a particular linguistic feature is determined by an aspect of culture and to simultaneously use it as evidence for that very aspect of culture. So, it is circular to claim that a language has evidentials because the culture values empirically based reasoning, and then to further claim that we know that the culture values empirically valued reasoning because it has evidentials. The way to avoid this is first to establish, using *nonlinguistic evidence,* particular values or meanings in a certain culture, such as the IEP. Next, using *noncultural evidence,* establish the meaning and structure of the relevant linguistic examples (examples would include standard arguments for constituency, displaced constituents, etc.). Finally, show how linking the two provides a conceptually and empirically (in terms of predictions, where possible, or explaining independent domains such as historical change) superior account of the facts that leave them unconnected. This is what I have attempted to do here.

Ethnogrammatical studies thus range from showing that, say, a language has honorifics because of a severe social structure, or a particular set of kinship terms because of its restrictions on marriage, to showing global, architectonic cultural constraints on grammar; for example, from taboos like the IEP.

Another issue is whether the researcher can successfully get the semantics right. The so-called translation fallacy is well known, but field linguists in particular must be ever-vigilant not to be confused by it. We must be on guard against the mistake of concluding that language X shares a category with language Y if the categories overlap in function. As Gordon (2004) argues, much of Pirahã is largely incommensurate with English, and so translation is simply a poor approximation of Pirahã intentions and meaning.

Cultural learning is discussed in a multitude of studies (e.g., the entire fields of cultural psychology [Heine 2011] and neuroanthropology [Lende and Downey 2012]). But perhaps the two most important mechanisms are (i) what D. Everett (2012a), going back to Aristotle, refers to as "the social instinct," and (ii) general cognition. Another way of referring to the social instinct is as the "interactional instinct" (Lee et al. 2009; Levinson 2006; Joaquim and

Schumann 2013). By general cognition I refer in particular to the general ability of the human brain to generalize and recognize patterns.

What might be the evolutionary utility of an interactional instinct? This instinct (however it is ultimately characterized) is the presumably unlearned need for humans to communicate, to interact with one another. Levinson (2006) makes a convincing case for the independence of interaction from language. The need to interact and the ability to interact are prior to language. The appeal of such an instinct is that it is a simple reflex that requires no learning curve (such as is required for the so-called language instinct, for example). The instinct is not the final product, of course, but it triggers movement in that direction and is arguably what distinguishes humans from other species that lack this social or interactional instinct. The social instinct is the "initiator" in that it provides the problem, while language and society provide the solutions. In this sense, language is the principal tool for building social cohesion through interaction. There might be ways around the need for an instinct like the "interactional instinct." But I appeal to it here as but one way to understand why humans have an unique need to communicate with their conspecifics. We could also call it a "social desire" or an "interactional need," based on human emotions and their intersection with human intelligence.

Many researchers (e.g., Tomasello [1999]) have made a case for qualitative differences between the interactions and social organizations of humans vs. other species. Clearly, though, since humans have bigger brains, an interactive instinct, and a transmitted linguistic history (passing along to subsequent generations both the content and the form; i.e., grammar), these differences will be both quantitatively and qualitatively different from nonhuman animals in communication. The idea of general learning (including such things as memory, motivation, emotion, heuristics, categorization, perception, and reasoning), heavily dependent as this is on the keen human ability to make tacit statistical generalizations, as a key to language differences between humans and other creatures has been defended many times in the literature. Kurzweil (2012) makes this case to a popular audience. But many researchers in Bayesian approaches to learning (e.g., Goldsmith 2015; Pearl 2013; Perfors et al. 2012; MacWhinney 2005) present much more technical and nuanced evidence, backed by extensive experimentation. Such claims in fact go far back, with a form of the argument to be found, at least implicitly, in Benedict (1934).

The effects of culture on the lexicon take on a greater significance these days when many linguists deny a strong bifurcation between syntax or grammar and the lexicon. In fact, if constructions (see Goldberg 1995) are lexical items that produce families of syntactic constructions, then the culture can affect the syntax of constructions, just as all linguists now agree it can affect the lexical items of any language. Our discussion here is not intended as a list of noncontroversial results. It is intended to provide evidence that culture profoundly affects grammar, and that understanding and studying this relationship between culture and grammar is not beyond our grasp.

Finally, the considerations above lead to the proposal of a simple formula for the development of grammar in our species:

COGNITION, CULTURE, AND COMMUNICATION → GRAMMAR

In other words, given human cognitive abilities, cultural/community shared experiences, and the social/interactional instinct, grammar emerges as a solution to the latter problem, facilitated by the first two abilities.

Summary

In this chapter we looked at the evolution of linguistic theory in North America away from the starting point that so distinguished it—the idea that language is a manifestation of culture and that culture exercises an architectonic effect throughout language—in syntax, phonology, morphology, and semantics. We then turned to consider in depth the analysis of Pirahã grammar; we concluded that only by recognizing the role of culture as a constraint and shaper of grammar as well as all other components of language, could we begin to understand languages, either as theoretical objects of study or as living partnerships between members of a speech community. From this perspective, we saw that Pirahã shows evidence for cultural effects in the syntax (lack of recursion) and in the phonology (the interaction between channels of discourse, prosody, and phonemic processes). Thus we concluded that culture exercises an architectonic effect on the grammar of the language and that in this sense language emerges from culture, society, functional constraints on linguistic form.

7

Gestures, Culture, and Homesigns

In the great open-air theater that is Rome, the characters talk with their hands as much as their mouths.

RACHEL DONADIO, "When Italians Chat, Hands and Fingers Do the Talking"

Just as dark matter affects grammar, so it is causually implicated in the components of grammar—words, gestures, phonology, syntax and so on. However, for many linguists and anthropologists, gestures are often omitted from discussions of grammar, being seen as somehow "paralinguistic" accouterments of speech. Certainly they tend to be overlooked in discussions of language and culture. It seems that many researchers see gestures as a separate, independent facet of human behavior. However, gesture researchers, from various theoretical perspectives, have developed research that shows the intimate connections between gestures, grammar, cognition, and culture. These connections are underwritten by dark matter. Thus a study of the manifestations of dark matter cannot be complete without a detailed analysis of the symbiosis between gestures and grammar.

Moreover, in recent publications, some researchers have argued that gestures show us something more—namely, that the dark matter behind gestures includes striking, highly language-specific components of gestures. This research, pioneered in the work of Susan Goldin-Meadow, examines the "spontaneous emergence" of gestures in children who have otherwise no access to linguistic input, as in the deaf children of hearing, non-signing parents. She calls these gestures "homesigns," and the gestural systems she studies might indeed crucial to our quest to tease apart native vs. cultural or a priori vs. a posteriori perspectives on the origins of (some) dark matter.

This chapter therefore looks in detail at the historical and evolving research into gestures and human language, from the ancients through the current and very important research of contemporary scientists such as David McNeill and his colleagues, as well as some of his critics and others (borrowing heavily from the work of Kendon [2004]). The chapter argues that without understanding gestures, we cannot understand grammar, the evolution of language, or the use of language. Gestures are vital for a fuller understanding of dark matter, its origins, and its broader role in human language, communication, and cognition.

Language Is Large

Language is holistic. Whatever grammar there is, language engages the whole person—intellect, emotions, hands, mouth, tongue, brain. And it likewise requires access to cultural information and dark matter. What I want to do in this chapter is twofold. First, I want to provide an overview and review of what I consider to be the most interesting research on gestures (some may disagree with my selection). We see through this that gestures are extremely complex. But second, I want to argue that this is all learnable and provides yet another case for the superiority of the dark matter concept we are pursuing here, as opposed to a priori, unlearned, innate dark matter.

If I am correct, then, the gestures that accompany all human speech reveal an intersection of culture, individual apperception, intentionality, and other components of dark matter as we have been discussing it until now. According to many gesture researchers, *dynamic* human gestures interact in complex ways with *static* human grammars to produce human language—neither grammar nor gesture is language apart from the other. Gestures, as principal components of the multimodal behavior of language, are themselves intricate in structure, meaning, and use. David McNeill (1992, 2000, 2005, 2012) and others' research on gestures make it clear that gesticulations are as analytically challenging and as intricate in design and function as syntax, phonology, morphology, discourse, and other aspects of language that some linguists consider "language proper." But gestures are not simply add-ons to language. There is no language without them.[1] And there are no gestures without dark matter. Although some gestures can be explained by their users (unspoken), most are beneath users' conscious awareness (ineffable). Moreover, though gestures themselves are crucial components of language, like words and phrases or intonation, they are shaped by the needs of the language they enhance and the cultures from which they emerge.

Kenneth Pike—the first person to ever discuss gestures in my linguistic training—saw them as evidence for his proposal that language should be studied in relation to a unified theory of human behavior:

> In a certain party game people start by singing a stanza which begins *Under the spreading chestnut tree* . . . Then they repeat the stanza to the same tune

> but replace the word spreading with a quick gesture in which the arms are extended rapidly outward, leaving a vocal silence during the length of time which would otherwise have been filled by singing. On the next repetition the word spreading gives place to the gesture, as before, and in addition the syllable chest is omitted and the gap is filled with a gesture of thumping the chest. On the succeeding repetition the head is slapped instead of the syllable nut being uttered . . . Finally, after further repetitions and replacements, there may be left only a few connecting words like the, and a sequence of gestures performed in unison to the original timing of the song. (Pike 1967, 25)

From this example, Pike raises a question and draws a conclusion. The conclusion is that gestures can replace speech (in this case, with "language-slotted gestures"; see below) in the song above because both speech and gestures are aspects not merely of language but of human behavior more generally.[2] Unfortunately, Pike oversimplifies by drawing this conclusion from a very particular type of gesture, but his basic point is valid: language and its components are human behavior guided by individual psychology and culture (i.e., dark matter).

Human behavior, including communication, is the working out of intentions. Language is the best tool for communicating those intentions (D. Everett 2012a, 160ff). Or, as Giorgolo (2010, 8ff) puts it, "Communication has evolved into a cooperative behavior guided by principles of rational interaction," all entailing tacit knowledge, values, and other components of dark matter.

The question Pike raises is an interesting one. I will call it "Pike's problem," one that should be addressed by a theory and typology of gestures. The problem is just this: Why is it that only sounds made by our mouth can be used in syllables and speech more generally? Why couldn't *slap* be [sla#], where [#] would represent the sound of someone slapping their chest? Why isn't this a possible word or syllable in anyone's language? As a student in Pike's introductory class, in which he asked this question, I thought it was interesting but did not appreciate adequately the degree to which the question impinges on the core of the linguistics enterprise.

Before proceeding to more detail on gestures, however, we might address the obvious question of whether speech itself is but a set of gestures. If so, then gesture research would encompass all of linguistics. And it is easy enough to see why speech could be a form of gesture—it involves movement of body parts (articulators to points of articulation) for the purpose of communication. However, there are nevertheless two principal motives for excluding the movements of the articulatory apparatus from the gesture continuum. First, speech gestures are generally unseen, happening behind the lips. Second, articulatory movements and manual gestures contrast functionally. Manual gestures add information that syntax and speech alone do not carry.[3] These then exclude both movements of the vocal apparatus and the static grammar of hand movements in sign language (though sign language, like non-signed speech, also has different types of gestures) from the theory of gestures we discuss here.

Gestures aim toward perlocutionary effects, such that if some information is not part of speech proper, the speaker will try to express it in gestural terms in order to help the hearer use or react to the information in the way intended. In this way, they are like intonation and other prosodies.

To more fully illustrate the need for a single theory of culture and language—indeed, all human behavior—Pike asked his audience to contemplate the following scene: Two men are watching other men move some heavy furniture down the stairs in their apartment building. One man passing on the stairway landing is huffing and puffing and concentrating solely on his heavy load. His wallet is hanging loosely from his back pocket, about to fall out. He clearly wouldn't notice if someone relieved him of this burden. The first observer looks at the second observer with raised eyebrows, looking at the wallet. The second one sees him and simply shakes his head to indicate no. What kind of communicative act is this? Is it language? Is it parasitic on language? Is it a communicative act of a different nature? Should such acts fall under a theory of gestures? Certainly there is a shared dark matter necessary for this kind of act, including gestures, language, grammar, and so on, to take place to begin with. A basic principle is that gestures enlarge to fill the communicative space shared with speech. As speech is diminished—as in Pike's example of the contemplation of robbing the overworked mover, or in the substitution song we mentioned earlier—gestures will take on a larger role to keep communication as effective as possible. This larger role is illustrated in delayed auditory feedback experiments (discussed below) or the use of gestures to communicate in the absence of speech while hunting or fishing. Other examples are discussed directly.

There is a broad popular interest in gestures. For example, in a much commented-on article in the *New York Times*, Rachel Donadio (2013) wrote about how "When Italians chat, hands and fingers do the talking." But all languages have gestures. So is there any scientific sense in which Italian gestures are noteworthy? For example, can gestures vary from language to language or from culture to culture? Or are they innately prespecified? Or could it be that all gestures are different in every utterance by every speaker? Italians' gestures are frequently caricatured on television and in movies. Are such caricatures based in fact?

PIONEERS OF GESTURE RESEARCH

What appears to draw attention to Italians' gestures is their extravagance. Even in the seventeenth century, Northern European Protestants disapproved of Italians' "flamboyant" gestures (Kendon 2004, 21ff). The first person to study these gestures from a scientific perspective was David Efron, a student of Franz Boas, who wrote the earliest modern anthropological linguistic study on cultural differences in gestures (focusing on gestures of recent Italian and Jewish immigrants) more than seventy years ago.

Efron's ([1942] 1972) study, *Gesture, Race, and Culture*, is at once a reaction against Nazi views of the racial bases of cognitive processes, a development of a model for recording and discussing gestures, and an exploration of the effects of culture on gesture. The core of Efron's contribution is his description of the gestures of unassimilated Southern Italians and Eastern European Jews ("traditional" Italians and Jews), recently emigrated to the United States and mainly living in New York City (though some of his subjects came from the Adirondacks, Saratoga, and the Catskills). According to Efron (330ff), Italians use gestures to signal and support content (e.g., a "deep" valley, a "tall" man, "no way"), of illustrating what was just said, with many *signs* ("emblems" in McNeill's sense) among the gestures. The Jewish immigrants of Efron's study, on the other hand, use gestures as logical connectives (to indicate changes of scene, argumental divisions, etc.). These conclusions underscore and support the Boasian perspective that language is shaped by culture.

The questions that Efron set out to investigate were (i) whether there were standardized group differences in gestural behavior between the two different "racial" groups and, if so, (ii) to determine what becomes of these gestural patterns in members and descendants of the same groups under the impact of social assimilation. He found a strong cultural effect—"Americanization" of gestures—in each group over time, minimizing their initially strong differences, thus illustrating the idea that at least some aspects of gestures are individual dark matter, group-developed cultural productions, another part of the social contract.

Working with the artist Stuyvesant Van Veen, Efron produced an effective methodology for studying and recording gestures, as well as a language for providing what Geertz (1973, 3–30) would have called "thick descriptions" of gestures (see chap. 4). Abstracting away from the long anti-Nazi-science section, the book continues to be of interest for contemporary scientists.[4] Efron's work was a pioneering one in many senses, but by no means the first work on gesture. Kendon (2004, chap. 3–5) provides a detailed overview of the history of gesture studies, going back to classical antiquity.

For example, Aristotle (384–322 BCE) discouraged the overuse of gestures in speech as manipulative and unbecoming. Cicero (106–43 BCE), on the other hand, argued that the use of gestures was important in oratory and encouraged their education. In the first century, Marcus Fabius Quintlianus (Quintilian, 35–after 96 CE) received a government grant (perhaps a linguistic first) for a book-length study of gesture. For Quintilian and most of the other classical (and some later) writers, however, gesture was not limited to the hands but included general orientation of the body and facial expressions, so-called body language. There are a couple of lessons to draw from these older findings: (i) communication is holistic, so we should not be surprised to learn that, for example, blind people gesture; (ii) gesture varies just as any kind of culturally articulated unconscious knowledge would, just as any other convention.

As the Renaissance rediscovered the work of Cicero and other classical scholars, Europeans became interested in gesture and rhetoric, as in the sixteenth-century work of Peter Ramus (1515–1572) of Paris on gesture in the style and delivery of oratory. Giovanni Bonifacio's (1547–1645) published the first book in Europe on gesture, *L'Arte de' Cenni* (1616). The first book in English on gesture was John Bulwer's (1606–1656) 1644 book *Chirologia: or the Naturall Language of the Hand.*

The sixteenth-century studies often looked a good deal like actual science, not merely prescriptive offerings for improving public discourse delivery. In the seventeenth century, some researchers moved on to question explicitly whether gesture might be developed to serve as a possible universal language. By the eighteenth century, some began to wonder whether gesture might have served as the original source for language (some researchers, such as Arbib [2005, 2012] and Corballis [2002], still think so), though most contemporary gesture researchers strongly discourage this view. Although these studies were to some degree misguided (as McNeill points out in his work), they do fit with the idea (also pointed out in D. Everett 2012a, 171ff, among other places) that speech as we know it almost certainly was evolutionarily secondary to the development of some sort of language, whether this were gestures and whistling or gestures and much less clear speech (lacking the vowels which are the hallmark of the speech of all humans—*i*, *a*, and *u*, for example). Gesture is fundamentally part of communication, and according to the logic and history of evolutionary development of language, communication precedes language.

By the medieval period (Kendon 2004, 328ff), thinkers from the early Renaissance were writing about gestural variation across European countries. In the Counter-Reformation period, there were attempts to reform gestures (due to the perceived inappropriateness of some).

One study that merits mention is the work of Abbé Charles-Michel de l'Épée (1712–1789), who wrote an eighteenth-century study of deaf education and sign language. L'Épée was ahead of his time because he was not interested in getting deaf people to use spoken language, but rather in helping both deaf and non-deaf people to understand and appreciate sign language.

Although interest in gesture research declined markedly in the nineteenth century, according to Kendon (2004, 17ff) there were nevertheless two fascinating studies published during this period. In 1832 Andrea de Forio published on gesture and cultural community in Naples, while Garrick Mallery published in 1881, *Sign Language among North American Indians Compared with That among Other Peoples and Deaf-Mutes*.[5] As has been common in North American linguistics and anthropological research, our general understanding of human language improves via studies of America's first nations.

After centuries of interest in gesture, from the nineteenth century, significant attention to gesture declined markedly. This was due in part, according to Kendon (2004, 64ff), to the rise of behaviorism and psychoanalysis. Both of these movements were interested in those aspects of human behavior that are beyond conscious control. Since gestures seem to be consciously controlled, these two fields had little interest in them. Yet this assumption of conscious control of gestures is incorrect. The average person has at best a vague knowledge that they are gesturing while talking and are not planning their gestures any more than they are their syllable structures. Gestures are controlled by dark matter, via emicization.

Another factor in the decline of interest and study of gestures was the growing reification of linguistics (Kendon 2004, 65; D. Everett 2013a; chap. 6 above). As methods developed for distributional analysis (e.g., Z. Harris 1951; Wells 1947; Bloomfield 1933) it became harder for some to see how to fit gesture into language, since their distributionalist methods could not readily account for gestures. Bloomfield (1933, vii) says little more of gesture than that it is "governed by social convention." Zellig Harris (1951)—arguably the leading Bloomfieldian (until his student, Chomsky)—was only interested in gesture to the degree that it was subject to the same kinds of distributional analysis as the rest of language (according to Kendon [2004, 67]). The contrast with Bloomfield, Harris, and others with the views of Sapir and Boas discussed in the next paragraph is striking. To my mind, for example, this contrast shows that Bloomfield really did represent a distinctly non-Boasian, structuralist (vs. descriptivist) tradition that today continues in some branches of linguistics, prolonging the structuralists' preoccupation with form over function (Sakel and Everett 2012, 152ff).

The culture-linguistic symbiosis pioneered by Sapir (1921, 1928), on the other hand, took a view of gestures that entailed a perspective of dark matter quite similar to mine. As Sapir said, "The unwritten code of gestured messages and responses is the *anonymous* work of an elaborate social structure" [emphasis mine]. Sapir seems to mean by "anonymous" something quite similar to my idea of dark matter. Sapir, Boas, and their students in fact understood language differently, as but a single manifestation of human behavior more generally, however specialized some aspects of language might be. Kenneth Pike, heavily influenced by Sapir (Pike 1978 and 1998; pers .comm.), was perhaps the most explicitly concerned of all linguists of the era in understanding language and gestures as components of human behavior more generally. Though Pike never developed a theory of gestures, he was a pioneer in prosodic studies, the functions of which often overlap with gestures (Pike [1943] 1945, 1945).

Moving beyond the work of Boas's student Efron, the marked return to and growth in gesture studies did not begin until the 1970s (when Efron's study was republished). At this time, researchers focused primarily on understanding how speech and gesture were integrated. Are they two distinct systems or parts of a single system? The interest of researchers in reviving gesture studies was sparked by a renewed interest in the evolution of language and the role of gestures in language development, the crucial role of visible body actions in interaction and communication, the universality of gestures, and understanding more precisely the relationship of gestures and speech as components of language more widely understood.

GESTURE EMICIZATION IN MCNEILL'S RESEARCH

In order to appreciate the necessity of emicization (and thus dark matter) in gesture acquisition and how gestures are governed by dark matter rather than (or in addition to) conscious knowledge, it is helpful to look at the classification of gestures. In all his work, McNeill looks at gestures in terms of their dimensions and their relationship to grammar and language. Relative to their relationship to language, McNeill places gestures along the continuum below:

The Gesture Continuum

Gesticulation Language-slotted gestures Pantomime Emblems Sign languages

Gesticulation is the core of the theory of gestures. It involves gestures that intersect grammatical structures at the "growth point" (see below) and which are practiced by all humans so far as we know, even the blind and others with different types of cognitive disorders, such as proprioceptive deficiencies. Gesticulation is in fact what the theory of McNeill, Kendon, and others is mainly about. Some gesticulations, like language-slotted gestures and pantomimes, are not conventional—they may vary widely and have no societally fixed form (though they are culturally influenced).

Language-slotted gestures are those that can actually replace a word—as, for example, in the substitution song mentioned above with which Pike begins his 1967 book. Or imagine that you tell someone, "He [use of foot in a kicking motion] the ball," where the gesture replaces the verb *kicked*; or "She [use of open hand across your face] me" (for "She slapped me"). These gestures occupy grammatical slots in an utterance and replace the grammatical units, usually words, that we otherwise expect there. They are improvised and used for particular effects in particular circumstances. They reveal the speakers' understanding of the positions, words, and structures of their syntax. As Pike said, they show that language is a form of individual behavior, heavily influenced by culture.

Pantomime is gesture that simulates an object or an action without speech. Like gesticulations and language-slotted gestures, pantomime is also not conventionalized, meaning that its forms may vary widely.

Emblems are conventionalized gestures that function as isolated "signs," such as the forefinger and the thumb rounding and touching at their tips to form the OK sign, or "the bird," the upraised, solitary middle finger.

Sign languages are gesture-based languages that use gesture in a static, rather than dynamic, way. That is, there are gestural morphemes, sentences, and other grammatical units that are tightly conventionalized. Sign languages replace spoken language. They do not enhance it or interact with it—one reason that McNeill argues that spoken languages did not begin, and could not have begun, as sign languages. On the other hand, sign languages themselves make use of gesticulation, in addition to the conventional signs that form the lexical items of the grammar.

Part of what McNeill intends by the term *dynamic* is action-based, imagery-creating performances that are not conventionalized. Using the Vygotskian terminology that influenced him (and me, among many others), he argues that language and gesture participate in a "dialectic interaction between unlike modes of thinking" (2005, 4). This bimodal language-forming dialectic (but let us not forget other modes of language, such as affective prosody, register alternations, facial expressions, body language, and so on) is not an add-on to language. It is as much part of the multimodal whole of language as syntax.

DYNAMIC VS. STATIC COGNITION

McNeill (2005, 162) claims that "dynamic cognition is environment-sensitive in its structure. Static cognition is synchronically insensitive to environment." I believe that he thus bridges a conceptual gap between different uses of the term *cognition* in modern cognitive theories. I myself have thought about this same distinction for some time. Before reading McNeill (1992), I had thought that D. Everett (1994) was perhaps the first paper to draw an explicit distinction between "dynamic" and "static" cognition. In that paper I argued that sentential grammars are "static" (learned and fixed) objects whereas the use of language in conversation and discourse was "dynamic" (131): "Discourse and sentence structures illustrate two types of cognition, *dynamic* vs. *static* and that . . . necessarily involve different theoretical constructs for their explanation. That is, they constitute distinct epistemological domains."[6] In that paper, I further defined—directly connected to what here we are calling dark matter—cognition as referring "to structures and processes which underlie reasoning, knowledge, and other higher-level, brain-caused behavior that cannot be accounted for in terms of neurons or arrangements of neurons in our present state of neurophysiological knowledge" (134).

I am pleased to learn that McNeill and I have been in agreement about the importance of these parameters for twenty years at least (though both of us could profitably have paid more attention to the dynamic aspect of even grammar and syntax that sociolinguists have discussed for so long). These parameters seem quite an important distinction to draw. Much of the debate about what is "cognitive" in the cognitive sciences could be elucidated or eliminated if researchers were to recognize this distinction. Formal linguistics and functional linguistics, for example, both study cognition but from different perspectives, if this distinction is correct—the static and the dynamic, respectively. The sociological and scientific consequences of the failure to recognize this distinction have long been with us. One wonders therefore how the history of linguistics and the cognitive sciences more generally might have been altered had gesture work impacted linguistic theories from the outset.

West Coast cognitive scientists (such as Lakoff and Fillmore at Berkeley and Langacker at UCSD) often focused on discourse-construction as an active process that engaged concepts that included, inter alia, "live vs. dead metaphors," "framing," "newsworthy," "foregrounding," and "backgrounding." They emphasized those aspects of language that varied by specific speakers' perceptions of context and communicative goals in real time. East Coast cognitive scientists (I am oversimplifying) at the time were concerned with greater knowledge of structure, formal semantic relationships, lexical meanings and structures, and so on. McNeill (2005, 17) sums up this approach, saying that "in this tradition, language is regarded as an object, not a process." He rightly links the concerns of the object-oriented vs. process-oriented perspectives to the early distinction between diachronic (process) vs. synchronic (static) studies introduced by Saussure.

Another way of conceptualizing the static vs. the dynamic perspectives on cognition is in the importance of variation in each. Static, object-focused linguistics traditions think of idiolects or dialects as (at least for a moment) stable or discrete entities. It is possible in such a tradition to talk, for example, of someone's "I(nternalized/Intensional)-language" as more than a convenient fiction, because a state of knowledge of grammar is seen as steady in some sense. But other linguists and cognitive scientists believe that the static view is a pernicious idealization. For these researchers—concerned with language as a process—variation is where the action is. From Pike's perspective, these two forms of cognition relate to particle vs. wave manifestations of behavior, yet both are equally emic, equally governed by internalized parameters that help shape our dark matter.[7] This distinction is all the more interesting to our discussion of dark matter because sentence grammars (from a Chomskyan perspective, at least) are formed from *static cognitive* capacities (e.g., learned constructions, rules, constraints, templates), whereas discourse principles (e.g., emphasizing the newsworthy, emphasis, intonation, use of idioms, ideophones) present *dynamic cognition*.[8]

As we move to consider gesture theory in more detail, the bedrock concept is the "growth point." This is the core unit of language in the theory of McNeill, and it must be understood if we are to understand his theory. The growth point is the moment of synchronization where gesture and speech coincide. In McNeill (2012, 24), a growth point is described as that point where "speech and gesture are (a) synchronized, (b) co-expressive, (c) jointly form a 'psychological predicate,' and (d) present the same idea in opposite semiotic modes." This description of the growth point contains several technical terms. By "co-expressive" McNeill means that for the symbols used simultaneously in gesture and speech "each symbol, in its own way, expresses the same idea" (1). By "psychological predicate" (terminology deriving from the work of Lev Vygotsky [1978]), McNeill intends the moment of the expression when "newsworthy content is *differentiated from a context*" (33). By "semiotic opposites" McNeill means that gestures are dynamic, created on the spur of the moment and, though influenced by culture and convention, are not themselves lexical or conventionalized units. Speech, on the other hand, contains grammar, which is highly conventionalized (grammatical rules, lexical items, etc.) and is thus "static communication."

In short, gesture studies leave us no alternative but to see language in dynamic, process-oriented terms. It is manufactured by speakers in real time, following a culturally articulated unconscious; it is dynamic; it is not merely application of uni-modal rules, but is a multimodal holistic event. Gestures are actions and processes par excellence. Nevertheless, they do have object properties too, showing the ultimate weakness of dichotomies. McNeill provides some of the best analysis of the object properties of gestures. Thus he defines a gesture unit as "the interval between successive rests of the limbs" (2005, 31). Like most units of human activity (Pike 1967, 82ff, 315ff), it is useful to break down gestures into margins (onset and coda) and nucleus. Thus, McNeill (and others) argues that gestures should likewise be analyzed in terms of *prestroke*, *stroke*, and *poststroke*. And just as onsets, codas, and nuclei in syllables may be long or short, gestures may be lengthened in their different constituents—what McNeill calls "holds." In the prestroke—which, like the other constituents of a gesture, may be "held" to better synchronize timing with the spoken speech—the hands move from their rest position in anticipation of the gesture. The stroke is the meaningful core movement of the hands. The poststroke is the beginning of the retraction of the gesture. The work on gesture is full of rich illustrations of these gestural constituents and how they at once synchronize with and add dynamically to speech.

These constituents and holds are strong indications that utterances are tacitly designed. As Kendon (2004, 5) says, "Gestures are part of the 'design' of an utterance." One of the clearest ways in which gestures show design are in the constituents of gestures (prestroke, stroke, and poststroke) and how they are often held to synchronize precisely with spoken speech. The question of how gestures are learned and controlled is every bit as interesting as those few aspects of language that linguistics as a discipline currently addresses.

Another crucial component of the dynamic theory of language and gestures that McNeill develops is the *catchment* (also referred to in places as a *cohesive*), which indicates that two temporally discontinuous portions of a discourse go together—repeating the same gesture indicates that the points with such gestures form a unit. In essence, a catchment is a way to mark continuity in the discourse through gestures (Givón 1983).[9] McNeill says that:

> [a] catchment is recognized when one or more gesture features occur in at least two (not necessarily consecutive) gestures. The logic is that recurrent images suggest a common discourse theme, and a discourse theme will produce gestures with recurring features . . . A catchment is a kind of thread of visuospatial imagery that runs through a discourse to reveal the larger discourse units that encompass the otherwise separate parts. (2005, 117)

In the notion of catchment, gesture theory makes one of its most important contributions to the understanding of dark matter: it underscores that speakers put the constituents of sentences to use in larger discourse functions that cannot be captured by focusing on static knowledge of sentence grammars alone. Thus catchments function simultaneously (each individual occurrence) at the level of the sentence and at the level of the discourse (the shared features of the catchment gestures), illustrating that sentences and their constituents are themselves constituents of discourses, once again reinforcing the Pikean ideas on behavior, language, and "hierarchy" (where the apex of the grammatical hierarchy is not the sentence, but rather conversations, which may contain monologic discourses as constituents[10]).

On the other hand, gestures are not linked to sentences and discourses merely by timing and visual features. They are also connected semantically via "lexical affiliates." The lexical affiliate concept was first introduced by Schegloff (1984; McNeill 2005, 305) and it refers to "the word or words deemed to correspond most closely to a gesture in meaning." Gestures generally precede the words that lexically correspond to them, thus marking the introduction of new meaning into the discourse. (An example might be a downward motion functionally preceding a word such as *downward*.) The dark matter control involved in linking gestures to their lexical affiliates is astounding in its subtlety and complexity. But this control reflects a knowledge of rhythm, highlighting, newsworthiness, and other ways of recognizing and attempting to communicate what is important.

To better understand the tacit relationship, or "unbreakable bond," between speech and gestures, there are numerous experiments that look at effects that result from real or imposed sensory deficits: *delayed auditory feedback* (DAF), blindness, and afferent disorder (proprioceptive deficits). In DAF experiments, the subject wears headphones and hears parts of their own speech on roughly a .2-second delay, close to the length of a standard syllable in English. This produces an auditory stuttering effect. The speaker tries to adjust by slowing down (though this doesn't help because the feedback also slows down) and by simplifying their grammar. Interestingly, the gestures produced by the speaker become more robust, more frequent, in effect taking more of the communication task upon themselves in these situations. And yet the gestures "do not lose synchrony with speech." (McNeill 1992, 273ff). In other words, gestures are tied to speech not by their own timing but by intent and meaning of the speaker—dark matter—and by their inextricable link to the content of what is being expressed.

And this inextricability is quite special. For example, in McNeill (2005,

234ff), the case of the subject "IW" is discussed. At age nineteen, IW suddenly lost all sense of touch and proprioception below the neck due to an infection. The experiments conducted by McNeill and his colleagues show that IW is unable to control instrumental movements when he cannot see his hands (though when he can see his hands, he has learned how to use this visual information to control them in a natural-appearing manner). What is fascinating is that IW, when speaking, uses a number of (what IW refers to as) "throwaway gestures" that are well coordinated, unplanned, nonvisually reliant, speech-connected gestures. McNeill concludes that at a minimum, this case provides evidence that speech gestures are different from other uses of the hands—even other gesturing uses of the hands. However, I am nonetheless unconvinced by McNeill's further conclusion that there is some innate thought-language-hand neural pathway in the brain. On the other hand, I *am* convinced that such a pathway arises developmentally, and that it is different from other pathways involving gestures and movement.

Finally, with regard to the special relationship between gestures and speech, McNeill (2012, 13) observes that not only do sighted people employ gestures when talking on the phone—showing that gestures are not simply something added for the benefit of an interlocutor—but also that the blind use gestures when speaking, indicating that we use gestures even when we cannot see them and thus that they are a vital constituent of normal speech.[11] Since the blind cannot have observed gestures in their speech community, their gestures will not match up exactly to those of the local sighted culture. But the blinds' use of gestures shows us that communication, as we stated earlier, is holistic, and that we use as much of our bodies as we are able when we are communicatively engaged. We need studies of how the blind first begin to use gestures. But to my mind the gestures of the blind simply follow from the use of our entire bodies in communication. We "feel" what we are saying in our limbs and faces, and so on. Thus DAF, the blinds' use of gestures, and the "throwaway" gestures emerging even in the context of IW's proprioceptive disorder suggest that the relationship between gestures and speech is, in McNeill's words, an "unbreakable bond."

Fascinatingly, however, though the bond may be unbreakable, it is culturally malleable. David Efron's ([1942] 1972) work may have provided the first modern study to examine the link between culture and gesture. But it is not the only one; there is a now a sizeable literature on such effects. To take one example, de Ruiter and Wilkins (1998) and Wilkins (1999) discuss the case of Arrernte, in which the connection, or "binding," of speech and gesture is overridden by culture and dark matter. According to de Ruiter and Wilkins, the Arrernte regularly perform gestures *after* the co-expressive speech. The cultural reason for this, the authors suggest, is that the Arrernte make much larger gestures physically than are found in many other cultures, using movements of the entire arm in gesturing. Thus, as the author interpret the phenomena, the larger gestures and space required by the Arrernte demand more planning time, favoring the performance of gestures following the relevant speech. A simpler analysis is suggested by McNeill (2005, 28ff), however; namely, that the Arrernte simply prefer the gestures to follow the speech. The lack of binding and different timing would simply be a cultural choice, a cultural value. Gestures for the Arrernte could then be interpreted similarly to the Turkana people of Kenya, also discussed by McNeill, in which gestures function in part to echo and reinforce speech, other potential cultural values, and functional enhancements of language communication. Whatever the analysis, one must appreciate the relevance and significance accorded to culture in McNeill's and other researchers modern analyses of gesture, following on in the Boasian tradition inaugurated by Efron.

EQUIPRIMORDIALITY

In his work McNeill introduces the vital term *equiprimordiality* into the discussion of gestures and their relationship to the evolution of human speech. By this he intends that gestures and speech were equally and simultaneously implicated in the evolution of language. To understand this, we must ask how the growth point and the imagery-language "dialectic" evolved. Here McNeill (2012, 65ff) relies on George Herbert Mead's (1974) seminal work of on the evolution of the mind as a social entity, with special attention to language and gestures. Mead's (1974, 47ff) claim on gestures is that they "become significant symbols when they implicitly arouse in an individual making them the same response which they explicitly arouse in other individuals" (this was probably written in the 1920s). McNeill's insight is to take Mead's conjecture and tie it in with Rizzolatti and Arbib's (1998) discussion of the involvement of mirror neurons in language. What McNeill claims is that Rizzolatti and Arbib missed a crucial step, which he refers to as "Mead's loop," wherein one's own gestures are responded to by one's own mirror neurons in the same way that these neurons respond to the actions of others, thus bringing one's own actions into the realm of the social and contributing crucially to the development of a theory of mind—being able to interpret the actions of others under the assumption that others have minds like we do and think according to similar processes. Thus McNeill at once links his research program and the evolution of language more generally to the brain and society in an interesting and unique way, also highlighting the ineffable cerebral, as well as social connections in the formation of language, culture, and dark matter.

According to McNeill (2012, 69), Mead's loop entails that "speech and gesture had to evolve together. . . . There could not have been gesture-first *or* speech-first[12]" [emphasis in the original]. This follows, he claims, because Mead's loop creates a "dual semiotic": "To create the dual semiotic of Mead's Loop, they [speech and gesture] had to be equiprimordial." Mead's loop made possible the dynamic aspects of speech as well as the analysis of otherwise holophrastic constructions into parts, such as words, phrases, sentences, morphemes, phonetic segments, and so on. McNeill explains this by claiming that

> semiotically, it [Mead's loop] brought the gesture's meaning into the mirror neuron area. Mirror neurons no longer were confined to the semiosis of actions. One's own gestures . . . entered, as if it were liberating action from action and opening it to imagery in gesture. Extended by metaphoricity, the significance of imagery is unlimited. So from this one change, the meaning potential of language moved away from only action and expanded vastly. (2012, 67)

I notice a couple of things in this quote relevant to our present concerns. First, the language focuses on action and meaning rather than structure, which sets it off from—while still complementing—a great deal of linguistic analysis. Second, Mead's loop and the growth point place compositionality in a somewhat different light in the evolution of human language. Most linguists (myself included), when asked what the great quantum leap in the evolution of language was, would have likely answered, "Compositionality." But if McNeill is partially right here, the growth point's evolution from Mead's Loop is more important than compositionality. In fact, in Everett (2012a) I allude to the possibility that compositionality relies on nonlanguage-specific cognitive abilities. Interestingly and unfortunately, this is completely ignored in most recent works on the evolution of language (e.g., Fitch [2010]; but see D. Everett [forthcoming] for integration of the growth point into the understanding of language evolution, including syntax, phonology, pragmatics, etc.). But what makes this important for our current concerns is the historicity of dark matter; connections are learned and passed down transgenerationally—in this case, most likely by example.

Once we get past this initial hurdle of how gestures become meaningful for humans, other notions arise to fine-tune the evolutionary story of the gesture-speech nexus. McNeill's theory (e.g., 1992, 311ff) takes a perspective similar to construction grammar (Goldberg 1995) in claiming that utterances—gesture/speech wholes—are initially "holophrastic." That is, they are used as single words or unanalyzable wholes. Through reuse and the aid of gestures to focus on specific components of the construction, they are later analyzed in more detail. This leads to the generation of syntactic constituents and rules reminiscent of the discovery methods of Z. Harris (1951), Longacre (1964), and others, i.e. distributional isolability and recombination).

As gestures and speech become signs in the social space, gestures take on one of two perspectives (McNeill 2005, 34). They represent either the viewpoint of the observer/speaker (OVPT) or the viewpoint of the person being talked about, or *character viewpoint* (CVPT). Thus as we practice language and culture we learn these things—different viewpoints, different ways of highlighting content and attributing ownership of content.

For example, McNeill gives an example of one person retelling what they say in a cartoon of Sylvester the Cat and Tweety Bird. When their hand movements are meant to duplicate or stand for Sylvester's movements, then their perspective is CVPT. But when their hand movements indicate their own perspective, then their perspective is OVPT.

Many researchers have speculated that gestures might have preceded speech in the evolution of human language. McNeill does not disagree entirely with this position. His reasons are similar to those suggested in Everett (2012a). Intentionality is a necessary prerequisite to language. And intentionality is shown not only in speech but also in gestures and other actions and states (e.g., anxiety, tail pointing in canines, focused attention in all species; see also D. Everett, forthcoming). One reason that gestures are used is because intentionality is focus, often involuntary. The orientation of our eyes, body, hands, and so on, vary according to the direction of our attention. This part does seem to be a very low-level biological fact, exploited by communication. The implications of McNeill's analysis of Mead's loop and the growth point are enormous. For one thing, if he is correct, then gesture could not have been the initial form of language. This is not to say that pre-linguistic creatures cannot express intentionality by pointing or gesturing in some way. It does mean that real linguistic communication must have always included *both* gestures and speech.

IMPLICATIONS FOR LANGUAGE EVOLUTION FROM GESTURES

Another interesting component of McNeill's theory of language evolution concerns his own take on recursion. Recursion is (see D. Everett 2010b) a tool for more tightly packing information into single utterances.[13] Thus he independently arrives at an important conclusion in recent debates on recursion by providing a model of language evolution and use in which recursion is useful but not essential, a very similar point to D. Everett (2005a, 2005b, 2008, 2009a, 2009b 2010a, 2010b, 2012a, 2012b, and many others).

Language evolution is of course vital to a theory of dark matter because a careful study of evolution helps us trace the origin of human knowledge of language. For example, does this knowledge wind back to Plato or to Aristotle's conception of knowledge—that is, remembered or learned? The answer, unsurprisingly, is that it goes to Aristotle. To see this, consider the role of gestures in language evolution as expounded by McNeill. Thus, although in recent years, Tomasello (1999, 2008), Corballis (2002), Hewes (1973), and Arbib (2005), among many others, have argued that "language evolved, not from the vocal calls of our primate ancestors, but rather from their manual and facial *gestures*" (Corballis 2002, ix), McNeill argues that there are two theory-busting problems with the "gesture-first" theory of language evolution. First, speech did not supplant gesture. Rather, as all the work of McNeill, his students, and many, many others show, the two form an integrated system. The gesture-first origin of language predicts asynchrony between gesture and speech, since they would be separate systems. But they are synchronous and parts of a single whole. Further, code switching between gestures and speech is common. Why, if speech evolved from gestures, would the two still have this give-and-take relationship? Moreover, if the gesture-first hypothesis is correct, then why, aside from languages of the deaf, is gesture never the primary "channel" for any language in the world?

The second major problem with the gesture-first theory is that if gestures could be substituted by speech, they would not then be of the right type to form a language. This follows because in the absence of language, the available communicative gestures would have to be pantomimes. But, as McNeill makes clear throughout his trilogy, *pantomime repels speech*. Pantomime does not accompany speech—it fills in missing values or gaps in speech. It is used in lieu of speech.

Also, as McNeill makes clear throughout his trilogy, speech is built on a stable grammar. The only gestures which provide stability are the conventionalized and grammaticized gestures in sign languages. In this case again, however, gestures are either used instead of or to supplant speech. Summing up, had sign language or other gestures—for example, pantomimes or language-slotted gestures—preceded speech, then there would have been no functional need for speech to develop. As McNeill (2012, 60ff) puts it: "First, gesture-first must claim that speech, when it emerged, supplanted gesture; second, the gestures of gesture-first would [initially] be pantomimes, that is, gestures that simulate actions and events; such gestures do not combine with co-expressive speech as gesticulations but rather fall into other slots on the Gesture Continuum, the language-slotted and pantomime."

One might attempt to reply that Pike's example shows that gestures can substitute for speech. But the gestures Pike discusses are language-slotted gestures, parasitic on speech, not the type of gesture to function in place of speech. On the other hand, Pike's example suggests another question, namely, whether there could be "gesture-slotted speech" corresponding to speech-slotted gestures (i.e., an output in which speech substitutes for what would normally be expressed by gestures). If speech evolved from gestures, after all, this is how it would have come about. And gesture-slotted speech is not hard to imagine. For example, consider someone bilingual in American Sign Language and English substituting a spoken word for each sign, one by one, in front of an audience. Yet such an event would not really exemplify gesture-slotted language, since it would be a translation between two independent languages, not speech taking the place of gestures in a single language. This is important for our thesis here for a couple of reasons: (i) the utilitarian nature of gestures offers us a clear route to understanding their genesis and spread; and (ii) their universality is as supportive of the Aristotelian view of knowledge as learned, over the Platonic conception of knowledge as always present, exactly because they are so useful. In fact, it is even more likely that they have been learned since they are so useful they would be rediscovered time and time again, and because to propose a Platonic view in light of their learnability is to complicate the story and hence is less parsimonious.

And as they stabilize by conventionalization, such gestures become sign languages. But these are all gestures replacing speech functions and thus for speech to develop from these would make little sense either functionally or logically.

However, in spite of my overall positive view of McNeill's reasoning about the absence of gesture-first languages, however, there seems to be something missing. If he were correct in his additional assertion or speculation that two now-extinct species of hominin had used either a gesture-first or a gesture-only language and that this is the first stage in the ontological development of modern language, then why would it be so surprising to think that Homo sapiens had also used gesture-first initially? I see no reason to believe that the path to language would have been different for either hominin species. In fact, I doubt seriously that pre-sapien species of Homo would have followed a different path, since, as D. Everett (2012a) argues, there are significant advantages of vocal vs. gesture communication.

Another question arising in connection to the equiprimordial relationship between speech and gesture is this: Is there a common, *specific* innate cerebral basis for language and gesture or gesticulation? McNeill (1992, 333ff) seems to think so, if I read him correctly. My own opinion is that no such evidence exists. In fact, the opposite seems to hold. For example, McNeill reviews evidence to show that the cortical proximity of speech and gesture is directly proportionate to the distance leftward where the gesture is located in Kendon's gesture hierarchy. This means that the closer a type of gesture is tied to speech, the greater the proximity of that type to speech in the brain. Yet this is not support for any innate pathway. There is nothing in cortical proximity that could not be accounted for *better* by hypothesizing merely that the two are learned together and initially experienced together. On the other hand, superficially stronger evidence for the neurological connectedness of speech and gesture emerges from studies of aphasia. In people with Broca's aphasia, meaningful gestures are produced in a choppy fashion. In Wernicke's aphasia, on the other hand, the gestures are meaningless but fluent—at least, according to Hewes (1973). Yet this proximity again does not support a nativist "bioprogram" (a word McNeill occasionally uses) of any sort. Setting aside my discussion in D. Everett (2012a) where I argue against the very existence of these two types of aphasia (based on the simple fact that these correspond to no language-specific parts of the brain), McNeill's data seem open to an explanation via the general principle of "adjacency" (sometimes also called "iconicity") noted across various studies in linguistics and other disciplines, namely, that the more two things affect one another, the closer they will be to one another. This idea has applications in the understanding of morphosyntactic constituents, vowel harmony, and neurological coordinates.

McNeill discusses a variety of different types of gestures. We have discussed catchments earlier, so now the other three—iconic gestures, metaphoric gestures, and beats, are introduced. Each reveals a distinct facet of the gesture-speech relationship and its relationship to cognition and culture. And each, like gestures and speech more generally, reveal a great deal of dark matter.

McNeill (1992, 12ff) describes iconic gestures as bearing "a close formal relationship to the semantic content of speech." Iconic gestures show that "what is depicted through gesture should be incorporated into a complete picture of a person's thought processes." These gestures depict or represent concrete objects to flesh out the imagery and meaning of speech—for example, making the motion and appropriate hand shapes of pulling back a bow when discussing shooting an arrow, a common occurrence in Amazonian communication.

Alongside iconic gestures, speakers also use metaphoric gestures. Metaphoric gestures are simultaneously metalinguistic (representing discourses or discourse genres, etc.) and cultural (based on what counts as a metaphor).

These gestures are abstractions. For example, McNeill (1992, 14) illustrates a speaker holding up both hands, palms facing toward each other, to represent a span of speech—a story told via a cartoon in the first example McNeill uses.

A third form of gesture is the beat: "Beats mark information that does not advance the plot line but provides the structure within which the plot line unfolds" (McNeill 1992, 15ff). These can signal departures from the main event line, such as a hand movement to accompany a "summing up" of what has been said in a discourse to this point. Beats can also be used to segment discourses or to accompany phonological emphasis in the speech stream. In D. Everett (1988) I discussed the case of my language teacher, Kaioá, using gestures to indicate stressed syllables.[14] Ladefoged, Ladefoged, and Everett (1997) also discuss training speakers to mark phonological beats in three different Amazonian languages. It turns out that this is not be as unusual as I thought it was at the time, since this is not an uncommon function of beat gestures, according to McNeill's lab, among others.

McNeill (2012, 77) further suggests that "an area of life where a syntactic ability could evolve is the cultural and social encounter." Here he cites the work of Freyd (1983) on "shareability"—the idea that structures and meanings must come to be shared among individuals if we are to say that they speak the same language (i.e., are utilizing the same outputs of conventionalization—another instance of the actuation problem). In particular, McNeill appeals to Freyd's "discreteness filter," an idea akin to the generative notion of *discrete* in the phrase *discrete infinity* (for criticisms of the latter, see D. Everett 2010a). The idea is that our utterances were initially holophrastic, noncompositional. Then, as humans began to learn a repertory of such utterances, these utterances changed via the GP, such that gestures would highlight some portions of the previously unanalyzable whole, leading to an analysis of the holophrastic into component parts—top-down parsing that eventually results in compositionality. This fascinates me because it presents a picture of how learning and emicization of the relationship between gestures and grammar can drive (dark matter) development of the very evolution of language.

This points to a stark difference between McNeill's theory and other theories of language evolution (such as Hauser et al. 2002). In McNeill's theory, the compositionality of syntax arises from actual language *use* via GPs, not from a sudden mysterious appearance of compositionality via recursion. In fact, in McNeill's theory (and in Kinsella's [2009] and D. Everett's [2012a], among many others), compositionality *precedes* recursion. And this is just as the dark matter model of language predicts—language emerges from a process of emicization, reanalysis, and re-emicization in order to better satisfy communication needs (Hopper 1988; MacWhinney 2006; Steels 2005; Rosenbaum 2014; etc.). The following quote expresses this well: "Contrary to traditions both philological and Biblical, language did not begin with a 'first word.' Words emerged from GPs. There was an emerging ability to differentiate news worthy points in contexts; a first 'psychological predicate' perhaps but not a first word" (McNeill 2012, 78). Ironically, by demonstrating how compositionality could have come about by use and thus entered all human languages from early human interactions, McNeill undermines the need to appeal to genetics or biology to account for this, instead supporting our account here. In the context of his discussion of compositionality, moreover, McNeill (2012, 78, 223) offers an extremely interesting discussion of how recursion itself might have entered grammar, one quite compatible with my own (D. Everett 2010a, 2010b, 2012a, 2012b). The story begins with the analyzability of holophrastic utterance via growth points.

Recursion would not have begun with gestures themselves. This is because gestures, unlike the eventually compositional static outputs of grammar, are gestalt units (though not all gesture researchers accept this). This is a fundamental difference between these dynamic units vs. static syntax. Gestures are wholes without meaningful parts. And the meaning of the whole is not derived from the meaning of the parts. Thus, although we can observe several submovements in the larger gesture, none of these smaller acts has any meaning apart from the gesture as a whole. Gestures are in this sense *anticompositional.*

But syntax became analyzable, and *following this*, recursion was able to play a role in the grammar. In this sense, recursion for McNeill, as for me (inter alia D. Everett 2005a, 2008, 2009a, 2010a, 2010b, 2012a, 2012b) is a nonessential, yet extremely useful component of language evolution (contra Hauser et al. 2002). Recursion is used to render the syntagmatic (string) paradigmatic (a slot), enabling speakers to pack more information into single utterances and, as I point out in D. Everett (2012b), making it easier to follow complex events via oral discourse. McNeill (2012, 223) cites Shelley's "The Masque of Anarchy" to illustrate the syntagmatic to paradigmatic shift:

> His big tears, for he wept full well,
> Turned to millstones as they fell.

He then points out: "The rhyming '-ell's, on the axis of combination, project a new semantic opposition." Due to the rhyming, the words are highlighted—potentially leading to their separate storage and analysis as words, *parts of utterances*, introducing compositionality into grammar. This opposition is between paradigmatic parts of larger sentences (syntagmemes) that themselves derive from syntagmemes. Thus having provided a plausible scenario for the evolution of syntax, McNeill turns to consider the resultant spread of static grammar.

McNeill (2012, 92) suggests that "if we take the Tower of Babel story as a parable of migration, it is not as far-fetched as one might suppose. The insight is that migration leads to encounters and breeds diversification; and the further the migration, the more the encounters, and the greater the diversification." Continuing with the Biblical metaphor, however, McNeill's speculation has to overcome the "Who did Cain marry?" problem. Many students of the Bible find it curious that Cain was the son of the first man, Adam, yet he was able to find a wife in a neighboring city (Genesis 4:1–5:5). Where did the inhabitants of that city come from? By the same token, one must legitimately inquire as to who encountered whom in the migrations from Africa. If the first encounter of human languages following the rising of ur-language produced linguistic changes because of distinct languages coming into contact, then how did the first language change into the second? There must be change without contact. And in our discussion of blowgun manufacture among the Banawás, I showed how cultural change can take place without contact, just as linguistic change could and does (Schönpflug 2008). And in fact, specialists in diachronic linguistics tell us that this is correct—that language change can be internal (e.g., sociolinguistic shift) or external (via language contact). But if change can occur without contact, does McNeill's hypothesis lose its force? No, because he also predicts that the trail of change should lead to greater and greater complexity the farther humans migrated from the geographical source of Ur-Language. The reasoning is that the earliest language would have had the simplest syntax, mapping meanings to temporal orders (i.e., iconically). But language contact would complicate that.

On the other hand, Thomason (2008) argues that contact does not necessarily make languages more complex. She presents a few cases where it can. But folks who work in language contact know that there are too many variables to say that a given language-contact situation will result in one of the languages becoming either more complex or more simple. Going further, theoretical linguists and typological linguists agree that there is no content to the claim "Language A is more complex than language B." There is just no widely accepted means to evaluate linguistic complexity. About all we can say is, "Phenomenon *x* in language A is more complex than the same phenomenon in language B." But even that is problematic because it assumes that we can say that this or that phenomenon in one language is the same as a phenomenon in another language, an idea I see little evidence for. That is, whether in smaller "societies of intimates" or larger "societies of strangers," diversity of pronunciation, grammar, and meaning all enter languages for any number of reasons that the diachronic linguistics literature is full of. So even if McNeill were correct (2012, 92) that initial syntax mapped meanings onto temporal orders of events—and frankly, I see no evidence that this was ever the case—why would contact unidirectionally increase complexity? In fact, as Trudgill (2011) argues, contact can either simplify languages or complicate them. I am therefore unable to accept McNeill's conjecture that languages with a greater history of contact and thus farther from Africa should be more complex than languages with a lesser history of contact (and how could one even measure that, in any case, apart from geography which need not necessarily entail greater contact?). It would also be necessary, were we to take McNeill's proposal here seriously, to discuss the possible confounding factor of the "serial-founder effect" common to migrating populations and how this might also impact language (see, inter alia, Slatkin and Excoffier 2012). All of this is important for our theory here because it shows (i) the utility of gestures in communication; and (ii) the connection of gestures to speech that slowly aggregates over time and hence need not be "assumed" because it can be learned so naturally.

Gestures, Apperceptions, and Culture Research

At this point of the discussion of gesture and dark matter, I want to consider again briefly the possible role of gestures in language evolution in other species, in particular with regard to the possible language of Neanderthals and *Denisovan hominins*. McNeill says this (contrary to what he says about Homo sapiens):

> Gesture-first may have existed in the two now extinct human lines, Neanderthals and Denisova hominin. It could have existed in our line and extinguished as well, but we have survived to evolve a new form of language, Mead's Loop based on speech-gesture equiprimordiality. This new language, as we have seen could not have emerged from gesture-first and was a second origin. (2012, 165)

From this, McNeill speculates further that these two species lacked language and failed to survive in connection with that deficiency. To me (D. Everett, forthcoming) we know too little about these species to warrant such claims.[15] One other speculation McNeill offers, a variant of "ontogeny recapitulates phylogeny," is worth additional thought. It is that while children apparently use gesture-first in their initial language acquisition, this dies out before two years of age (McNeill 2012, 165). Then, at three or four years of age, GPs emerge, "suggesting that gesture-first had existed once phylogenetically but went extinct and was followed by a new form of language in which speech and gesture imagery merged into the unified packages inhabited by thought and being that we ourselves have now" (ibid.). Although, again, the significance of children's acquisitional stages is somewhat conjectural, it is interesting enough to merit further investigation. It is a very positive indication of the quality and originality of his thinking, that even where McNeill goes out on very thin empirical limbs, his suggestions are interesting and worth considering further. Nevertheless, even this is equally supportive of the more parsimonious idea that we communicate holistically, using our entire bodies, not just mouths, hands, and brains (but see Iverson and Goldin-Meadow 1997). This resonance with our communicative actions brings an entire body—facial expressions, gestures, words, body orientation—into the act (Kita 2000). Indeed, an alternative to McNeill's ontogeny-recapitulates-phlylogeny view immediately suggests itself. This is that the need or desire or instinct to communicate precedes words—obviously—and shows up initially in body movements (e.g., gestures) prior to learning lexical items. This makes sense, again, if communication is a holistic effort of the individual as a whole, not merely their mouths.

Other Research on Gestures

Though McNeill's work is exceptionally fecund—in particular as it relates to language more broadly, culture, and dark matter—I want to transition now to briefly discuss other work on gestures, beginning with the work of Gianluca Giorgolo, of Carleton University and Oxford University. Giorgolo's work is sophisticated and formal. As he says in the abstract of Giorgolo (2010):

> The paper presents a formal framework to model the fusion of gesture meaning with the meaning of the co-occurring verbal fragment. The framework is based on the formalization of two simple concepts, intersectivity and iconicity, which form the core of most descriptive accounts of the interaction of the two modalities. The formalization is presented as an extension of a well-known framework for the analysis of meaning in natural language. We claim that a proper formalization of these two concepts is sufficient to provide a principled explanation of gestures accompanying different types of linguistic expressions. This supports my gesture-in-culture understanding. The formalization also aims at providing a general mechanism (iconicity) by which the meaning of gestures is extracted from their formal appearance.

Giorgolo's work is clearly within the general framework of formal linguistics, and thus ought to have a growing influence in the years to come on the integration of gesture into formal syntax and semantics studies. Many of the ways in which mainstream linguistics ought to incorporate gesture research are already being pioneered by Giorgolo, though it won't be easy, as Giorgolo's work underscores the complexity of our tacit knowledge of *how to communicate*.

Another prominent researcher whose work has been heavily influenced by McNeill is his former student, Justine Cassell, at Carnegie Mellon University. Cassell's work on computational communication is ground-breaking. This research, summarized as follows, includes:[16]

> developing the Embodied Conversational Agent (ECA), a virtual human capable of interacting with humans using both language and nonverbal behavior. More recently Cassell has investigated the role that the ECA can play in children's lives, as a Story Listening System (SLS): peer support for learning language and literacy skills. And Cassell has also employed linguistic and psychological analyses to look at the effects of online conversation among a particularly diverse group of young people on their self-esteem, self-efficacy, and sense of community.
>
> Once machines have human-like capabilities, can they be used to evoke the best communicative skills that humans are capable of, the richest learning? This is the goal of Cassell's research: to develop technologies that evoke from humans the most human and humane of our capabilities, and to study their effects on our evolving world.

Cassell's work ties into dark matter by showing how even machines can learn nonverbal behavior connected to communication. This fascinating result shows that learning of gestures certainly need not be innate. However, it does not escape the general constraints on communication and thinking by machines that we have noted several times above.

The final researcher I would like to mention who is doing research on the multimodal nature of human language is Dr. Jennifer Green. Green's (2014) work is important to our discussion because it shows that communication ranges into the environment, such that humans can use various communicative strategies exploiting their enveloping ecology (see also L. Green 2013; Kohn 2013; Descola 2013). Green, at the University of Melbourne, works on Aboriginal sand stories. Her summary is worth citing at length because of the originality and significance of her focus of multimodal research for linguistics more broadly.

> Sand stories from Central Australia are a traditional form of Aboriginal women's verbal art that incorporates speech, song, sign, gesture and drawing. Small leaves and other objects may be used to represent story characters. This detailed study of Arandic sand stories takes a multimodal approach to

> the analysis of the stories and shows how the expressive elements used in the stories are orchestrated together.
>
> Speakers of the Arandic languages of Central Australia have a range of semiotic resources or "systems" in their communicative repertoire. These include everyday language, spoken auxiliary languages, such as those used to encode respect for certain kin, sign language, the esoteric language of songs, and symbolic or graphic conventions used in sand stories and in various forms of Aboriginal art. Spontaneous gesture is also part of this complexity. In everyday communication it is the norm for several of these systems to coexist and be interdependent. The performance of Arandic sand stories (called tyepety in some Arandic languages) is a traditional form of visual storytelling in which co-speech graphics form an essential part. A skilled narrator of these stories incorporates multiple semiotic systems and uses the potentials within these systems to great creative effect. Speech, sign, gesture and drawing are employed, in sequence and in unison. As well as drawing on the ground, narrators may also use a variety of objects to establish a visual field in front of them, somewhat like a miniature stage-set. Leaves or sticks are used to represent story characters, and other small items which come to hand may be used to symbolize objects that are part of everyday life, such as shelters, shades, windbreaks and fire pits. The use of the ground for illustrative and explanatory purposes is pervasive in the environment of Central Australia where there is ample inscribable ground, and this attention to the surface of the ground arises partly from a cultural preoccupation with observing the information encoded on its surface. (J. Green 2014, i, 1ff)

Giorgolo, Cassell, and Green are on the cutting edge of gesture and multimodal aspects of language research that one hopes will gain momentum. All of this research demonstrates, as clearly as one could ever hope, how misguided it was in the early days of linguistic theory to set the sentence as the "start symbol" of the grammar, rendering it in effect the exclusive empirical domain of the majority of formal linguistics research. Grammar itself is multimodal, from the conversation down to the morpheme. This research also strongly reinforces the view of language as primarily communicative in function, using various subtools, of which sentences are but one. It also shows that without a multimodal perspective, sentence grammars are extremely limited in their contribution to our understanding of either interaction or cognition. Earlier, we referred to this as the "reification" of linguistics and discussed many ways in which the otherwise innocuous idealization of restricting analytical focus to sentences has arguably retarded progress in the understanding of human languages.

Cognitive scientists, anthropologists, philosophers, linguists, and others should be grateful for the careful, painstaking, long-term research into the multimodal nature and origin of human language—research that hopefully will not continue to be ignored in debates on the evolution, use, and structuring of human languages.

Perhaps the greatest lesson from gesture research is that it settles—to my mind, at least—the debate of whether language evolved for mental life or for communication. If McNeill is correct about the role of the growth point in the evolution of syntax, for example, then the preeminence of communication as the key to language evolution (over expression of thought) is unavoidable. In this regard, McNeill's arguments are essential. They make the case that language is both static and dynamic and that therefore compositional meaning is not sufficient to provide humans with language. His theoretical understanding of the role of the growth point, the theory of mind, and Mead's loop in human language are or ought to be transformational for the discipline. This is strong support for the interactional instinct as well.

Another way in which gesture studies come to bear on fundamental issues of human language and cognition is their relevance to what has come to be known as "embodied cognition" (Gibbs 2005; Chemero 2011; Lakoff and Nuñez 2001; etc.; see also C. Everett 2013b for very different ways in which language can affect cognition). For example, in a recent report on research at the University of Chicago, McNeill's home institution, it is reported that the use of gestures—that is, embodying cognition—can contribute to cognitive acquisition of concepts as difficult as mathematics (Ingmire 2014).

What we learn from gestures in normal human languages is that they vary from culture to culture in their forms and meanings, but they are found apparently in all cultures. There are important reasons for their universal appearance since the consonant-vowel speech stream, word order, and other grammatical devices need help to get across the informational richness and nuances of communicative intentions. Prosody—the use of pitch, loudness, and intensity—is one way to help out, as we have seen. Gestures are a complementary form.

So far, we have seen nothing in grammar, gestures, or other aspects of language that would lead us to believe that anything needs to be attributed to the genome of Homo sapiens that is specific to language. Cultural learning, statistical learning, and individual apperceptional learning complemented with human episodic memory seem up to the task, especially when considered with the arguments of D. Everett (2012a) and this essay. Nevertheless, the literature is rife with claims to the contrary, namely, that there are phenomena that can only be explained if language is acquired at least partially based on language-specific biases in the newborn learner. One of the sets of studies that has attracted a great deal of attention in this regard is the work of Goldin-

Meadow on "homesigns," gestural languages that emerge from the deaf children of non-signing parents or who otherwise, Goldin-Meadow claims, have no access to linguistic input.

HOMESIGNS

One thing is clear from all claims of the emergence (what Goldin-Meadow [2015] calls "resilience") of language features in communities that are claimed to otherwise lack language—from Nicaraguan Sign Language to Al-Sayyid Bedouin Sign Language to creole languages—is that they begin simple and then become more complex over time with more social interaction. Often it takes at least three generations to develop a complexity roughly like many older languages. Thus, even if homesigns and the like are evidence for nativist or Bastian-like knowledge of language, such knowledge is very limited, perhaps no more than vague biases or solution spaces (which is one way to interpret, for example, the work of Berlin and Kay [1969] on the development of color terminology from the biological bases of color perception). More important, what marks the work of Goldin-Meadow and many others is what I consider to be an over-charitable interpretation of the linguistic aspects of the signs and a less charitable view of the cultural input the child receives, as well as the nature of the task the child is facing. Absent a serious consideration of either the task or the input, such claims of nativism are severely weakened.

For example, Goldin-Meadow (2015) argues that homesigners develop symbols for objects, word-ordering constraints, part-whole relationships of gestures (i.e., compositionality, where a single gesture, she claims, can be broken down into separate parts, just as a word can be broken into morphemes); that gestures can fill slots in larger structures; that there is evidence for hierarchical structuring of homesigners' utterances; that homesigners use gestures to mark different modes, such as negation and interrogation; that homesigners use their gestures communicatively; and that some categories in the world are grammaticalized according to patterns we see in all or many human languages (e.g., homesigners' use of gestures for size and shape but not gestures for hardness, texture, temperature, weight). She discusses other characteristics of homesigners' gestures, but these are sufficient for showing the potential problems of her approach.

First, note that all of these characteristics evolved, so at some point humans learned them before natural selection could have used them for determining relative fitness. Second, none of these characteristics are specific to language. Symbols—unless we are using this term in the strict Peircian sense, as we have seen, are used in one way or another by several species. But even if the reference is to Peirce's theory of the symbol as a conventionalized connection between form and meaning, then much like Saussure's signs, there is no reason to believe that this cannot be learned easily by humans. In fact, on one interpretation, this is all that Goldin-Meadow's results on symbols show us—namely, that children readily adopt symbols. The object is a form with a meaning. As the child learns the object and desires to communicate, this desire to communicate perhaps the most striking characteristic of our species—whether an interactional instinct or an emotional urge—the child will iconically represent the object, and the meaning of the object in the particular culture comes along for the ride. Children participate in their parents' lives and try to communicate, even if without language, as Helen Keller's remarkable odyssey shows us. With an ability to see or hear or feel, the child *can* receive input from the environment, from its caretakers, and in fact *will*, with most caretakers and in most environments. Learning the use of the object and the salience of the object to their parents and environment, children communicate about objects, as most other species (at least, mammals) do. Whole objects, as perceivable in a particular space in time, are most salient and learned relatively easily by dogs, humans, and other creatures. Humans try to represent their objects, unlike other animals, because humans strive to communicate.

The fact that some features of the objects stand out to children is likewise unsurprising, though the particular reason that shape and size win out over many other features, if Goldin-Meadow (2015) is correct, is not yet clear. She ascribes it to the child's native endowment. But I would suggest looking first at the way that objects are used, presented, structured, and valued in the examples of the child's caretakers. Furniture, dishes, houses, tools, and so on, are far more easily arranged and far more prevalent in the environment of caretakers' salient objects than other features. At least that could be tested, and there is no suggestion that any such tests were contemplated.

With regard to the claim that homesigners' speech is organized hierarchically, there are two caveats. First, as I have argued in my own work (D. Everett 2005a, 2005b, 2008, 2009a, 2012a, 2012b, and others), hierarchy and parataxis are difficult to tease apart. Either may be misinterpreted as the other, and they are often confused due to what might be seen as theoretical. For example, in Pirahã utterances, we might say, "The man is here. He is tall." Or "I spoke. You are coming." And these could be interpreted as "The man who is tall is here" or "I said that you are coming." But the analysis is likely much simpler, with the syntax lacking hierarchical structure. In none of Goldin-Meadow's examples purporting to show hierarchical structure in homesigners' utterances, did I see convincing evidence for hierarchy. The second caveat is that hierarchy is a natural solution independent of language, and thus if one finds hierarchical organization in some language structures, this is not evidence that there is an innate linguistic bias. As information demands grow due to increasing societal complexity, hierarchy is the most efficient solution to information organization, across many domains (Simon 1962). Atoms, universes, and many other complex objects of nature are organized hierarchically. Hierarchies are found in automobiles, canine behaviors, and computer filing systems. It is a naturally occurring and observable solution. In fact, for any action that involves selectional constraints, such that you must do *x* before you do *y*, there is hierarchy. There is absolutely nothing special to language.

The ordering that homesigners are claimed to impose on their structures is mundane. First, they have no alternative but to put their symbols in some order. And since the main ingredients of any utterance are the thing being reported about and what happened to it—what has long been called *theme-rheme* or *topic-comment*—the topic will usually precede the comment (it is most effective communicatively to begin with shared or old information before giving new information, likely because it exploits short-term memory). And within the comment, where the new information is placed, a large number of languages in the world prefer to place the patient or object before the predicating element. So if someone somewhere is eating a fruit, this can be described as either *fruit-eat* or *eat-fruit*, with most languages choosing the former (e.g., German, Pirahã, Japanese, and thousands of others). E. Gibson et al. (2013, 1079) have argued that word orders may be heavily influenced by strategies to deal with "noise corrupting the linguistic signal." They predict in particular that subject-verb-object (SVO) will be more common than subject-object-verb (SOV) order in the absence of case-marking. Now, on the one hand this does not prohibit SOV languages without case marking (and in fact, there are many of these crosslinguistically). On the other hand, it is not clear whether there is anything remotely like case marking or argument marking in homesigning. And yet, the basic problem that communicators have to solve is communicating new information about shared information. So the object/patient is expected, ceterus paribus, to be adjacent to the verb, at least when both communicate new information—which is generally so—and the subject/agent farther away. And there are only two choices: object-verb or verb-object. Thus nothing very serious should be read into finding one of these orders among homesigners.

Nor should it be surprising that once an order is chosen, it is easier to stay with it than to order objects and verbs randomly. Nor is there anything in the genetic endowment that should be applied about information structure. Topic-comment is a natural communicative arrangement. But Goldin-

Meadow neglects to discuss the information-theoretic possible constraints on ordering, and hence misses a potential alternative explanation for her facts.

Homesigning clearly illustrates the desire of all members of our species to communicate (Aristotle's "social instinct"). And it shows a range of common solutions to the problem of how to communicate. But not only do we have no convincing syntactic analysis of the facts, but evidence suggests (Andrén 2010; Duncan 2002, 2006; Zlatev, forthcoming) that in fact gestures are sufficiently motivated by communicational needs that it makes little sense to attribute them to the genes as language-specific biological endowment.

Summary

This chapter has given an overview and detailed discussion of gestures and their relationship to dark matter, grammar, communication, and the origin of language. Though we looked primarily at the pioneering and cutting-edge work of David McNeill, we also surveyed other approaches to gestures. We learned that, many claims to the contrary notwithstanding, gestures provide no strong evidence of innate dark matter but, on the contrary, show how culture and individual psychology can produce gestures as part of language, just as they do with syntax and other components of human language.

8

Dark Matter Confrontations in Translation

Even more important than what takes place inside the translator's brain is what takes place in the total cultural framework in which the communication occurs.

EUGENE NIDA, *Toward a Science of Translating*

In the previous chapter, we examined ideas about the origin of language and the multimodal, hoslitic nature of linguistic communication. In this chapter we want to look at how distinct languages-culture pairs can be mapped on to one another—that is, how translation works and what makes it possible. Evidence is presented that translation is never completely successful, due to the way that language is shaped and controlled by dark matter. The chapter also offers a discussion of methodology for what Quine referred to as "radical translation."

In the introduction to this book I mentioned problems that I encountered in translation when I went to the Pirahãs to translate the Bible. Those problems included things like the fact that the Pirahãs have no concept of God—certainly no "Supreme Being"—that Pirahãs do not like for any individual to tell another individual how to live; that Pirahãs do not feel spiritually lost, and so on. The introduction also discussed some of the mistaken assumptions—dark matter—of American missionaries and perhaps others who translate for more traditional societies. These included things like the fact that although most American missionaries believe that God has "prepared" every culture to understand the "gospel" (the "good news"; i.e., to understand that God's son, Jesus, died on a cross for their sins), the Pirahãs find the concepts of savior, sin, and salvation incomprehensible; that in spite of American missionaries' belief that people like the Pirahãs are afraid of a dark, threatening evil spiritual world and that many of them will be overjoyed at the missionary's arrival with the news that Jesus has freed them from this fear, the Pirahãs fear nothing and were uninterested in the missionary message; that American missionaries believe all languages will be able to convey all the concepts necessary to express the full New Testament message. It is the job of the translator to find the appropriate words and phrases in the target language and then to match them with the appropriate Greek, Aramaic, or Hebrew concepts. This view of translation is false, as we will see in what follows.

But, as stated from the outset, more important than even the differences between the translator and his or her audience is the fact that their profound differences are unspoken. Both are guided through any encounters by the invisible hands of conflicting cultural values.

Negotiating a Cultural Semantics

Throughout our lives we work with one social group or another to negotiate the meanings of words, actions, expressions, omissions, and the other facets of the meanings arising from our lives-in-the-world. Meaning is never the possession or creation of a hearer and a speaker alone, but also is created from the social situation, which itself emerges from a particular union of cultures in time and space (Foley 1997, 2007; Pike 1967; Geertz 1973, among others). It is the negotiation to find fit between dark matters—and it often fails exactly because the matters are *dark*.

Translation (by which term I also subsume what some call "interpretation") is the occupation of vast numbers of our species. Shamans, preachers, imams, swamis, and priests regularly interpret the will or words of god/the gods to the laity. US Supreme Court justices translate the US Constitution. Teachers translate other minds' overt and dark matter for students. Contractors translate blueprints into concrete, steel, and wood. Scientists translate observations into theories. We all map meanings onto other meanings and forms onto other forms, working each minute of the day to express the etic in the emic.

Translation is our principal tool for understanding and being understood. And we all translate in one of two senses. First, we translate others' actions, words, postures, dress, intonation, gestures, facial expressions, driving, table etiquette, and so on—all that is visible about them and interpretable by our dark matter, for ourselves. Let us call this *self-directed translation*. We have evolved to be good enough at this to remove ourselves from danger, to dominate others, to navigate through the social worlds that envelop us. We do this in part by seeing etics and imposing emics. We assume that we are experiencing the other's emic perspective because it is the only way that we, without special training or experience, know how to interpret the world around us. For example, when someone raises his eyebrows after you make a remark, is he being funny or sarcastic or seriously doubting what you said?[1] If someone puts her shoes on, is she going out or is she just cold? (My dog is good at telling the difference, by looking at what shoes I have put on and whether I tied them or not.) And how do we interpret language in our environment, whether directed at us or not?

But there is another, also common, form of translation: *other-directed translation*. In this mode of translation, our goal is enable one party to understand another by learning the emic perspectives of each of them—that is, coming to know the dark matter of two differently emicized others, retelling the emic of one as the emic of another.

Failure to translate well or at all—misunderstanding others—is perhaps more common than accurate translation—understanding others; and this failure is caused by a variety of factors. Perhaps the most common is to interpret the etic actions of one person as the emic actions of another (your own, for example), and therefore not attempting to understand the interlocutor's dark matter.[2] Another is assuming that the other's etic actions are in fact *their* emic intentions. For example, if someone raises his voice, you may assume he's angry because to you that etic action indexes anger emically; or an Indian in Delhi "waggles" her head subtly and you misinterpret acquiescence as disapproval.

Another reason for translation failure is incomplete emicization—thinking that you have an insider's perspective but not, in fact, being quite there. You have lived in a culture for several months and think you can now semi-reliably interpret, say, body postures and hand gestures, yet you still are capable of embarrassing gaffes. Quine argued (1960) that because words are associated with their objects only by means of stimuli and responses, and because we can never say for certain what stimulus another is responding to, we can never fully translate the other; there always exists an indeterminacy—even if we share a significant number of cultural values, for example. Our apperceptions, constructed selves, and dark matter will be differently formed.

Bakhtin (1984) recognizes other complications in translation; namely, that we and our interlocutors speak simultaneously with different voices (heteroglossia) and that we are always hearing multiple voices (polyphony). For example, I may incorporate into my speech the speech of others, heteroglossia—knowingly or unknowingly. And I will always be sensitive to the linguistic environment around me, speaking words under social constraints. So as we translate one voice into another, we cannot be sure that there is only one voice or two that we are hearing, or that we are speaking with ourselves (from, say, our different roles in society, different knowledge structures for different roles). In fact, Bakhtin would no doubt argue that translation is always many-to-many and never one-to-one.

When we take upon ourselves other-directed translation, our success is also based on our conception of how we go about establishing emic to emic links. What translation model should we use? Should translation be literal, dynamic, free, or something else? We discuss each of these as we proceed, since each of these types of translation impose different demands on dark matter.

A literal translation is, roughly, a word-for-word translation. Most literary translators, United Nations translators, and other secular translators would recognize immediately that a literal translation in this sense is neither possible nor even desirable. It is not possible because the lexicons and grammars of different languages are never in one-to-one correspondence. Literal translation is also undesirable because it fails to link substance and style, form and content, appropriately in the target language (see the discussion below of Susan Sontag's thoughts on these matters).

TRANSLATION CONTROVERSIES

Take, for example, the translation of the word *virgin* that is found in the Bible in relation to the birth of the Messiah:

> Therefore the Lord himself will give you a sign: The virgin will conceive and give birth to a son, and will call him Immanuel. (Isaiah 7:14)

In this translation, the New International Version, Isaiah is said to prophesy that "Immanuel" (literally, "God with us")—that is, the Messiah—is to be born from a woman that had no sexual relations. This translation is crucial to the New Testament story because in the Christian interpretations of Matthew 1:22–23, the "virgin birth" is said to have been uniquely fulfilled by Mary's "immaculate conception" (New International Version). All this took place to fulfill what the Lord had said through the prophet: "The virgin will conceive and give birth to a son, and they will call him Immanuel."

But in fact this is not what Isaiah said. The prophet Isaiah used the Hebrew word *almah* הָמְלַע, which means "young woman." This can mean either a young woman who is a virgin or simply a young, though sexually initiated, woman.

This ambiguity is interesting because upon the subtle translation choice that might resolve the ambiguity rests an entire theology of Jesus and his mother, Mary. According to some theologians, the Hebrew word *almah* is ambiguous between "young girl" and "virgin." Many Jewish scholars disagree with the Christian (and Muslim) interpretation that Isaiah prophesied that the Messiah would be born of a virgin. But for some Christian scholars, the issue was settled in the Septuagint, LXX (Ἡ μετάφρασις τῶν Ἑβδομήκοντα), or the "70," a translation of the Old Testament into Greek in the second century BCE. Since the translation of the Hebrew Bible into Greek was done by Jewish scholars, it is assumed that they would know what the original cultural interpretation of Isaiah 7:14 would be. Conservative theologians say that the inherent ambiguity is resolved because the Greek New Testament used the word *parthenos* (παρθένος), which is claimed to unambiguously mean "virgin," a young maiden who has not had sexual relations.

It is then claimed that since the LXX translation uses the word *parthenos* for the verse of the prophet Isaiah, that this *must* mean that Isaiah intended to communicate that the Messiah would be born of a virgin. Again, this is because the same Christian theologians claim that *parthenos* means only virgin.

But this is false. Such pontification confuses dark matter and word meaning. What we have here is simply a word that means "young woman," along with the dark matter expectation that most young women are virgins in that culture. Like *almah*, *parthenos* is indeed also ambiguous between "young woman" and "virgin." For example, the same LXX version, in Genesis 34:2–4, translated the rape of Dinah, the daughter of the patriarch Jacob, by Shechem and refers to her, *after her rape*, as *parthenos*. Perhaps the idea that parthenos means exclusively "virgin" comes from the building for Athena in Athens, the *parthenon* (παρθενών), or "chamber of the young woman or virgin." Athena was most assuredly considered to be a virgin in Greek mythology, and her temple's name is usually interpreted as representing her virginity. But this doesn't entail a fact such that a temple dedicated to her with the name *parthenon* can only mean "chamber of a virgin." The cultural and dark matter emic knowledge that Athena was a virgin leads to this interpretation of the name *parthenon*—not its literal meaning per se.

Conservative Christian scholars have values, knowledge structures, and roles that give them a vested interest in establishing that Isaiah had foretold that the Messiah (i.e., Jesus in Christian theology), would have an immaculate conception, that he would not be the product of a semen-fertilized ovum. Therefore, they find themselves in the uncomfortable position of having to argue, based on their values, against the commonplace usages of two words from Biblical Greek and Biblical Hebrew.

What is important for the present discussion is not whether Mary was a virgin, the mother of the Messiah, and so on. Rather, we see in this example that translation is a conflict of dark matters, a culture-plus-individual-dark-matter-controlled, emic-to-emic mapping. It is not possible to do it, except under the most rudimentary circumstances, mechanically. It is a cultural-psychological endeavor always. One's cultural values and memory-apperceptional constructed self constrain one's translation preferences, goals, and interests. Word-for-word or literal translations—in literature, operating manuals, and so on—in this regard can be seen as either undesirable or desirable, depending on goals.

To see this from another angle, consider the controversial Christian concept of baptism. The word *baptize* is not even a translation but rather a transliteration from the original Greek. (Transliterations arise because on some occasions the translator believes that he or she needs to use a single word for a new concept yet also believes that there is not one available in the target language. Therefore, he or she may employ transliteration rather than translation, the conversion via sound-for-sound or letter-for-letter mapping of a word from one language to another.) In the Koiné (common) Greek dialect of the New Testament, the word *baptize* usually meant "to immerse." However, the word *baptizo* (βαπτίζω) can also mean "to dip repeatedly." It is used to describe a particular kind of ceremony. Thus for some Christians, one can be baptized only if one is immersed in water. For others, the word allows for sprinkling, since they interpret it merely as ceremonial wetting in some fashion. The translator of the New Testament from Greek to Latin (St. Jerome) simply chose not to create a conflict by translating *baptizo* as "immerse"—choosing, perhaps wisely, to avoid translating it at all, instead transliterating it. Thus, in addition to translation, the decision to transliterate is also a culturally or dark-matter motivated choice.

TRANSLATION DIFFICULTIES MORE GENERALLY

Translation is fraught with issues in which the constructed self's dark matter or cultural values can affect the type of translation or choices in the translation process (for perhaps the best recent treatment, see Becker 2000). For example, consider two kinds of overlap we find across languages: overlapping meanings and overlapping forms. Take first the case of overlapping forms, especially where these forms come from overlapping histories. Nida (1964, 160) calls these "false friends." As an example, the words *exquisite* (English), *exquisito* (Spanish), and *esquisito* (Portuguese) serve nicely. All three words derive from the Latin construction that originates as *ex quaerere*, which means "to seek out"—something sought out. In its development it took on the sense of "rare," "hard to find." The word has the sense of "rare" in all three languages. However, in Spanish and English it means "beautiful," whereas in Portuguese it means "weird" or "odd" or "strange" and so on—extremes of rarity. In cases like this, although the form of the three words overlaps, has a common etymology, and even shares a meaning at a very general level, in everyday usage, the Portuguese term is very different.

There are also cases where forms are distinct but meanings are similar. For example, the Portuguese word *ja*, "already," is distinct in form from the Pirahã word *soxóá*, but they overlap in meaning. For example, to say, "I am leaving now" in Portuguese, one would say, "*Eu ja vou.*" To say the same thing in Pirahã, one says "*Ti soxóá kahápií*" (for both languages, literally, "I already go.") However, to say that something has been used already in Pirahã—for instance, a used article of clothing—one would say "*baósaí soxóái*" ("cloth already thing,", or "used clothes"). This usage is not found for any sense of the Portuguese *ja*. Thus, although it can be easy to jump to the conclusion that two words have exactly the same meaning, they can in fact have very different meanings, in spite of overlapping in one or more contexts.

There are many other areas of translation difficulty, including phenomena such as idioms, litotes, metonymy, synecdoche, metaphor, simile, and analogy. These can be very difficult to understand or even to see unless the translator has been able to achieve emic perspectives from both the target and source languages and cultures. The lesson is simply that there is no one-to-one meaning, form, or other correspondence between two languages or cultures. As Quine (1960, 51–52) says, translation is marked by indeterminacy—thus the need to become an insider in both groups.

In translation we draw upon all the emicized and random dark matter we possess to map the etic and emic of one knowledge domain, person, or society onto another. What enables this? Should it be hard or easy? In 1978, when I first read Quine's (1960, 29ff) "*gavagai* problem" from his *Word and Object* discussion of the indeterminacy of translation, I was returning from my first field trip among the Pirahãs. The phrase I heard most frequently during that research period was *tíi ʔóogabagaí*, "I almost begin to want (that)," where the most prominent phonological stretch were the morphemes *gabagaí* "frustrated initiation"—very similar to *gavagai* and, initially at least, equally inscrutable. And during this trip I had understood almost nothing, had little idea about how to arrive at any understanding of the language, and felt that learning to speak the language would forever remain beyond my ability. Quine's example was strikingly familiar to me.

Entering into a new culture and language precisely in order to learn them, with the larger aim of translating meanings from one culture and language (source) into the new one (target), can be one of the most daunting intellectual, personal, emotional tasks imaginable—especially when no language or culture is shared in common. Rereading Quine's passage on exactly the problems of this sort of field situation many years later, I am struck by the contrast between the mistakes that never got made and the kinds of mistakes that frequently occurred.

On the side of mistakes never made, however, Quine's *gavagai* problem is one. In my field research on more than twenty languages—many of which involved monolingual situations (see D. Everett 2001; Sakel and Everett 2012), whenever I pointed at an object or asked "What's that?" I always got an answer for an entire object. Seeing me point at a bird, no one ever responded "feathers." When asked about a manatee, no one ever answered "manatee soul." On inquiring about a child, I always got "child," "boy," or "girl," never "short hair."

Why is that? According to Quine, the answers I never got might be as common as the answers that I in actuality received. In other words, Quine would say that I have no idea whether I simply missed crucial answers, whether I failed to elicit or recognize them, or not. The missing answers certainly could make sense in some of the cultures I have worked with. And in some contexts, these answers would be quite reasonable, as if I had touched a child's head and pointed to their hair, or held a bird and grasped one of its feathers. I believe that the absence of these Quinean answers results from the fact that when one person points toward a thing, all people (that I have worked with, at least) assume that what is being asked is the name of *the entire object*. In fact, over the years, as I have conducted many "monolingual demonstrations,"[3] I have never encountered the *gavagai* problem. Objects have a relative salience—whole objects (see the discussion in the previous chapter of whole objects in homesigns). This is perhaps a result of evolved perception. Perhaps animals perceive wholes before parts. If we are being threatened by a wolf, we are being threatened by the entire wolf, not merely its ears or paws or even teeth. And it is likely that the wolf sees a person-object when looking at us. We would not last very long in the wild if we saw ears without understanding that ears are part of something else, *more important than its parts*, that could turn out to be foe, friend, or food. Initial focus always seems to be on the whole. Perhaps this is due to biological values of hunger satisfaction, self-preservation, or the like. In any case, it seems to be what happens transculturally.[4]

On the other hand, other kinds of mistakes *are* common. These involve easily confusable, equally plausible perceptions of situations. For example, in the Pirahã word list collected by the nineteenth-century explorer Karl von Martius, in 1821, most of the words and translations he provides are impressively accurate. Von Martius even attempted to represent tones and nasalization, before the existence of systematic phonetic representations of these features. However, he made a single translation mistake: for the form *ʔabaáti*, "remain," his translation is its antonym, "come." He was likely with the Pirahãs as they were going to the jungle, but did not understand that the Pirahãs apparently did not want him to come with them. Aside from the humor of this nearly two-hundred-year-old misunderstanding, the nature of the problem is easy enough to comprehend—of the two a priori, equally plausible interpretations of the situation in which von Martius found himself, he assumed that he was not being left behind. As one whom the Pirahãs have also asked to stay behind on occasion, in the jungle, while they went on ahead to hunt for hours, I can agree with von Martius that this is unexpected. These kinds of elicitation errors are common, whereas the ones underscored by Quine are not.

GENRES OF TRANSLATION

When one is translating, the first decision to make is to decide what kind of translation one wishes to produce. One type of translation is known as "free translation." In the lines below extracted from a Pirahã text (the full text appears in chap. 6), the first line is the vernacular, the second line the literal translation, and the third line the free translation:

1. *Ti aogií aipipaábahoagaí. Gíxai. Hai.*
I Braz.woman began to dream You. Hmm.
I dreamed about Alfredo's wife [aside to Sheldon, "you probably know her"].
2. *Ti xaí* Xaogií *ai xaagá. Xapipaábahoagaí.*
I thus. Braz.woman there be began to dream
I was thus. The Brazilian woman was there. I began to dream.
3. *Xao gáxaiaiao. Xapipaába. Xao hi igía abaáti.*
she spoke dreamt Braz.woman she with remain.
[Casimiro] dreamt. The Brazilian woman spoke. "Stay with the Brazilian woman."

The literal translation (the second line, non-italic, non-bold) is barely coherent from the perspective of English, since it merely uses the order and words of Pirahã, translated without trying to convey purpose, intent, and so on. The third line aims to express the meaning in natural English, even though this natural English might require words, orders, complex or simply structures, and the like, that are not found in the vernacular.

CULTURAL FORM AND TRANSLATION

Because of these difficulties, religions take different views on translation of their scriptures into other languages.[5] Interestingly, of the two major proselytizing religions, Christianity and Islam, the Muslim view of translation is superficially the more scientifically well founded. In Islam, the prevailing doctrine is that the Quran cannot be translated. One can only make "interpretations" of it in different languages.

As Wikipedia puts it:[6]

> Translations into other languages are necessarily the work of humans and so, according to Muslims, no longer possess the uniquely sacred character of the Arabic original. Since these translations necessarily subtly change the meaning, they are often called "interpretations" or "translation[s] of the meanings" (with "meanings" being ambiguous between the meanings of the various passages and the multiple possible meanings with which each word taken in isolation can be associated, and with the latter connotation amounting to an acknowledgement that the so-called translation is but one possible interpretation and is not claimed to be the full equivalent of the original).

The good part about this idea is that it recognizes the inseparability of form and content—the Quran is not simply a content that wears different forms. No work is. It is a style, a form, cloaked in content as much as the other way around. This is poetry. It is all true art.

Interestingly, the Islamic view overlaps with Susan Sontag's (2013) perspective, in at least a couple of respects. In particular her agreement with Islamic views on translation in statements like (21ff): "Style and content are indissoluble . . . the strongly individual style of each important writer is an organic aspect of his work." And also, "The notion of a style-less, transparent art is one of the most tenacious fantasies of modern culture." "In almost every case, our manner of appearing *is* our manner of being. The mask is the face." And, finally, "It is not only that styles belong to a time and a place; and that our perception of the style of a given work of art is always changed with an awareness of the work's historicity, its place in a chronology . . . the visibility of styles is itself a product of historical consciousness."

Sontag has captured a deep insight here, one that still eludes many linguists, philosophers, anthropologists, and others; namely, that our living in a particular place, time, and society produces a complex network of ideas, values, styles, conventions—a particular dark matter that is responsible for creating an identity that is simultaneously form and function. Our form, or style, can include word choices, chronological or logical structuring of the stories we tell, our body-fat percentage, or—as we saw with the Dutch—even our height. This is also why there is no completely objective or completely skeptical life. None of us ever escapes the wings or the shackles of our dark matter, of the influences of the other and the symbiosis of style and substance.

Another type of translation is what Nida (1964, 166) refers to as "dynamic equivalence" translation: "In such a translation the focus of attention is directed, not so much toward the source message, as toward the receptor response." Dynamic translations try to produce the same perlocutionary effect in the target-language audience as was produced in the source-language audience. As unrealistic as such a goal might be, it is an interesting one and places even greater responsibility on the translator to know both the source and target cultures and languages well—to begin and end with each of the two emic systems involved and, however unlikely, to have one's dark matter encompass the nuances of both systems. If the original audience was disgusted, so should the target audience be. If the passage means *x* to the source audience, it must mean *x* and not *x* + *1* or "more or less *x*" to the target audience. If the original audience believed in the miracle, so should the target audience. If the original audience felt the need to be "saved," then so should the target audience. If the original audience cried at a particular verse of a poem so should the target audience, and so on. Using this method—or better, philosophy—of translation, one is not concerned so much with the individual words, sentence structures, and the like. One must rather understand how and why things were expressed in certain ways in the source language, and how and why to express exactly those things in the target language.

These modes of translation—as well as talk of the emic, of dark matter, and of culture—bring us to the question of meaning, or as Ogden and Richards ([1923] 1989) put it, *The Meaning of Meaning*. Basically, meaning is the referent in the real world, the action desired, the theme or the object of a proposition (Soames 2010) and so on. The information that the meaning conveys is the change in the (informational, emotional, legal, physical, etc.) state in the hearer produced by the words or actions of the speaker or actor. Getting at meaning requires knowledge of language, which in turn requires knowledge of culture.

Knowledge of language is not merely a knowledge of all sentences, but the structure of that knowledge—for example, the mapping between sentences. Another way of thinking about this (one that I would strongly urge) is that language is not a set of any units, such as sentences or words, but rather an understanding of storytelling—how to tell stories, what they should be about, how they should be structured, and so on. Beyond this, language is knowledge of the social group, of the culture from its narrowest community boundaries to its widest extent. This is clearly reminiscent of Bakhtin's (1982, 284) view: "Any understanding of live speech, a live utterance, is inherently responsive . . . Any utterance is a link in the chain of communication." According to Bakhtin's concepts of heteroglossia and polyphony, everything we say is a mixture of things we have heard others say and the interaction with them in dialogue.

Translation, like culture itself, is a dance, an action-reaction pairing in which two beings adjust themselves according to the necessity of the environment—for example, maintaining the same speed while dancing. Thus not everything—perhaps not much at all, even—in human interactions is the result of mental representation, but is rather acting together. This brings us to a deeper and more foundational question we have avoided so far; namely, whether translation is even possible. It is a truism among linguists that all languages are capable of expressing the same things. But if language is formed symbiotically with culture, does it necessarily follow that translation of anything from one language to any other language can be done? What does translation entail?

No Universal Intertranslatability

Consider the possibility of translating the following sentence into Pirahã (thanks to Geoffrey K. Pullum [pers. comm.] for suggesting this example):

> My cousin totally ran out of cash during his second year at art college in California and had to draw $10K from the Bank of Mom and Dad to pay his tuition.

What does this sentence show? First, it shows a syntactic device common in English that is lacking in Pirahã: embedding. This means that in Pirahã, this single English sentence would have to be expressed by more than one sentence, since without embedding one sentence cannot be placed inside another. (The absence of embedding in Pirahã results from a lack of recursion in Pirahã; D. Everett 2014b; Futrell et al., forthcoming). The absence of embedding need not be fatal to translating the content, though it does demonstrate that translation from English to Pirahã cannot be done by matching forms, thus losing some of the translation's accuracy (according to Sontag, though not necessarily so according to the dynamic translation philosophy).

Neither, however, can a translation match the English concepts with Pirahã concepts. There are no words in Pirahã for *mother* or *father*, only a generic word for *parent*, which, as we have seen, has a wider meaning than biological parent, referring also to someone one is hoping to get something from, or someone who holds power over you in a particular situation. There are no quantifier words, hence "totally" cannot be translated. Further, Pirahã culture has no money, though they do have a vague concept of it from watching Brazilians. They have no numbers, so *one dollar* or *ten thousand dollars* would not be translatable. They have no concept or word for *art*. They have no word for *year*, per se, though they could say "water," referring to a high-water/low-water cycle (one year). They have no concept or word for *bank*. Thus this entire sentence is ineffable in Pirahã, in both form and content. Therefore it is not possible in practice to translate from all languages to all languages, QED. It might be possible, though, if speakers and hearers were to come to share sufficient dark matter, via prolonged cultural contact, that they became bicultural. But that is not the case here, thus underscoring the importance not only of dark matter to translation, but of translation as a test for degree of dark matter overlap, of knowledge of a language, of cultural understanding, and so on.

Whether translation is actually possible in a complete or interesting sense depends on the objectives of the translation. A linguist translating a language for other linguists will have different objectives than a missionary translating the Bible or a company translating an operating manual. The important point is that translation is a cultural activity, putting one culture into terms of another culture, not just transferring one language into another. Language is only part of the translation process. This is shown by the texts we examined in earlier chapters and by the example above. The implications of this are profound: not all things can be said in all languages. This is the implication of the notion of dark matter in grammar and cognition we are considering here.

Another way in which translation interacts with dark matter revolves around the notion of relevance. Dark matter shares in common with Sperber and Wilson's (1995) *relevance theory* the idea that we need assistance to process the vast amount of input the world bombards us with. Relevance theory says that we attend to what is relevant, where relevance is determined by context, topic, interlocutors' assumptions, culture, and so on.

The idea of relevance is motivated by a "bottleneck" theory of cognition—namely, that since our attentional resources are limited, we use some principle of relevance to *license our disregard* for the majority of the input and to place our focus on a narrow band of the environmental stimulus. (Broadbent [1958], Treisman [1991], and others made the point that we can switch our attention to previously unattended information if that becomes relevant, such as overhearing our name in an across-the-room conversation, "sensing" someone staring at us, seeing flames, being attacked, or seeing the main speaker walk on stage.)

Sperber and Wilson's work was inspired by the writings of philosopher Paul Grice (1991) on maxims of conversation, constituents of the larger *cooperative principle*. This principle describes our linguistic exchanges as making our contribution, such as is required or desired at a particular stage in conversation by the accepted purpose or direction of the talk exchange we are engaged in.

The four conversational maxims that manifest the cooperative principle are as follows.

1. MAXIM OF QUALITY

Do not say what you believe to be false. Do not say that for which you lack evidence.

2. MAXIM OF QUANTITY.

Make your contribution no more nor less information than is required.

For example:

Q: "How are you?"
A_1: "Uh-huh." (underinformative)
A_2: "I was constipated yesterday and not sure if I should eat more cheese today, given that my lactose intolerance isn't going away . . ." (overinformative)

3. MAXIM OF RELATION

Be relevant.

This is harder to control for and study, because culture is the foundation of relevancy. The hearer will assume that the speaker is saying something relevant, and the speaker will assume that the hearer will look for the relevance; shared cultural values are crucial. The relevancy detected/assumed/constructed need not be literally there, since the culture-specific knowledge and cultural nature of the exchange will fill in the gaps. For example:

Q: "Have your parents arrived?"
A: "Mother is ill."

The hearer will infer that the mother's illness is relevant to the question of whether the parents have arrived. If one speaker asserts in this context that his mother is ill, then since this must be relevant to the question, and the parents must not have arrived. Or:

Q: "Have your parents arrived?"
A: "There were fish in the river."

This is a stretch. How could fish in a river have anything to do with whether one's parents arrived? They do not, literally. But the hearer will again assume that there is something relevant in the answer to the question and so think along the lines of "Perhaps there is a river his parents needed to drive over with a bridge of some sort that cannot be employed when there is also a high concentration of fish in that section of the river."

A conversation is another form of dance. Each dancer assumes that his or her partner is trying to dance the same dance. Relevance and cooperation, however, are culturally influenced and individually determined (Henrich and Henrich 2007). There is relevance in the sense of the culture as we are using that term here, as well as relevance to each partner as self-constructed, and thus what they each choose to attend to at any particular time, how they choose to interact, and so on.

MAXIM OF MANNER

Be clear, avoid obscurity, avoid ambiguity, and say no more than is required.

These are not grammar rules. They are expected by the hearer as means to more easily interpret what the speaker means. For example, if someone asks you, "Do you believe that abortion should be illegal?" the expected answer is "yes" or "no." Thus if, as candidates do in political debates, you begin your answer with "My campaign has always been clear that some issues are complex," then the hearer will take this as a poor answer, one that for some reason flouts the maxim. (All the maxims may be flouted.) This doesn't mean that the offending answer is meaningless, just that the speaker intends to say something other than what was expected.

Translation is also an important test of the understanding of dark matter because the ability or inability to translate tells us about shared concepts. Concepts are crucial to our hermeneutics and translations. They are often held up as the ultimate example of knowing-that vs. knowing-how. One of the most influential and original discussions of concepts in recent years is found in the work of Susan Carey, in particular her 2009 book, *The Origin of Concepts*. Her theses in this book are (i) that humans are born with rich conceptual representational structures (back to Bastian); (ii) that human perceptual abilities help us to identify innate concepts; (iii) that our innate representational structures/systems never change, and therefore, there will be cross-cultural conceptual universals; and finally, (iv) that we have mechanisms for transcending *core cognition*. We can create culturally influenced representations that are incompatible with the fixed set we are born with.

Carey's theses contrast with Brandom's. Again, Brandom, in all his recent work, argues that concepts are not given but "earned," by deriving, demonstrating, and learning their meanings via inference. And only Brandom ultimately provides a satisfying account of shared concepts—concepts built by interaction and individual apperception, varying tremendously from culture to culture and family to family.

The theory of dark matter developed throughout the different divisions of the preceding discussion takes a skeptical view of nativist perspectives of concepts (e.g., Carey's). Some values and emotional centers of our brain have a biological source, though the expressions and conceptualizations of these—the value "avoid pain," "avoid death," "be happy," and others—have local cultural interpretations.

Before innate concepts are proposed, we must survey world cultures, account for differences, assess proposals of universals, and understand the range and typology of learned concepts, for even Carey's theory of concepts recognizes that many concepts (e.g., "US president") must be learned. However, once we have understood how, why, and which concepts are learned transculturally, what is left for nativism, aside from standard poverty of stimulus arguments? And what, after all, does "poverty of stimulus" mean in practice, other than that we cannot think of a stimulus responsible for a particular concept, action, or other learning? As many have said in the past, when looked at carefully, the expression "poverty of stimulus" is interchangeable in practice with "poverty of imagination."

Problems in Translation

A chapter on translation should perhaps include more examples of the difficulties of what Quine (1960) referred to as "radical translation." Recall that Quine's discussion of radical translation and the "indeterminacy" of translation flow from Quine's concern with a linguist confronting a linguistic community that speaks a language unrelated to any that the linguist is familiar with—what Pike called a "monolingual setting." Quine (based on my reading of Quine 1985) came up with the example, I believe, via his contact with Kenneth L. Pike (my first linguistics professor). The idea is roughly that we have no way to exactly know what the referents of a given term are. This section, then, is an expansion on the earlier question of intertranslatability.

One of the first questions visitor to the Pirahãs ask me is "How do you say 'hello' in Pirahã?" Or "How do I say 'thank you'?" These were once my questions too. How *does* one greet others among the Pirahãs? Or take leave, express gratitude, and so on? Well the answer is simple—generally one does not do these things. Such uses of language are called "phatic" language, and Pirahã, among other groups, simply lacks this type of language. One can say "I am going now" for "Good-bye" or "You gave me this" rather than "Thank you," and so on. But normally one doesn't say anything at all. One leaves. One arrives. One accepts. One gives. No special declaration is needed culturally—after all, these actions are obvious. This means, however, that there is no way to accurately translate *thanks*, *bye*, *hi*, and so on, into Pirahã. So as I once translated portions of the New Testament into Pirahã, I might translate *thank you* as "I accept this. I will pay you back." Yet, though Pirahãs might say this kind of thing, they simply use no formulaic phatic language. This is a translation problem of "category vs. no category."

Another example of category problems is what I call the *partially matching category*. When I first attempted to translate some Pirahã words into English, I used what to my then Christian mind were the closest English equivalents. This confused my readers and, ultimately, the Pirahãs as they received my attempts to translate. For example, one of the hardest terms for me to understand was *kaoáíbógí*, which literally means "fast mouth."

In the evenings, sometimes throughout the night, one can hear a falsetto male voice coming from the jungle into the village. It may give advice on where to fish, what foreigners to avoid, or how to spend one's next day, among other things. It can talk about what living underground is like for the dead. It may tell stories about jaguars. Or the speaker may walk into the village naked and start acting vulgar and saying vulgar things. The people will say, animatedly, that "a fast mouth is here." And when I reply, "But that looks just like so-and-so," they respond, "Yes, fast mouths look just like Pirahãs. But you can see that they are not Pirahãs: They are naked. Pirahãs are not. They talk high [in a falsetto]. Pirahãs do not talk high."

Kaoaibogis can live under the ground, under the water, inside tree trunks, everywhere. They do not have blood. Thus I originally interpreted these jungle entities as "spirits." But this is incorrect. First, kaoaibogis are material. Second, they play no role in a larger (e.g., religious), belief system—no more than a jaguar or a fish does, in any case. Their function in Pirahã society is for drama, for humor, for scaring others, for correction of children, for advice. They are real and physical. They are similar to people but not identical to people. They are similar to people lacking normal cultural prohibitions. Beliefs about them—what they do; how they live; whether they are good, bad, or neither—vary among the Pirahãs. Yet there is no sense or referent of *kaoáíbógí* that makes it translatable as "spirit" in English. They are entities that do not map well into a North American ontology or the English language.

Kaoaibogis do not map neatly to fact or fiction. The Pirahãs enjoy them and see them, but they are liminal creatures, neither mundane nor otherworldly. They can be seen, heard, smelled, touched, and so on. Yet they also are funny, even when trying to be serious, as though the people were embarrassed by the person "playing with the jungle entity." At the same time, all Pirahãs swear that kaoaibogis are real. The person they resemble (what a nonbeliever might refer to as the "actor") will refuse to admit that he (never a woman, even when the kaoaibogi dresses like, talks like, and is claimed to be female) was present when the kaoaibogi was present and will add that he never saw the kaoaibogi.

The "fast mouth" label is just one for a series of nonhuman, humanoid entities. There are also "big tooth" (*xaitoii*) and "jaguar" (*kagaihiai*), among others. A *kagaihiai* is not a real jaguar, it just looks like (and can act, kill, and eat like, etc.) one. And so on. These creatures are all just among the multitude of types of jungle entities encountered by the Pirahãs, and like other entities, each individual may have a personal name. It isn't clear to me whether they change individual names as the Pirahãs do, but I suspect that they do.

With this background, we see that *kaoáíbógí* has/have no simple one-word or one-phrase translation into English. These entities are *neither* fact nor fiction in our sense, neither friend nor foe. In a sense, to quote Philippe Descola (2013), they are "beyond nature and culture." We ought not to translate these terms as spirits, sprites, demons, possessed people, or any other off-the-shelf English word. In fact, the point is not that there is no one-word translation, but that there is no translation, period. At best we can describe them, but the Pirahãs' conception of and classification of sentient entities in their world fails to match our own (and vice versa, of course). Likewise, the Pirahãs have no concept of "god" and thus cannot, without much learning about, say, Brazilian culture, even approach an accurate understanding of what we mean by a supreme deity. Previous missionaries tried to translate *God* as *baixi hiooxio*, "up-high father." I also attempted to use this as a translation for *God*. But the term, like the concept it is trying to communicate, turns out to have no purchase among the Pirahãs. They have no supreme being, no creator, nor any savior among their beliefs about the world.

On the other hand, Pirahãs do have beliefs about the organization of nature. For example, they believe that the universe that they can see is structured into layers. There is the layer we walk on and breathe in, our biosphere, which they call *xoí*. If you ask a Pirahã how to say *mato* (the Portuguese word for "jungle") in Pirahã, they will say *xoí*. But *xoí* refers to place of living, jungle, environment, personal space, and so on. For example, to tell someone to be motionless, as in a canoe when you are about to shoot an arrow standing up at a fish swimming just below the surface, you say "*Xoí kabao xaabaati!*" "Biosphere" is the closest English translation, but it is inadequate because the English word lacks a center, as the jungle is the center of the *xoí* for the Pirahãs. Such an exploration of Pirahã nature words begins to point us toward the inadequacy of the view of translation that simply matches up words and concepts. Culture and nature form each other, as Descola argues, and thus each triad of culture-language-nature is unique. Translation cannot be literal because a literal translation would be gibberish. Translation requires a certain liberty to communicate the aspects of meaning relevant to a specific, situational, culture-language-nature mapping.

As another example, consider the Pirahã word *bigí*, which appears to mean "sky." But it also means "ground." How can the same word mean both "sky" and "ground"? Because the Pirahãs believe that the sky and the ground are each natural divisions, differentially permeable barriers, between our *xoí* and some other world's *xoí*—above the sky or beneath the ground. The word *bigí* is thus only effectively translated as we understand the Pirahã cosmology. There is no easy, unique translation for this term either. And the list goes on.

Moreover, there are also no translations from Pirahã into English for numerous kinship terms that other languages take for granted, such as *mother, father, grandparent, niece, uncle, brother, sister, cousin*, and so forth. They (as I point out in D. Everett 2005a and 2008) have the simplest kinship system ever documented and have only the words *xoogii*, "big" (used for "parent" or "grandparent" or "older sibling"); *baixi*, "parent or someone with power over you in a given situation"; *xahaigi*, "anyone of ego's generation"; *hoisi* or *hoaagi*, "son"; and *kaai*, "daughter."

Pirahãs also lack all number words (a well-documented fact, however surprising, in M. Frank et al. [2008]; D. Everett [2005a]; C. Everett and Madora [2012]). Thus it is impossible to translate *one, two, a million*, or any other precise numerical concept into Pirahã. They also lack quantifiers (e.g., *all, each, every, some, many*), so translations of quantifiers, like numbers, can at best be approximated (D. Everett 2005a). And they also do not talk about "duty" or "salvation" or "sacrifice" or "atonement" and so on. I am not claiming that it would be impossible to translate these concepts into Pirahã, but the task is more than linguistic—it would require connecting cultural concepts from first-century Jewish culture to contemporary Pirahã culture, mediated through contemporary American culture, mediated through millennia of theology.

But in addition to the lack of terms for easy translation, there is an even more difficult aspect of translation that I encountered in working with the Pirahãs: all speakers are (or at least were on my last visit) monolingual. When a Pirahã-language helper gives you a word or a phrase, only rarely are they able to translate it into Portuguese. Someone might tell me something and the verb form they employ could have seven or more suffixes. Then someone else (or even the original speaker) might repeat what was said, and the verb used could have the same root but only three suffixes, no suffixes, or very different suffixes than the first utterance. For stories it is impossible to get more than the briefest summary of the gist of the story in Portuguese, and that only from my most "bilingual" language teachers. So how does one translate stories and phrases and verbs from a language one barely speaks and which has no speakers who speak any language you understand (my situation in my early days among the Pirahãs)?

My method was to use paraphrase. I purchased several small, inexpensive, but reasonable quality cassette recorders, in addition to a more expensive, high-quality digital recorder, with a professional-quality headset microphone. I used the higher-quality gear for recording texts. Then I would record the text from the digital recorder to a smaller analog cassette recorder. (The cassette was for playback to the Pirahãs, though not for analysis—that was the job of the digital device.) Then I would play back the cassette recording of a text for a Pirahã who had not been involved in the original telling of the text (either telling it or overhearing it told). Then, with a second recorder running, I would ask him to explain the text to me. As the text was explained by this second language teacher, very often the second speaker simply repeated the first speaker. But regularly, with varying frequency, the second speaker would change things the first speaker said, inserting value judgments about the content or form of the first iteration of the text, with perhaps some remarks about its storyteller as well (e.g., "He speaks poorly"; "He doesn't know."). He might use different verb forms, different verbs and nouns, different intonation and gestures, and so on. Often, following the second speaker, I would work with a third speaker, asking them to "tell me what they said," referring to the first two speakers. I would play a line by the first speaker, and then the comments by the second speaker on that line, finally recording the comments of the third speaker.

Armed with these commentaries and the original text, I would return to the high-quality recording of the original and transcribe and translate it (to the best of my ability). I transcribed it all in detail—phonemes, intonation, accent, morpheme breaks, sentence boundaries, and so on. Then I would read my transcription to yet other Pirahãs, sometimes serially, sometimes in a group, offering my interpretation and soliciting their comments, criticism, and corrections of my pronunciation, interpretation, and so on. Thus did they supply the overt material in the recordings, their translations of their own dark matter, via their interpretations of the texts. In this way I learned culture, covert→overt language, and individual personalities through a laborious process of dialogue.

Summary

This chapter examined the notion of a culture-based semantics, based on emic understanding of a given language's semantic fields. By way of illustration, we examined the effects of different approaches to meaning and culture via an examination of translation controversies in biblical studies. We then saw how different approaches were manifest in genres of translation. In particular, we considered and accepted the idea of meaning negotiation as a type of dance in which each interlocutor matches the meaning moves of his or her partner in communication. We were led to conclude by the role of culture in the grounding and imputation of meaning, that not everything—perhaps not much at all, can be said to be intertranslatable crossculturally or crosslinguistically. One way in which culture enters in to translation and discourse, we argued, was through the notion of relevance as developed in relevance theory.

We concluded with a discussion of the methodology of translation in what Quine (1960) labels "radical translation."

PART THREE

Implications

9

Beyond Instincts

> We believe that civilization has been created under the pressure of the exigencies of life at the cost of satisfaction of the instincts.
>
> FREUD, *Introductory Lectures*

No "Man in a Can"

The idea that there is a psychic unity of mankind via a set of innate, phylogenetic concepts and/or instincts that unfold ontogenetically and largely set the parameters of our identity, of our "human nature," is what I refer to as the "man in a can" view of cognitive science. Not only do I find this view simplistic, anti-intuitive, anti-anthropological, and anti-evolutionary, I think that the research results obtained under man-in-a-can assumptions—for example, by linguists and psychologists working on universal grammar, or psychologists working in evolutionary psychology—are unimpressive, an unfortunate cul-de-sac in the progression of knowledge about our species.

In this chapter, therefore, I lay out a case against instincts, looking at and rejecting several proposals on instincts—especially those that entail "native concepts"—proposed in the literature, all of them relatively popular with scientists and/or the general public.

We started this book with some questions, which included the following: If you give a lecture, how might you know from people's faces whether they are understanding you? When you use a concept, why do you believe that you understand it? Why do you like the music that you like? How do you know that the cry you heard is from your own child? How can people tell without looking whether someone is running upstairs or downstairs? How do you know what your mother looks like? What does tofu taste like? Why do you say "red, white, and blue" instead of "white, blue, and red"? I made the case that apperceptions, values, violable value hierarchies, and knowledge structures of the enveloping culture in conjunction with the idiopsychology of each individual lead to the formation of dark matter, and that this dark matter is the answer to each of these questions, depending on the history of the individual.

Evolution and the Minimization of Instincts

If the concepts of culture, apperception, and dark matter we have been constructing to this point are on the right track, then they leave little role for instincts. Perhaps a better way to put this is that the theory of dark matter *implies* a minimization of instincts. It is not that instincts are incompatible with culture or dark matter. Rather, they just become less relevant to the discussion. For example, if we indeed learn from all around us and participate in culturing and languaging, such that by these activities we also construct ourselves, then a great deal of what we want to account for is accounted for (W. Prinz 2012, 2013). The question is, why then reserve other aspects of the self for instincts, rather than pressing on and searching for evidence that our higher cognitive activities are learned? A closely related question is, how scientific is talk about instincts in the first place? (The answer here is "not very.") Of course, the concept of instincts is common enough in the literature on animal behavior, evolutionary psychology, and Chomskyan linguistics, among other fields. Instincts are of interest in the context of the present discussion because to the degree that they are claimed to represent innate *knowledge*, they would support the Platonic-Bastian tradition of dark matter, over the Aristotelian perspective I am urging. Considered from the vantage point of the conception of knowledge here, however, the existence of innate dark matter would represent an additional source of knowledge for our species, in addition to apperception and culturing. Since we know that postpartum acquisition of knowledge takes place and that general learning is responsible for a great deal of how people come to construct a sense of self and group identity, one could reasonably argue that instincts would complicate the picture of human development, going against the inherent cognitive and cerebral plasticity of the species, and that appeal to epistemological nativism should be limited by Occam's razor unless very strong evidence exists for them. Part of the purpose of the present section is to argue that no strong evidence exists—and that a huge gap exists between the postulation of (epistemic) instincts and convincing arguments for them. As a preview, here are some of the things that bother me about proposals that important aspects of human knowledge are innate (e.g., morality, language): (i) the nonlinear relationship of genotype to phenotype; (ii) failure to link "instincts" to environment—today's instincts are often tomorrow's learning, once we learn more about the environmental pressures to acquire certain knowledge; (iii) problematic definitions of innateness; (iv) failure to rule out learning before proposing an instinct; (v) the unclear content of what is left over for instincts after acquired dark matter is accounted for; and (vi) lack of an evolutionary account for the origin of an instinct.

Knowing a person's genome tells us very little about how that person's genes are going to interact with their environment. As we saw in chapter 2 with the discussion of Dutch height, roughly the same genes can produce shorter-than-average people or the tallest people on the planet, depending on their interaction with the environment. To belabor this point for a moment, there is never a period in the development of any individual, from their gamete stage to adulthood, when they are not being affected by their environment. It would be misguided, therefore, to think that newborns of any species begin to learn from their environment only when they are born. Their cells have been thoroughly bathed in their environment before their parents mated—a bath whose properties are determined by their parents' behavior, ecological surroundings, and so on. The effects of the environment on development are so numerous, unstudied, and untested in this sense that we currently have no wholly reliable basis for distinguishing environment from innate predispositions or instincts.

Another reason for doubting the usefulness of terms like *instinct* and *innate* is that many things we believe to be instinctual can change radically when the environment changes radically, even aspects of the environment that we might not have thought relevant. For example, in 2004 a group of scientists led by Kerry Walton (Walton, Benavides, et al. 2005) carried out experiments on the ability of rats to right themselves in a low-gravity environment. What they discovered was that the self-righting routine (the way in which they come to their feet) that many had thought to be instinctual was ineffective in low gravity. But the rats didn't simply fail to self-right. They "invented" a new strategy that worked while they were weightless. They showed behavioral flexibility where none had previously been expected. As the authors put it (ibid., 593), "Postnatal motor system development is appropriate to the gravitational field in which the animal is reared." But then it is unlikely that it is innate, rather than "negotiated" as the body resonates (J. J. Gibson 1966, 1979) with its environment. Likely, the innate aspect should these be limited to the musculature and physiology, providing a non-concept-based ability to self-right, one that normally develops in earth-based gravity.

Consider further the relevance of innateness (also referred to as nativism) for linguistics. In the 1960s linguists, psychologists, philosophers, and others became excited and fascinated with Chomsky's bold and brilliant revival of Cartesian rationalism in the form of the hypothesis known most commonly as universal grammar. In the intervening decades, hundreds if not thousands of researchers inspired by Chomsky have conducted research on human language under nativist—usually Chomskyan nativist—assumptions. However, as we take stock of the tens of thousands of man-hours invested in this endeavor, I am not aware of a single psychological analysis in which UG is causally implicated. There is not a single analysis I have seen in which the solution could not have been reached without UG. There are in fact no predictions and no analysis of any linguistic fact in which UG is crucial. When we survey the literature on instincts that has arisen since Chomsky's earliest work on universal grammar, we find a strong desire to find "design" instead of "evolution," especially in the context of the nature of human cognition. Though some refer to this as "designer appeal," I prefer "Bastian appeal."

We find it in Chomskyan linguistics—the work responsible more than any other for reviving the moribund philosophical school of rationalism, in particular the dualistic rationalism of Descartes, which separated the mind from the body—but also in theories inspired by it, such as evolutionary psychology, innate morality, natural semantic metatheory, and many others. It is the desire to find the Platonic conceptual a priori, to think that somehow we are the product of our genes rather than, in conjunction with our societies, the shapers of our phenotype.

To deny instincts in one domain does not entail denial of the obvious fact that our genes impose strong limitations on us. There obviously are things such innate characteristics—eye color, adipose cell concentration, blood type, height, and so on. The question here, however, is whether there are Bastian-like innate conceptual structures. Carruthers, Laurence, and Stich believe that there are:

> Though they [evolutionary psychologists] are broadly sympathetic with the sociobiologists' attempts to give evolutionary explanations of cultural phenomena, evolutionary psychologists maintain that sociobiology's focus on behavior and its neglect of psychology are misguided. When genes influence behavior, they argue, they do so by building brains with a bevy of specialized mental modules. Behavior is the result of the interaction between these mental modules and the environment. (2007, 9)

Moreover, Tooby and Cosmides (1992, 91) go so far as to claim that there is a single human metaculture—a cultural UG, in other words—that is "evoked" by experience. Carruthers, Laurence, and Stich (2007, 11) claim this means that "as in the case of Chomskian parameters, the information required to deal with the problem at hand is innate, and the environment serves only to trigger the appropriate package of information." This includes, for example, food-sharing within band-level groups.

The HM (human metaculture) of Cosmides and Tooby is perhaps the most radical form of Bastianism or Platonistic a priori knowledge that has been proposed since the *Meno* (and largely falsified as a claim about *human* nature, if animals with quite different cognitive organization turn out to have culture—see Laland et al. [2009] and Whitehead and Rendell [2014]). In what follows I want to argue that all forms of innate conceptualism—Platonic a priori knowledge, all Bastianisms—are detriments to understanding, passé, and deeply flawed. We are prepared to move beyond these primitive notions of design and instinct on to a more empirical, scientific understanding of human behavior, knowledge, and the dark matter arising at the intersection of culture and the individual.

To begin our discussion, I want to examine what I take to be the philosopher Ludwig Wittgenstein's transition from the a priori of Plato to the conventional of Aristotle. Since Cosmides, Tooby, the editors of the three-volume *Innate Mind*, and most other nativists take Chomsky's work on language as their inspiration, I offer a discussion of why instinct is in fact of little use even in discussing language, using Wittgenstein as a starting point. Then I argue that it is also of little use in understanding either culture or dark matter more generally. Finally, I circle back in the next chapter to show how the learning of dark matter entails the construction of self and a relationship to culture and language that obviates the need for a construct such as human nature.

I have argued in preceding discussions that dark matter helps us understand the relationship between the individual and his or her culture, by linking culturing, languaging, interpreting, and remembering. The case made is that the Aristotelian line of thought favoring the social over the innate, timeless knowledge of Platonism or rationalism, is the most compatible with the facts and discussion here. I further argue that empiricism is in many ways preferable to rationalism as a foundation for understanding human cognition. What I propose to do in this next part of the discussion is to examine the standard model of the mind and instincts within nativist cognitive science and then to move on to different perspectives, leading to my own views. I want to begin with "Wittgenstein's shift."

NO SYNTACTIC INSTINCT

Ludwig Wittgenstein (1889–1951) was one greatest figures of twentieth-century philosophy. He was initially inspired to think about language from the work of Bertrand Russell and Gottlob Frege, among others. However, as he began to find his own voice, it became clear to all—Russell in particular, early on—that what he had to say was unique. Wittgenstein is in part revered because he did not allow his past states, whether from his personal life or from his earlier philosophical commitments, to govern his future direction or present ideas.

For example, on a personal level, although he was born into one of the wealthiest families in Vienna, he was determined not to allow his material fortune to adversely affect his intellectual objectives. Therefore he simply gave away his inheritance to his siblings and dedicate himself full time to philosophy. Later he even abandoned his teaching post at Cambridge University in order to undertake a Thoreau-like existence, so as to better focus on his thinking (not his writing, necessarily, because he published so very little relative to many modern philosophers). In his thinking and writing, Wittgenstein was no less prepared to prune or destroy the past to serve the present.

Wittgenstein's first book was the *Tractatus Logico-Philosophicus* (translated into English in 1922 from the 1921 German original), in which he develops a Platonic, formal view of language as set of propositions representing states of affairs. As he says ([1922] 1998, 3:3): "Only the proposition has sense; only in the context of a proposition has a name meaning." The form and extent of language are given by logic, according to the early Wittgenstein of the *Tractatus*. Logic (i.e., the a priori) underlies all of language; meaning arises only in the context of referential items (names) linked in a logically well-formed proposition. Anything outside of this—that is, logically ill-formed propositions, or propositions containing names that lack (empirically verifiable) referents (e.g., unicorns, God, Santa Claus, Truth)—are nonsense. This view of language in effect makes language the arbiter of the world (in some exegeses of Wittgenstein, at least). The importance of the notions of truth and the a priori in Wittgenstein's early writings are what lead me to label this phase his "Platonic era." This version of his philosophy led to the school of logical positivism, the school that advocated the construction of a purely logical language for thought, a self-contained language in which the ways it is used are dictated by its form or orthogonal to it.

Later, as his ideas evolved in a very different direction, Wittgenstein criticized the *Tractatus* view of language as "dogmatic"—the idea that there is a unique, true interpretation of every proposition.[1] His antidogmatic view began to permeate his thought, and he grew interested more in the function of language than its form (my interpretation), making what I would refer to as his "Aristotelian shift" from form to usage.

The thesis of his posthumously published book, *Philosophical Investigations* ([1953] 2009), is perhaps best captured by his statement that "for a large class of cases—though not for all—in which we employ the word meaning it can be defined thus: the meaning of a word is its use in the language game. And the 'meaning' of a name is sometimes explained by pointing to its bearer" (43). Thus did the school of "ordinary language philosophy" (Ayer 1940; Austin 1975; Searle 1970—the "Oxford School," etc.) emerge from the

Investigations much as logical positivism arose from the *Tractatus*. The views of the *Investigations* are prefigured in his 1933 *Blue Book* (Wittgenstein [1958] 1965, 4): "If we had to name anything which is the life of the sign, we should have to say that it was its *use*" [emphasis in the original]. Wittgenstein transformed the philosophical study of language from deductive philosophy to inductive science. He insisted that to know the meaning of a sentence, we have to *look* at its usage, not analyze its logical form—a message that some have ignored, many preferring his earlier views to his later conception.

Perhaps the most famous expression from Wittgenstein's later work on language was "language games." He argued that there is no core meaning of any word, but rather a "network" of uses. In other words, language is "languaging," as culture is "culturing." This is what Pike labeled the "dynamic" and "field" perspectives on language, whereas the earlier Wittgenstein took what Pike would have described as a "static" perspective on meaning. Additionally, Wittgenstein took languaging to be a social activity, thus ruling out a "private language." Grammar is an activity, not an instinct, thus leading to the strange statement that "I can know what someone else is thinking, but not what I am thinking" ([1953] 2009, 222). This aphorism makes sense only if language is inherently social and meanings emerge from social use.

Wittgenstein's influence on the study of meaning and language continues robustly through the present century, through philosophers such as Quine (1960), Austin (1975), Searle (1970), Rorty, (1981), Brandom (1998), Grice (1991), and Sperber and Wilson (1995), among others, all of them agreeing with the perspective that language is a tool, or in the words of the philosopher Marcelo Dascal (2002), a "cognitive technology." Some aspects of this work are prefigured in earlier writers, of course, especially the work of the pragmatists (e.g., James 1907; Peirce 1992, 1998). These writings are important for the thesis here that dark matter arises in the individual by means of their interpretations of their experiences, being, and saying in society, as well as through their memory-based construction of a concept of themselves.

This line of reasoning about language is on the one hand orthogonal to the question of whether language is innate. After all, it is possible that the genes provide grammatical and cultural constraints that serve as the parameters within which usage shapes meaning and form. Many philosophers, even those sympathetic to Wittgenstein's view of meaning, find their work compatible with universal grammar certainly and thus would agree with this rough characterization. But, on the other hand, once the genie is let out of the bottle, once we agree that usage can be responsible for meanings and forms, why not probe the extent of this responsibility? How much is left for innate knowledge such as UG to do?

In D. Everett (2012a) I argue that not much at all, perhaps nothing, would be left. Information flow, word order, sentence size, vocabulary, how to code concepts—for instance, either as word affixes as in many American Indian languages, or as words in languages as different as Mandarin and English—help create language forms and can be ascribed economically and perspicaciously to cultural history, values, practices, knowledge structures, and so on, as we saw earlier in our discussion of the emergence of grammar. This reminds me of a quote by Becker (2000, 3) on the challenges of translation: "If you take away grammar and lexicon from a language, what is left? . . . Everything!"

This Bastian hypothesis can be tested only if specific knowledge of grammar, via statements of UG, are clear in a testable manner, along with their falsifiability conditions—and, crucially, only if the biology being appealed to can also be explicated (Lenneberg [1967] is only the barest beginning). I (D. Everett 2012a) and various others have pointed out myriad objections to specific "invariant properties," lack of falsifiability, and so on. Intriguingly, aside from the widespread perspective that proposals of UG lack falsifiability (some defenders argue strenuously that this is false, but I find their arguments unconvincing), there is in fact no analysis claiming to be based on UG that would change if language derived from function or some other non-UG foundation. The analyses lack a causally engaged biology. Generally, biology is not even used in explanations. And so, unless it is causally implicated in specific analyses, UG is an incantation; it lacks any significance whatsoever.

The possibilities for the emergence or origin of current languages, UG notwithstanding, are exhausted by the following:[2]

1. Language similarities are the result of monogenesis—all languages began in Africa, and so they are all daughter languages of the original mother language; hence there will be other physical, external property or process similarities among languages.
2. Language is a priori: it emerges from the genes, the physics of the brain, or some combination of these.
3. Language forms and meanings develop together symbiotically and have a number of relationships, including iconicity (i.e., the more complex a concept, the more complex its linguistic expression). So the preposition *to* is shorter than the preposition *around* because the latter expresses more information.
4. Language is a mathematical system and has no more connection to biology than mathematics. Two plus two equals four whether you are human or vegetable. By this reasoning, phrases are endocentric in all possible worlds.[3]

5. Language, culture, biology, and so on, all impact one another, and thus teasing them apart is hard.
6. We have no idea and we do not care terribly about the "cause" of language—we just want to know how it works.
7. We will never know the answer to any of this without more field research on the seven thousand or so languages in the world, about which we still know relatively little.
8. Independent principles (such as physics, phonetics, and semantics) guarantee a degree of organization without the need to appeal to either innate or cultural knowledge.

None of these possibilities are implausible. None of them entirely excludes the others. To rule out any of these from consideration without strong empirical motivation would be a bad move. Therefore any researcher—Chomsky, myself, any other—must ask themself: "Does my model exclude some of these or others without warrant?" It was the purpose of an earlier paper of mine (D. Everett 2010a) to underscore this: "The Shrinking Chomskyan Corner in Linguistics." Proponents of UG have painted themselves into a corner by ruling out other possibilities without evidence, though this is not a necessary implication of research in UG per se (no more than it would be for any other theory).

For my money, the best hypotheses are numbers 5, 7, and 8. No one in my lifetime will likely know much about 1 or 2. Numbers 4 and 6 together have produced important results, such as the work of Katz (1972) and Postal (2009) on Platonic linguistics. Another example comes from formal models such as head-driven phrase structure grammar and several works by Geoffrey Pullum on formal semantics and syntax—both of these possibilities emerge from mathematical linguistics. Number 7 is where we find most fieldworkers as they confront the most complex task in all of linguistics: figuring out how little-studied languages work.[4] In none of this does UG offer great enlightenment.

If there are any instincts crucial for language, they will bear little resemblance to what Pinker (1995) refers to as "the language instinct." From what we currently know about language(s), the only candidate for an "instinct" is what some (Lee et al. 2009; Joaquim and Schumann 2013) refer to as the "interactional instinct," or what I (D. Everett 2012a and above) referred to as the "social instinct" (see below for a detailed discussion of a purported "phonology instinct").

To paraphrase Chomsky, *it is perfectly safe to attribute knowledge to the genes or instincts, so long as we realize that there is no substance to this assertion.*[5] Or, as Blumberg (2006, 205) puts it, "Nativists and evolutionary psychologists have draped themselves in the blanket of science, but, when all is said and done, they are merely telling bedtime stories for adults."

NO PHONOLOGY INSTINCT

It has been recognized at least since Sapir's 1908 PhD dissertation on Takelma, directed by Franz Boas, that phonology—the study of how speakers organize and perceive their sounds—is an interesting source of insight into human psychology. Sapir's main Takelma teacher, Tony Tillohash, helped Sapir recognize the psychological reality of the phoneme. This work further affected Sapir's sense of the connection between psychology, cognition, and culture, both for the aspects of this relationship that can be seen overtly and via those aspects that are worked into the dark matter of speakers.

Therefore, it is only natural that contemporary phonologists and psychologists also probe native-speaker intuitions and behaviors to discover more of the interesting and profound connections between sound systems and dark matter. One of the best studies I am aware of in recent years is Cutler's (2012) profound research on speech recognition, *Native Listening: Language Experience and the Recognition of Spoken Words*. However, there are a number of phonologists who seem to want to probe even deeper, for the ever-appealing "instincts"—innate content—regarding sound systems that all Homo sapiens are, ex hypothesi, born with. Obviously, since it is a thesis of this book that there is neither need nor convincing evidence for inborn knowledge in Homo sapiens, some of these more detailed and better-argued claims should be addressed here.

Rather than restate the arguments of D. Everett (2012a) against UG, though, let's briefly explore two other nativist proposals on human language meanings and forms. One is the work of Iris Berent (2013a, 2013b) on knowledge of sound systems. The other—addressed in the next section—is Wierzbicka's (1996) theory of universal semantic knowledge, her *natural semantics metalanguage* (NSM). I want to examine these claims for innate knowledge of language before leaving this part of our discussion, because where Chomsky's work centers on syntactic knowledge, these proposals on phonological knowledge and semantic knowledge cover the remainder of language from a nativist perspective.

First, let's take Berent's argument. Though the comments that follow are mostly negative, Berent's research is worth the relatively large space below that is dedicated to rebutting it. This is because it is one of the best-articulated arguments for linguistic nativism, with a large amount of experimentation and a detailed, painstakingly built case. Even if my criticisms are all on the mark, her work is worth reading and a milestone in the history of the nativist program. Arguably there are no arguments for nativist syntax as detailed and carefully laid out as her arguments for nativist phonology. Berent's claim is that phonotactics (the organization of segments into syllables, seen in the fact, e.g., that [bli] is a possible syllable of English while [lbi] is not) is an "instinct," a grammaticalization[6] of a functional principle of sound organization that has entered universal grammar and in which the original functional principles are no longer directly relevant for the resultant instinct.[7]

In Berent's (2013a) *The Phonological Mind*, we find a sustained argument on behalf of the proposition that there is innate phonological knowledge, centering around preferences for sounds and sound sequences and signs and sign sequences in spoken and signed languages. My criticisms here are limited to a small portion of Berent's monograph, in particular those she focuses on in Berent (2013b). As we see, there are many serious problems with Berent's concept of a phonological mind, the most important of which is the "origin problem." Where did the phonological knowledge come from? Without an account of the *evolution of an instinct*, proposing such nativist hypotheses are pure speculation. Rather, at best, we can take nonevolutionary evidence for an instinct as *explanada* rather than *explanans*. But other problems are also serious—in particular, the overinterpretation of observations (as "wonders"); the use of a falsified proposal as the basis for an instinct; a failure to conduct phonetic studies of the sound sequences she studies; and a failure to independently study the bases of phonotactics, simply accepting one "principle" as given and conducting experiments that are not only unconvincing but also somewhat circular. Berent concludes that her experimental results from English, Spanish, French, and Korean support her proposal that there is a universal sonority sequencing generalization (SSG) inborn in all Homo sapiens.

To understand her arguments, we must first understand the terms she uses, beginning with "sonority." Sonority is just is the property of one sound being inherently louder than another sound. For example, when the vowel [a] is produced in any language the mouth is open wider than for other vowels, and like other vowels, [a] offers very little impedance to the flow of air out of our lungs and mouths. This makes [a] the loudest sound, relatively speaking, of all phonemes of English. A sound with less inherent loudness (e.g., [k]) is said to be less sonorous. Several of Berent's experiments demonstrate that speakers of all the languages she tested—children and adults—prefer words organized according to the SSG. The idea behind the SSG is that the least loud

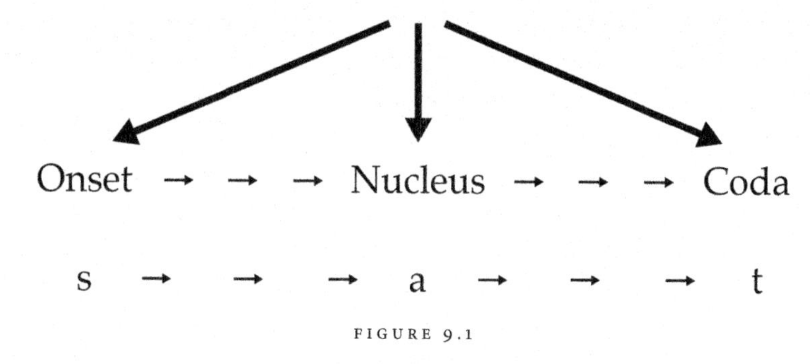

FIGURE 9.1

(sonorous) segments are found at the far edges of syllables, while the loudest segments are found closer to the nucleus of the syllable.

To see what is meant more clearly, consider a single-syllable word (monosyllable) such as *sat*, whose structure would be as shown in figure 9.1. Since [a] is the most sonorous element, it is in the nucleus position. [s] and [t] are at the margins—onset and coda—as they should be. Now take the hypothetical syllables mentioned earlier, [bli] and [lbi]. Both [bli] and [lbi] have what phonologists refer to as "complex onsets" (see figs. 9.2 and 9.3): multiple phonemes in a single onset (the same can happen with codas as with *pant*, in which [n] and [t] form a complex coda). Now, according to the SSG, since [b] is less sonorous than [l], it should come first in the onset. This means that [bli] is a well-formed syllable. In other words, it rises in sonority to the nucleus and falls from the nucleus to the end.

Such preferences emerge even when the speakers' native languages otherwise allow grammatical strings that appear to violate the SSG. Since the SSG is so important to the work on a phonological instinct, we need to take a closer look at it. To make it concrete, let's consider one proposal regarding the so-called sonority hierarchy (as we will see, not only do many phoneticians consider this hierarchy to be a spurious observation, but it is also inadequate to account for many phonotactic generalizations, suggesting that not sonority but some other principle is behind Berent's experimental results).[8] This hierarchy is illustrated below (Selkirk 1984, 116), from most sonorant on left to least on right:

[a] > [e o] > [i u] > [r] > [l] > [m n ŋ] > [z v ð] > [s f θ] > [b d g] > [p t k]

The hierarchy has often been proposed as the basis for the SSG, which might also be thought of as organizing syllables left to right into a crescendo, peak, and decrescendo of sonority, going from the least sonorant (least inherently loud) to the most sonorant (most inherently loud) and back down, in inverse

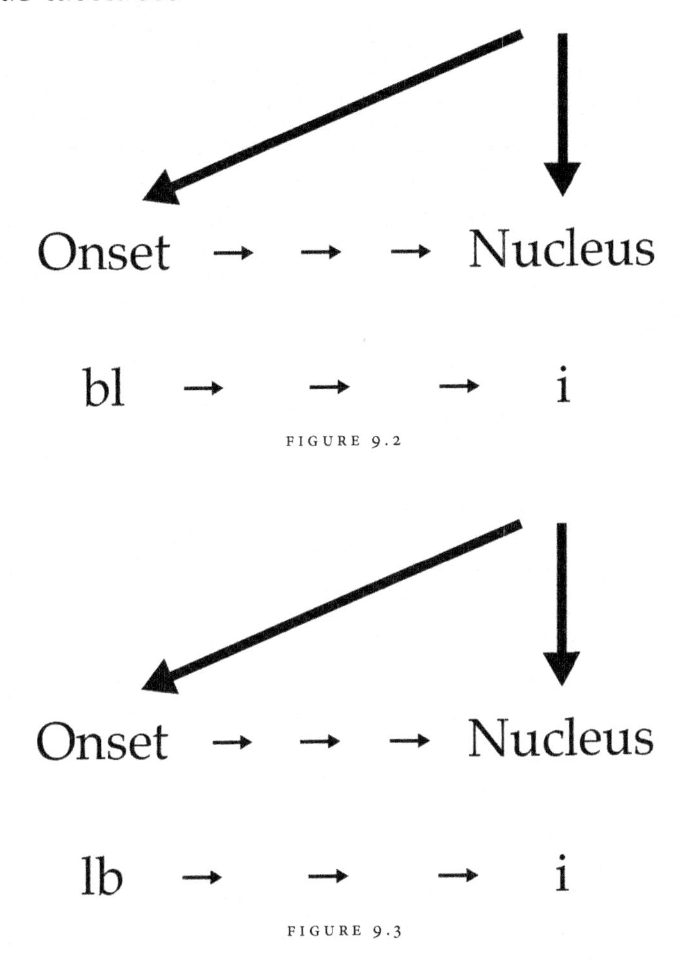

FIGURE 9.2

FIGURE 9.3

order, to the least sonorant (in fact, I was once a proponent of the SSG myself; see D. Everett 1995 for a sustained attempt to demonstrate the efficacy of this hierarchy in organizing Banawá syllable structure).

Without reviewing all of her experimental results (which all roughly show the same thing—preference in subjects for the SSG in some conditions), consider the following evidence that Berent brings to bear:

> The evidence comes from perceptual illusions. Syllables with ill-formed onsets (e.g., lba) tend to be systematically misidentified (e.g., as leba)—the worse formed the syllable, the more likely the misidentification. Thus, misidentification is most likely in lba followed by bda, and is least likely in bna. Crucially, the sensitivity to syllable structure occurs even when such onsets are unattested in participants' languages, and it is evident in adults [64, 67–70, 73] and young children. (2013b, 322)

Again, as we have seen, a licit syllable should build from least sonorant to most sonorant and then back down to least sonorant, across its onset, nucleus, and coda. This means that while [a] is the ideal syllable nucleus for English, a voiceless stop like [p, t, k] would be the least desirable (though in many languages—e.g., Berber—this hierarchy is violated regularly). Thus a syllable like [pap] would respect the hierarchy, but there should be no syllable like [opa] (though of course there is a perfectly fine *bisyllabic* German word *opa* "grandpa"). For the latter word, the SSG would only permit this to be syllabified as two syllables, [o] and [pa], with each vowel its own syllable nucleus. This is because both [o] and [i] are more sonorous than [p], so [p] must be either the coda or the onset of a syllable in which one of these two vowels is the nucleus.[9] Moreover, according to the SSG, a syllable like [psap] should be favored over a syllable like [spap]. This gets us to the obvious question of why "misidentification" by Korean speakers is least likely in *bna* (even though Korean itself lacks such sequences); because, according to Berent, all humans are born with an SSG instinct.

I do not think anything of the kind follows. My criticism of Berent's conclusions take the following form. First, I argue that there is no SSG either phonetically, grammatically, or even functionally. There is simply nothing there to have an instinct of. Second, *even if some other, better (though yet undiscovered) principle than the SSG were appealed to, the arguments for a phonology instinct do not follow*, as seen in my suggested alternative explanations of her results. Third, I offer detailed objections to every conclusion she draws from her work, concluding that there is no such thing as the "phonological mind."

Let's address first the reasons behind the claim that the SSG is not an explanation for phonotactics. The reasons are three: (i) there is no phonetic or functional basis for the generalization; (ii) the SSG that Berent appeals to is too weak—it fails to capture important, near-universal phonotactic generalizations; (iii) the generalization is too strong—it rules out commonly observed patterns in natural languages (e.g., English) that violate it. But then, if the SSG has no empirical basis in phonetics or phonology and is simply a spurious observation, it is unavailable for grammaticalization (i.e., to be incorporated as a grammatical principle) and cannot serve as the basis for the evolution of an instinct (though, of course, some other concept or principle might be; see below). One might reply that if the SSG is unable to explain all phonotactic constraints, that doesn't mean that we should throw it out. Perhaps we can simply supplement the SSG with other principles. But, as we see, why accept a disjointed set of "principles" to account for something that may have an easier account based more solidly in phonetics and perception? Before we can see this, though, let's look at the SSG in more detail.

The ideas of sonority and sonority sequencing have been around for centuries. Ohala (1992) claims that the first reference to a sonority hierarchy was in 1765. Certainly there are references to this in the nineteenth and early twentieth centuries. As Ohala observes, however, references to the SSG as an explanation for syllable structure are circular, descriptively inadequate, and not well integrated with other phonetic and phonological phenomena.

According to Ohala, both the SSG and the syllable itself are theoretical constructs that lack universal acceptance. There is certainly no complete phonetic understanding of either—a fact that facilitates circularity in discussing them. If we take a sequence such as *alba*, most phonologists would argue that the word has two syllables, and that the syllable boundary must fall between /l/ and /b/, because the syllable break a.lba would produce the syllable [a], which is fine, but also the syllable [lba] which violates the SSG ([l] is more sonorous than [b] and thus should be closer to the nucleus than [b]). On the other hand, if the syllable boundary is al.ba, then both syllables respect the SSG: [al] because [a] is a valid nucleus and [l] a valid coda, and [ba] because [b] is a valid onset and [a] is a valid nucleus. The fact that [l] and [b] are in separate syllables by this analysis means that there is no SSG violation, which there is in [a.lba]. Therefore SSG guides the parsing (analysis) of syllables. However, this is severely circular if the sequences parsed by the SSG then are used again as evidence for the SSG.

The SSG is also descriptively inadequate because it is at once too weak and too strong. For example, most languages strongly disprefer sequences such as /ji/, /wu/, and so on, or, as Ohala (1992, 321) puts it, "offglides with lowered F2 and F3 are disfavored after consonants with lowered F2 and F3.[10,11] Ohala's generalization here is vital for phonotactics crosslinguistically; yet it falls outside the SSG, since the SSG allows all such sequences. This means that if a single generalization or principle, of the type Ohala explores in his article, can be found that accounts for the SSG's empirical range plus these other data, it is to be preferred. Moreover, the SSG would then hardly be the basis for an instinct, and Berent's experiments would be merely skirting the edges of the real generalization. As we see, this is indeed what seems to be happening in her work. The SSG simply has no way of allowing a [dw] sequence, as in *dwarf*, or [tw] in *twin*, while prohibiting [bw]. Yet [dw] and [tw] are much more common than [bw], according to Ohala (though this sequence is observed in some loanwords, e.g., *bwana*), facts entirely missed by the SSG.

Unfortunately, Berent neither notices the problem that such sequences raise for the SSG "instinct," nor experimentally tests the SSG based on a firm understanding of the relevant phonetics. Rather, she assumes that since the SSG is "grammaticalized" and now an instinct, the phonetics no longer matter. But this is entirely circular. Here Berent's lack of phonetic experience and background in phonological analysis seem to have led her to accept the SSG based on the work of a few phonologists, without carefully investigating its empirical adequacy. This is not a problem in some senses—after all, her results still show speakers do prefer some sequences and disprefer others—but it is a crucial shortcoming, as we see below, when it comes to imputing these behaviors to "core knowledge" (knowledge that all humans are hypothesized to be born with) that would have to have evolved. It hardly needs mentioning, however, that a spurious observation of a few phonologists is not likely to be an instinct.

To take another obvious problem for the SSG, sequences involving syllable-initial sibilants are common crosslinguistically, even though they violate the SSG. Thus the SSG encounters problems in accounting for English words like *spark* or *start*. Since [t], [k], [p]—the voiceless stops—are not as loud/sonorous as [s], they should come first in the complex onset of the syllable. According to the SSG, that is, [psark] and [tsart], should be grammatical words of English (false), while [spark] and [start] should be ungrammatical (also false). Thus the SSG is too strong (incorrectly prohibits [spark]) and too weak (incorrectly predicts [psark]) to offer an account of English phonotactics. Joining these observations to our earlier ones, we see not only that the SSG allows illicit sequences such as /ji/ while prohibiting perfectly fine sequences such as /sp/, but it simply is not up to the task of English phonotactics more generally. And although many phonologists have noted such exceptions, there is no way to handle them except via ancillary hypotheses (think "epicycles") if the basis of one's theory of phonotactics is the SSG.

I conclude that Berent's phonology instinct cannot be based on the SSG, because the latter doesn't exist. She might claim instead that the instinct she is after is based on a related principle or that the SSG was never intended to account for all of phonotactics, only a smaller subset, and that phonotactics more broadly require a set of principles. Or we might suggest that the principles behind phonotactics are not phonological at all, but phonetics, having to do with relative formant relationships, along the lines adumbrated by Ohala (1992). But while such alternatives might better fit the facts she is invested in, a new principle or set of principles cannot rescue her proposal. This is because the evidence she provides for an instinct fails no matter what principle she might appeal to. To see why, let's consider what Berent (2013b, 320) infelicitously refers to as "the seven wonders of phonology." She takes all of these as evidence for "phonological core knowledge." I see them all as red herrings that merely underscore faulty reasoning than as any evidence for a phonological mind or an instinct of any kind. These "wonders" are:

1. Phonology employs algebraic rules.
2. Phonology shows universal constraints or rules (e.g., the SSG).
3. Phonology shows shared design of all phonological systems.
4. Phonology provides useful scaffolding for other human abilities;
5. Phonology shows regenesis—phonological systems (e.g., sign languages) created de novo always draw on the same principles; they never emerge ex nihilo.
6. Phonological constraints such as the SSG show early ontogenetic onset.
7. Phonology shows a unique design, unlike other cognitive domains.

I believe that every one of these "wonders" is insignificant and tells us nothing about language, offering no evidence whatsoever, individually or together, for a "phonological mind." Let's consider each in turn.

"Algebraic rules" are nothing more than the standard rules that linguists have used since Panini (fourth century BCE). For example, Berent uses an example of such a rule that she refers to as the "AAB rule" in Semitic phonologies. Thus, in Semitic languages, consonants and vowels mark the morphosyntactic functions of words, using different spacings and sequences (internal to the word) of Cs or Vs based on conjugation, or *binyanim*—the order of consonants and intercalated vowels. An example of such variables are illustrated below:

Modern Hebrew

CaCaC katav	"write"
niCCaC niršam	"register"
hiCCiC himšix	"continue"
CiCeC limed	"teach"
hitCaCeC hitlabeš	"get dressed"

In other languages, such functions would most frequently be marked by suffixes, infixes, prefixes, and so on. So, clearly, taking only this single, common example, variables are indeed found in phonological rules.

Now, in Berent's AAB rule (more precisely, it should be stated, as a constraint "*AAB," where * indicates that the sequence *AAB* is ungrammatical) is designed to capture the generalization that the initial consonants of a word cannot be the same. Thus a word like **sisum* would be ungrammatical, because the first two consonants are /s/ and /s/, violating the constraint. The constraint is algebraic because A and B are variables ranging across different phonological features (though A must be a consonant). But calling this an algebraic rule and using this as evidence for an instinct makes little sense. Such rules are regularly learned and operate in almost every are of human cognition. For example, one could adopt a constraint on dining seating arrangements of the type *G_1G_1X,—that is, the first two chairs at a dinner table cannot be occupied by people of the same gender (G), even though between the chairs there could be flower vases and the like. Humans learn to generalize across instances, using variables frequently. Absolutely nothing follows from this regarding instincts.

Universality is appealed to by Berent as further evidence for a phonology instinct. But as any linguist can affirm (especially in light of controversies over how to determine whether something is universal or not in modern linguistic theory), there are many definitions, uses, and abuses of the term *universality* in linguistics. For example, some linguists, such as Joseph Greenberg (1966) and N. Evans and Levinson (2009), argue that for something to be meaningfully universal, it actually has to be observable in every language. That is, a universal is a concrete entity. If it is not found in all languages, it is not universal. That is simple enough. But some linguists, such as Berent, Jackendoff (2003), and Chomsky (1986), prefer a more abstract conception of *universal* such that for something to be universal, it need only be available to human linguistic cognition. This set of universal affordances is referred to as the "toolbox." I have argued against this approach in many places, for being imprecise and often circular (in particular D. Everett 2012a, 2012b). But in any case, Berent clearly follows the notion of "universal" advocated by Chomsky and Jackendoff, among many others. Such universals need not be observed in all languages. Thus Berent would claim that the SSG is universal, not because it is obeyed in all its particulars in every language—like me, she would recognize that English allows violations of the SSG—but because her experiments with speakers of various languages show that they have preferences and so on that seem to be guided by knowledge of the SSG, even when their own native languages do not follow the SSG or have a simple syllable structure that is by definition unable to guide their behavior in experiments. If a Korean speaker, for example, shows preference for or perceptual illusions with some onset clusters and not others—in spite of the fact that there are no such clusters in Korean (and thus he or she could not have learned them, presumably), then this shows the universality of the SSG (as part of the linguistic toolbox). But there is a huge and unjustifiable leap taken in reasoning from this type of behavior to the presence of innate constraints on syllable structure. For example, there are phonetic reasons why Korean (or any) speakers prefer or more easily perceive, let us say, [bna] sequences rather than [lba], even though neither sequence is found in Korean. One simple explanation that comes to mind (and highlighted by phoneticians, though overlooked by many phonologists), is that the sequence [bna] is easier to perceive than [lba] because the interconsonantal transition in the onset of the former syllable produces better acoustic cues than in the second. Berent tries to rule out this kind of interpretation by arguing that the same restrictions show up in reading. But reading performance is irrelevant here for a couple of reasons. First, we know too little about the relationship between speaking and reading cognitively to draw any firm conclusions about similarity or dissimilarities in their performance to use as a comparison, in spite of a growing body of research on this topic. Second, in looking at new words, speakers often try to create the phonology in their heads, and so this "silent pronunciation" could guide such speakers' choices. Everyone (modulo pathology) has roughly the same ears matched to roughly the same vocal apparatus. Thus, although phonologies can grammaticalize violations of functionally preferable phonotactic constraints, one would expect that in experiments that clearly dissociate the experimental data from the speaker's own language, the functionality of the structures (e.g., being auditorily easier to distinguish) will emerge as decisive factors, accounting for speakers' reactions to nonnative sequences that respect or violate sonority sequencing, and so on. In fact, there is a name for this, though with a somewhat different emphasis, in optimality theoretic phonology (Prince and Smolensky [1993] 2004; McCarthy and Prince 1994): the "emergence of the unmarked." So there is nothing special I can see about the universality of these preferences. First, as we have seen, the SSG is not the principle implicated here, because there is no such principle. It is a spurious generalization. Second, local phonologies may build on cultural preferences to produce violations of preferable phonetic sequences, but the hearers are not slaves to these preferences. Let us say that a language has a word like *lbap*. In spite of this, the phonetic prediction would be that in an experimental situation, the speakers would likely prefer *blap* and reject *lbap*, since the former is easier to distinguish clearly in a semantically or pragmatically or culturally neutral environment. In other words, when asked to make judgments in an experiment about abstract sequences, it is unsurprising that the superiority of the functionality of some structures emerges as decisive. Such motivations reflect the fact that the ear and the vocal apparatus evolved together. Therefore, what Berent takes to be a grammatical and cognitive universal is neither, but rather a fact about perceptual ability, showing absolutely nothing about a phonology instinct.

Next, Berent talks about "shared design." This is just the idea that all known phonological systems derive from similar phonological features. But this is not a "wonder" of any sort. There is nothing inherently instinctual in building new phonological systems from the same vocal apparatus and auditory system, using in particular the more phonetically grounded components of segmental sequencing.

Another purported "wonder" is what Berent refers to as "scaffolding." This is nothing more than the idea that our phonologies are reused. They serve double duty—in grammar and as a basis for our reading and writing (and other related skills). This is of course false in much of some writing systems (e.g., Epi-Olmec hieroglyphics, where speaking and writing are based on nearly non-overlapping principles). In fact, "reuse" is expected in cognitive or biological systems to avoid unnecessary duplication of effort. Not only is it a crucial feature of brain functioning (Anderson 2014), but it is common among humans to reuse technology—the use of cutting instruments for a variety of purposes, from opening cans to carving ivory, for instance. Therefore, reuse is a common strategy of cognition, evolution, resource management, and on and on, and is thus orthogonal to the question of instincts.

Next, Berent talks about "regenesis," the appearance of the same (apparently) phonological principles in new languages, in particular when principles of spoken phonology (e.g., the SSG, according to Berent) show up in signed systems. The claim is that the SSG emerges when humans generate a new phonological system de novo. But even here, assuming we can replace the invalid SSG with a valid principle, we must use caution in imputing "principles" to others as innate knowledge. We have just seen, after all, how the particular phonetic preference Berent calls the SSG could occur without instincts.

But even if we take her claims and results and face value, "regenesis" is still a red herring. In spoken languages, the notion simply obscures the larger generalization or set of generalizations that *people always prefer the best-sounding sequences perceptually*, even when cultural effects in their native languages override these. Berent again attempts to counter this with research on sequences of signs in signed languages. Yet there is no sound-based principle in common between signed and spoken languages—by definition, since one lacks sounds altogether and the other lacks signs (see chap. 7). Both will of course find it useful to organize word-internal signs or sounds to maximize their perceptability, but no one has ever successfully demonstrated that signed languages have "phonology" in the same sense as spoken languages. In fact, I have long maintained that, in spite of broadly similar organizational principles, sign organization in visual vs. spoken languages is grounded in entirely different sets of features (for example, where is the correlate of the feature "high tone" or F2 transition in signed languages?), and thus that talking of them both as having "phonologies" is nothing more than misleading metaphor.

Another "wonder" Berent appeals to, to show that phonology is an instinct, is the common poverty of the stimulus argument or what she refers to as "early onset." Children show the operation of sophisticated linguistic behaviors early on—so early, in fact, that a particular researcher might not be able to imagine how it might have been learned, jumping to the conclusion that it must not have been learned but emerges from the child's innate endowment. Yet all Berent shows in discussing early onset is the completely unremarkable fact that children rapidly learn and prefer those sound sequences that their auditory and articulatory apparatuses have together evolved to recognize and produce most easily. This commonality is not linguistic per se. It is physical, like not trying to pick up a ton of bricks with only the strength in one's arms,. or more appropriately, in not using sounds that people cannot hear (e.g., with frequency that only dogs can hear).

Finally, Berent argues for "core phonological knowledge" based on what she terms "unique design." This means that phonology has its own unique properties. But this shows nothing about innate endowment. Burrito making has its own unique features, as does mathematics, both eminently learnable (like phonology). And Berent's discussion fails to explain why these unique features could not have been learned, nor why there would be any evolutionary advantage such that natural selection would favor them.

Summing up to this point, Berent has established neither that speakers are following sonority organization that is embedded in their "core knowledge," nor that her account is superior to more intuitively plausible phonetic principles. Nor are any of her "seven wonders of phonology" remotely wondrous.

And yet, in spite of all of my objections up to this point, there is a far more serious obstacle to accepting the idea of a phonological mind. This is what Blumberg (2006) refers to as the problem of "origins," which we have mentioned and which is discussed at length in several recent books (Buller 2006; R. Richardson 2007; Blumberg 2006, among others)—an obstacle Berent ignores entirely, and an all too common omission from proponents of behavioral nativism. Put another way, how could this core knowledge have evolved? More seriously, relative to the SSG, how could an instinct based on any related principle have evolved? As we have seen, to answer the origins problem, Berent would need to explain (as Tinbergen [1963], among others, discusses at length) the survival pressures, population pressures, environment, and so on, at the time of the evolution of a valid phonotactic constraint; if the trait appears as a mutation in one mind, what leads to its genetic spread to others in a population—what was its fitness advantage? In fact, the question doesn't even make sense regarding the SSG, since there is no such principle. But even if a better-justified generalization could be found, coming up with any plausible story of the origin of the principle is a huge challenge, as are definitions of *innate* and *instinct*, and the entire line of reasoning—as we saw in chapter 1—based on innate knowledge, inborn dark matter.

I therefore reject Berent's proposals for a phonological mind. Moreover, I believe that such proposals underscore the problem of psychology (Geertz 1973) that experiments, however clear their results, are no more useful than the quality of their foundational hypotheses. As a psychologist rather than a linguist, Berent seems to have muddled the questions.

NO SEMANTIC INSTINCT

Similar issues face other nativist accounts of language, such as the *natural semantic metalanguage* (NSM) of Anna Wierzbicka (1996) of the Australian National University and her followers. The idea of NSM is that there is a universal set of concepts that are found in all human languages. These are so-called "irreducible" semantic primitives. The proposed list includes things like:

Substantives: *I*, *you*, *someone*, etc.
Relationals: *kind*, *part*, etc.
Determiners: *this*, *the same*, etc.
Evaluators: *good*, *bad*, etc.
Quantifiers: *one*, *two*, *all*, *each*, *every*, etc.
Logical concepts: *not*, *maybe*, *if*, *can*, etc.

Wierzbicka believes (John Colapinto, pers. comm., 2007) that such primitives support the psychic unity of man, thus explicitly connecting (however unknowingly) her research program to Bastian's. But, aside from her Bastian bias, what is the empirical support for NSM? And where might these primitives sources derive from? Looking into such questions, as we saw with Berent's theory, Blumberg's (2006, 33) observation comes to mind: "When we ask questions about origins, we defeat designer-thinking." In other words, if we asked a proponent of instincts, "Where did the instinct come from?" or "Exactly how did the instinct arise in evolutionary terms?" we would likely be told, "No one knows." (And how well did an author consider alternatives to nativism before positing an innate principle?)

Before discussing this issue, however, we should note that the empirical evidence for NSM is weak. For example, in D. Everett (2005a, 2012a, 2012b) it is made clear that Pirahã data are counterexamples for many of these NSM primitives. Pirahã lacks numbers, quantifiers, Boolean operators (conjunctions, disjunctions, conditional markers, etc.), many of the substantives in the NSM list, and others (see also C. Everett and Madora 2012 and M. Frank et al. 2008 for follow-up corroborating studies of number).

Second, setting aside the empirical weakness of the theory, Wierzbicka, like Berent, offers no account of the sources of the primitives the theory proposes. This is a severe weakness. For example, if one wanted to claim that NSM primitives are innate, one would have to argue why a nativist account is warranted when the primitives would otherwise be of obvious utility in communication. Because of their utility, these "primitives" would appear independently in many, if not most, languages; however, they are implemented in particular languages. For example, numbers and quantifiers are so useful that some would argue (erroneously) that without them, there can be no human language (Davidson 1967). But at least the utility of numbers and quantifiers make it completely unsurprising to learn that these categories are found in many or most of the world's languages. Therefore the desire to establish them as a priori, universal knowledge is unwarranted, especially since the Pirahã data show that they are not all always useful in every culture—hardly a surprising fact to the anthropologist.

Moreover, while the work of Berent and Wierzbicka is undoubtedly correct in many areas, they both state their problem roughly in the following way: "Here are some (near) universal facts about aspects of language. These also show up with infants and in experiments. We see no way for them to be learned. Thus they are innate dark matter."

Before explaining my objections to nativist accounts more generally, however, I want to consider a final example of a nativist "module" of the mind—namely, the proposal that there is a moral instinct or innate basis for morality (apart from emotions and survival or other biological values in the sense defined earlier).

The idea of innate morality emerges in the work of Paul Bloom (2013), Marc Hauser (2006), and others. Both Bloom and Hauser are outspoken proponents for nativist epistemologies. Bloom even goes so far as to claim that the greatest misconception people have about morality is that "morality is a human invention" (S. Harris 2013). To support such claims, Bloom turns to research with human infants, claiming that "the powerful capacities that we and other researchers find in babies are strong evidence for the contribution of biology." He argues that "moral capacities are not acquired through learning" (ibid.). Bloom and his colleagues' research is highlighted in the media and is outlined in his popular book, *Just Babies: The Origins of Good and Evil* (2013), as well as in numerous articles for peer-reviewed scientific journals. The reasons behind the public appeal of Bloom's work seem to boil down to three: (i) designer bias; (ii) Ivy League bias; and (iii) simple answers for complex questions.[12]

THE CULTURAL APPEAL OF INSTINCTS

Designer bias we discussed earlier as the strong appeal to the general public and to many scientists of the notion that humans are the way they are for reasons beyond their control—because our genes tightly constrain us. There are hugely popular books on art instinct, religious instinct, moral instinct, language instinct, and so on. And these works are popular because they strike a chord with the general public. Many people—scientists and laypeople alike—reasonably prefer a simpler story over a messier story. Instincts are simple and clear to state, to try to understand. But the idea that human knowledge, culture, identity, and selves may result from a combination of general properties of the brain interacting with the world in numerous ways is just too complicated to have the same appeal. Instincts thus play a simplifying role. In this sense, they are similar to the idea that pyramids in Mexico and Egypt were both designed by space aliens.

The "Ivy League" or status bias, on the other hand, is helpful when issues are complex or controversial. This bias provides a simpler path to deciding who is right—just assume that the person at the most prestigious university is correct. This kind of bias is fairly strong in US culture, as well as in many other highly literate societies. This allows us to avoid weighing issues based on reasoning, providing a much quicker "emulate the famous or authority figure" alternative. This is much like what we do when we choose clothes based on what famous people wear, imitate the mannerisms of a well-known personality, or wear sports jerseys with the numbers of our heroes/sports favorites. Vicarious thinking takes less effort than actual thinking. By and large, as Boyd and Richerson (2005) argue, status bias is a rational, low-risk, effort-saving strategy.

There is nothing wrong with such biases per se. Taking "so-and-so's word for it" can save time, especially when the individual in question is eminent in his or her field. And, after all, Harvard, Yale, Princeton, and the other Ivy League schools are at the top of the university rankings because their faculty are homogenously outstanding. This is one reason people of certain cultures come by such biases. The idea that instincts override culture is often (though by no means always) associated with names from Ivy League schools, thus supporting the idea of such instincts among the public, along with the designer bias and the simple-vs.-complexity bias.

The other bias—one that has a long, respectable philosophical tradition behind it—is that when faced with two explanations of the same phenomenon, prefer the simpler one. In actual scientific practice, this doesn't mean "choose the conceptually simper theory," necessarily, but rather, choose the *shorter*—in strings of symbols—answer. Popularly applied, however, for the general public this bias means "choose the easier argument to follow, the argument with fewer variables, etc." This helps to explain part of the popular appeal of these ideas. But pointing it out is not a criticism of the theory, merely underscoring a widespread cultural value that contributes to the theory's popularity.

No Morality Instinct

There are criticisms to make, however, of innate ideas, especially with regard to morality (Lieberman [2013] and Patricia Churchland [2012] being two of the most incisive). Let's begin by evaluating what I call the "Monomorality idea"—the notion that morality is innate—by taking a look at this theory's methodology. My criticisms apply not only to Bloom's work, but also to Hauser's (2006) *Moral Minds* and most other research on the epistemology of babies.

Working with babies is very difficult. Babies cannot talk. They cannot move well. They are subject to the "Clever Hans" effect—that is, picking up subtle cues from the experimenter about what response is desired. And, most crucially and surprisingly, most studies of baby cognition are based on a single technique with variations. As Blumberg (2006, 167) says, "Amazingly, virtually every claim by nativists regarding the remarkable, unexpected competencies and core knowledge of infants is based on experiments using the eyes as portals to the mind. Yet few people are aware of how the resurgence of nativism rests on little more than a bold decision to play fast and loose with an experimental procedure that has been used since the 1960s to answer legitimate questions about infant perception."

Here's how the method works (and virtually every experimenter on infant cognition employs this method): The idea is that infants shift focus to new things from familiar things, by looking away from the latter toward the former. So an infant looking at *x* for a period of time will pay less and less attention to it the longer they are exposed to it, whether *x* is a color, person, event, and so on. If *y* is then shown, the infant will shift its gaze to *y* from *x*. This is the infant's "novelty preference." One can show how problematic this is fairly easily. For example, enter a room where a baby is sitting with its mother, fire six very loud blanks at the mother, and have the mother fall to the ground screaming, then faking death. The infant would most likely shift attention and stare! Does this mean there is a "do not murder" preference? (Ethics aside, of course.)

This methodology does indeed unambiguously show infants' perceptual ability to distinguish *x*'s from *y*'s. So if *x* is "green" and *y* is "red," and if the infants in our opinion have had no chance to learn the difference between green and red, are we not then justified in concluding that these colors are distinguished innately by the child? As Blumberg (2005, 168) puts it, yes, *so long as we are unconcerned with either parsimony or falsifiability*. Clearfield and Mix (2001) show that such experiments in fact miss numerous variables. One is whether the infant is acting upon information that is different than the information the researcher is thinking about. One principle example that Clearfield and Mix raise is that in numerical studies, there are times when infants focus on the length of what they are seeing as opposed to the specific numbers or amounts that they are seeing. They argue that in fact, the children are attending to different stimuli than contemplated by the experiments, and thus that these experiments are flawed. When a large part of an entire field rests on the idea that it can perfectly interpret infant gazes to understand infant morality, numerical cognition, and so on, the conclusions so derived are questionable at best.

Beyond the methodology, there are other problems. I have been arguing that there are various sets of values, ranked in different ways, across different societies and across individuals in the same society. Bloom's work gets at one or two of these based largely on emotions—what I refer to earlier as "biologically based" values. For example, there simply is not a sufficient number of cross-linguistic studies of morality to support the conclusions. In some societies (D. Richardson 2005), dishonesty is claimed to be valued over honesty in many situations. In other societies, theft is generally not a problem. In other societies, marital infidelity is insignificant. Thus while some taboos, such as incest, may be universal, the variation across societies and within others is sufficient to render any nativist account anemic and practically of little value. Consider, for example, what it means to be "bad." Say that this were an innate concept. What would it mean in Wittgenstein's sense of meaning as use? How would one come to understand the nuances of the meanings entailed without understanding the contexts in which the concept is used? Or perhaps what is innate is a small schema that includes food theft as the model where every extension must be learned. Under what evolutionary scenario could that have evolved? That is, what were the populations like, the availability of food, the nature of theft at the time of evolution that could account for this as a behavior of twenty-first-century infants? And where would this be encoded in the brain?

How does one know that five-month-old infants or even one-month-old infants have not experienced and disliked selfishness or the withholding of something desired from them—such as their mother's breast? Why appeal to a vague nativism until experience has been ruled out? The dark matter theory of mind maintains that the sources of our knowledge are apperceptions, reasoning in a general sense (though likely Bayesian), and culture. Until these three broad possible sources of knowledge have been ruled out, until the semantics of the moral judgment have been sorted out, until the evolutionary story has been properly worked out, claiming that infant glances in one direction or another are evidence of some sort of innate knowledge is unconvincing.

The proponents of semantic universals, innate phonological knowledge, and moral instincts—like other nativist approaches to human epistemology—all fail to systematically evaluate alternatives to their nativist accounts. More important, their theories crash upon three massive rocks: Occam's razor (they are unnecessary unless proven to be so); empirical results (they are wrong or untested); and evolutionary theory (they are just-so stories in the absence of serious evolutionary origin accounts).

With regard to the first problem, Occam's razor, or "do not multiply entities beyond necessity" (a summary of the views of William of Ockham, first presented with this name in 1852 by Sir William Hamilton), the problem is clear. Everyone agrees that we need learning apart from instinct—learning to find square roots, learning to like fine whiskey, learning to raise cattle, and so on. If learning can be extended to all human knowledge, that is, if every kind of knowledge could be learned (including grammatical structures, skills, and concepts) then we would obviously not need instincts. Instincts make sense only if we have clearly shown that an animal brings *unlearnable* knowledge or skills from their genome. If it cannot be clearly and uncontroversially demonstrated that learning of a particular skill or other form of knowledge is impossible, then proposing an instinct is proposing an unnecessary entity, a violation of Occam's razor. And yet arguments for a "poverty of stimulus" or instincts, especially barring origin arguments, are nothing more than arguments from silence. They raise intriguing problems but settle nothing. Those who find a more compelling case for learning as opposed to knowledge-in-the-genes approaches will perhaps be forgiven for being underwhelmed by nativism for this reason alone.

But this is not the only reason for rejecting nativism. Beyond Occam's blade, proposals for nativist epistemologies are unconvincing empirically. There are never exceptionless cases, for example. If you are claiming that some knowledge is innate and yet we find evidence to the contrary, then this is an empirical problem. D. Everett (2012a) is chockablock with counterexamples to nativist claims about language. Of course, there are possible answers for any perceived empirical shortcoming raised. The question any such answer raises, though, is how many ancillary hypotheses are required. We have seen some of the empirical problems for nativist linguistics and morality above. The literature is full of more (e.g., Buller 2006; R. Richardson 2007; Gopnik 2010; V. Evans 2014; Blumberg 2006; Lieberman 2013). The works just mentioned all raise many more varied and detailed empirical problems for a wide range of nativist accounts.

But even if Occam's razor and empirical shortcomings were not problems for any nativist account of human knowledge, the evolutionary question, the question of origins, is fatal. I return to this directly, following a more detailed discussion of Chomsky's universal grammar.

No Universal Grammar

I want to summarize our discussion to this point on universal grammar before moving on. It is interesting that in sixty-plus years of generative linguistics, not a single analysis ever proposed causally implicates UG. By this I do not mean that there has been no analysis where the linguist assumes that what he or she is proposing is universal because it is part of UG. Rather, I refer to an analysis of a phenomenon that could not be proposed without assuming that grammar is innate. There are many theories of what UG would look like, but no predictions and no analysis of any syntactic structure in which UG is crucial.[13]

Before beginning a detailed criticism of universal grammar, however, we should note that there are several definitions of it. Here is one from Chomsky in an e-mail to me on April 8, 2006: "UG is the true theory of the genetic component that underlies acquisition and use of language." This seems innocuous enough. In fact, it could easily be translated as meaning that the cerebral cortex itself is universal grammar. But more than this is meant, if one searches for other definitions. For example, in an interview with Wiktor Osiatyński, Chomsky offers a less biological and more epistemological, or Bastian, view:

> I think the most important work that is going on has to do with the search for very general and abstract features of what is sometimes called universal grammar: general properties of language that reflect a kind of biological necessity rather than logical necessity; that is, properties of language that are not logically necessary for such a system but which are essential invariant properties of human language and are known without learning. We know these properties but we don't learn them. We simply use our knowledge of these properties as the basis for learning. (Chomsky 1984, 96)

In fact, however, the concepts of "innate" and "instinct" are both imprecise and overused with little scientific or philosophical rigor. Philosopher Matteo

Mameli, in a series of papers, has revealed the widespread imprecision in these terms, as have Buller (2006), R. Richardson (2007), and Blumberg (2006), among others, from a more empirical perspective. I want to examine their criticisms here and then explain why I think that instinct-based accounts of human behavior are inferior to the dark matter account I have been developing until this point.

Problems with Instincts: Definitions and Origins

Instinct is a difficult concept to pin down. Animals seem to be born able to do things that they had no opportunity to learn—baby turtles finding the sea; dogs pointing or herding; ducks imprinting on their mothers; human infants' production of most speech sounds in the worlds' languages early in their lives, followed by a great winnowing process as they home in on their mothers' speech sounds; and so on.

Often it is believed that animals have instincts in far greater numbers than humans. In Darwin's day (Blumberg 2006, xiii), instincts were "God's way of manifesting rationally intelligent behavior in otherwise irrational animals." And as Blumberg goes on to say, "The term *instinct* is often merely a convenience for referring to complex, species-typical behaviors that seem to mysteriously emerge out of nowhere" (ibid, xiv). In any case, the strongest reason for retiring the terms *instinct* and *innate* from scientific thought is the devil of the details, which shows them to be unnecessary at best.

But there is also a definitional problem—that is, how to express with clarity what the term *innate* actually means operationally. For example, here is a partial list of possible definitions of *innate*:

> A trait is innate if it is not acquired; a trait is innate if it is present at birth; a trait is innate if it reliably appears during a particular, well-defined stage of life; a trait is innate if it is genetically determined; a trait is innate if it is genetically influenced; a trait is innate if it is genetically encoded; a trait is innate if its development doesn't involve the extraction of information from the environment; a trait is innate if it is not environmentally induced; a trait is innate if it is not possible to produce an alternative trait by means of environmental manipulations; a trait is innate if all environmental manipulations capable of producing an alternative trait are abnormal; a trait is innate if all environmental manipulations capable of producing an alternative trait are statistically abnormal; a trait is innate if all environmental manipulations capable of producing an alternative trait are evolutionarily abnormal; a trait is innate if it is highly heritable; a trait is innate if it is not learned; a trait is innate if (i) the trait is psychologically primitive and (ii) the trait results from normal

> development; a trait is innate if it is generatively entrenched in the design of an adaptive feature; a trait is innate if it is environmentally canalized, in the sense that it is insensitive to some range of environmental variation; a trait is innate if it is species-typical; a trait is innate if it is pre-functional; etc. (Mameli 2008; see also Mameli and Bateson 2006)

Mameli and Bateson (2006) show in painstaking detail that *all* of these definitions are inadequate.

But let's suppose that, in spite of this daunting problem, we can find a workable definition of instinct or innate. Even then, however, we would still not be in a position to use these terms. The reason is that we cannot attribute something to the human genotype convincingly without some evolutionary account of how it might have gotten there. And such an account would have to offer a scenario by which the trait could have been selected. To do this we would need information about the extent and character of variation in ancestral forms as well as differential survivorship and reproduction of those forms. To know how something was selected, however, we need to know something about the ecology under which the selection took place, such as an answer to questions like what were/are the ecological factors that explain the innate trait, either in the biological or social or other abiotic environment? Next, to use instinct or innate we would need to know how the traits could be passed on to subsequent generations. There should be a correlation between phenotypic traits of parents and offspring greater than chance. Then we would need to know about the population during the time of selection. Any evolutionary biologist also knows that we must have information concerning population pressures, gene flow, and the environment leading to the diffusion of the trait. We do not know the answers to these questions. We are in no position at present to know the answers. And we will *never* be able to know some of the answers. Therefore, there simply is very little utility to the terms instinct and innate.

Alison Gopnik, a professor of psychology at the University of California at Berkeley whose research highlights the child's learning abilities, as opposed to native instincts, offered a similar assessment of nativism:

> It's commonplace, in both scientific and popular writing to talk about innate human traits, "hard-wired" behaviors or "genes for" everything from alcoholism to intelligence. Sometimes these traits are supposed to be general features of human cognition—sometimes they are supposed to be individual features of particular people. The nature/nurture distinction continues to dominate thinking about development. But its time for innateness to go. (Gopnik 2014)

She offers an apt illustration:

> [Researchers] took two different but genetically identical strains of mice which normally develop different degrees of intelligence and cross-fostered them—the smart mice mothers raised the dumb mice pups. The result was that the dumb mice developed problem-solving abilities similar to those of the smart ones and this was even passed on to the next generation. So were the mice innately dumb or innately smart? The very question doesn't make sense. (ibid.)

And finally:

> The increasingly influential Bayesian models of human learning, models that have come to dominate recent accounts of human cognition, also challenge the idea of innateness in a different way. At least since Chomsky, there have been debates about whether we have innate knowledge. The Bayesian picture characterizes knowledge in terms of a set of potential hypotheses about the world. We initially believe that some hypotheses are less probable and others are more so. As we collect new evidence we can rationally update the probability of these hypotheses. We can discard what initially looked very likely and eventually accepting ideas that started out as longshots. (ibid.)

To sum up, the evidence against evolutionary psychology and massive modularity includes not only our earlier points, but also the following:

1. Evolutionary psychology is incompatible with cerebral and mental plasticity.
2. EP confuses arguments between innatism and modularity—that if something is innate, it is an argument for modularity or highly specific epistemological cognition, as opposed to very general perceptual or emotional biases.
3. EP fails to offer any serious account for profound culturally influenced variability in so-called module-directed behaviors (or even much field research to check cross-cultural variability), such as in non-Indo-European cultures with little contact with the outside world.
4. There is no physical portion of the brain dedicated to any module—not even of the five senses—that cannot be used for something else instead.[14]
5. Most of EP's thesis of "massive modularity" is based on the reasoning used in arguments for a language instinct and linguistic module, as argued by Chomsky, Fodor, Pinker, and others. But as I show in D. Everett (2012a) and as we saw above, there is no compelling evidence for any such instinct or module. There is not a single component of language that cannot be explained by either (i) the nature of the communicative task; (ii) the effect of culture; or (iii) general biological constraints on humans (see also Dascal 2012; Enfield 2013a; D. Everett 2013a).

People often raise the issue of "solution space" as an argument for UG and other innate knowledge. If UG is correct, then the task of the child faced with learning its native language is considerably easier. If a child has to search at random for generalizations, the story goes, without innate guidance, it might never learn its first language. Of course, though solution spaces are important, so is motivation. UG has nothing to say about motivation, assuming that all children learn their first language equally well, which is not at all clear (D. Everett 2012a). But the solution space could emerge from images (see V. Evans 2014 among many other works on this), from cultural meanings, from mathematical properties of language, from very basic, easily acquirable input for a Bayesian general learning mechanism common to many tasks humans face, and so on. Nativism by no means is the only perspective that can explain child knowledge acquisition. Goldsmith (2015) offers a detailed assessment of solution space needs and comes down on the side of what he calls "the new empiricism" rather than nativism—that solution spaces need not be hardwired.

Having discussed some of the problems with proposals of innate knowledge in different domains, I want to discuss what would be required to argue for instincts from an evolutionary perspective. Many of the points discussed below are from R. Richardson (2007). They are essential to any theory of innate knowledge, for if there is no potential explanation of how traits could have evolved, proposals that such traits are innate are weakened.

1. *Selection:* We need information about the extent and character of variation in ancestral forms, as well as differential survivorship and reproduction of those forms.
2. *Ecology:* What are the ecological factors that explain the innate trait, either in the biological or social or other abiotic environment?
3. *Heritability:* The mechanisms for genetic transmission must be understood clearly. There should be a correlation between phenotypic traits of parents and offspring greater than chance.
4. *Population structure:* There must be information concerning population structure, gene flow, and the environment leading to the diffusion of the trait.
5. *Trait polarity:* Which traits are primitive and which are derived? As is well known, adaptive features are themselves not necessarily adaptations (e.g., the usefulness of skull sutures for human parturition—these did not evolve for that, however useful).

Other questions that require answers before accepting instincts or nativistic accounts of traits anywhere in nature include the following (also from R. Richardson 2007):

6. How often and in what circumstances is a species-typical trait developmentally and post-developmentally environmentally canalized (insensitive to environment)?
7. How often and in what circumstances is a species-typical trait generatively entrenched (resisting disturbance because it is closely connected to other traits)?
8. How often and in what circumstances is a species-typical trait the result of stable developmental sequences?
9. How often and in what circumstances are developmental and post-developmental environmental canalization both present?
10. How often and in what circumstances does natural selection result in developmental or post-developmental canalization?
11. How often and in what circumstances does natural selection result in generative entrenchment?
12. How often and in what circumstances does natural selection result in stable developmental sequences?
13. How often and in what circumstances does natural selection result in adaptive plasticity?
14. How often and in what circumstances can the mechanisms responsible for adaptive plasticity result in developmental or post-developmental canalization?
15. How often and in what circumstances can the mechanisms responsible for adaptive plasticity result in generative entrenchment?
16. How often and in what circumstances can the mechanisms responsible for adaptive plasticity result in stable developmental sequences?
17. How often and in what circumstances is a species-typical trait due to mechanisms for adaptive plasticity?
18. How often and in what circumstances are stable developmental sequences produced by evolved canalizing mechanisms (as opposed to, say, developmental constraints)?

There is no attempt to answer these or any of the other questions raised in this chapter in any account of nativist knowledge. Therefore, instincts add nothing to our attempts to understand how humans come by their unspoken or ineffable knowledge. We must move beyond instincts to an empirical epistemology, based on culture and the growth of dark matter of the mind. Once we do abandon nativists "bed-time stories," as Blumberg (2006, 205) calls them, we are able to move beyond yet another obstacle to the understanding of our species: the idea of "human nature."

Before concluding this essay on dark matter, therefore, I want to say in the next chapter why it is time to move on, beyond the idea of human nature, leaving such an idea to religions and superstitions, where it belongs.

In this chapter we considered and rejected numerous claims for instincts, from phonology and semantics to morality. To be sure, the proposed instincts we addressed require much more experimentation and research to conclusively demonstrate that the noninstinctual alternatives are superior across the board, but the very fact that we are able to account for the core proposals even moderately well should raise concerns that the instincts are unneeded, based on parsimony alone.

There is of course a huge body of work in favor of instincts. But only a couple of research programs seem to escape the kinds of criticisms that have been raised in this chapter. Those are the work by Charles Yang (2015), Lisa Pearl (2013), and their associates. Unsurprisingly, these researchers, while clearly favoring inborn abilities, do not quite arrive at claims for innate conceptual knowledge. Each in effect makes the reasonable case that the child learning its grammar and the like needs a solution space—it needs to be able search for answers or solutions to the problems of analysis it faces (e.g., settling on a grammar for the language it is learning) in a feasible space that will enable it to find the answers in reasonable time (in particular, in the time we actually observe among young language learners).

Having said that, however, there are nonnativist proposals on the solution space problem, among other issues raised by nativists. For example, the competition model of Brian MacWhinney (2004) and the new empiricism of Goldsmith (2015) provide promising avenues of research.

Thus, although intriguing proposals have been and continue to be made for inborn conceptual knowledge—"instincts," according to some popular usage—none of these overcome the joint problems of parsimony, origins, and empirical adequacy.

10

Beyond Human Nature

The very advantage that [great artists and others distinguished by creative gifts] enjoy consists precisely in the permeability of the partition separating the conscious and the unconscious.

CARL JUNG, *Psychology of the Unconscious*

Whether there is such a thing as human nature is an important question. As this chapter makes clear, looked at from the biological perspective or the behavioral perspective, there clearly is a human nature. But this is not the sense of human nature that many people have in mind. Another view of human nature is that there is innate knowledge that all humans are born with that provides our species with a psychic unity. In this chapter, we consider and reject this latter notion of human nature, while maintaining the former.

What Would a Human Nature Be?

Line up a salmon, an eagle, an ape, a random human, and a robot. Could you pick out the human? Probably. So does this means humans have a nature? At one level, obviously we are a species and obviously we are as distinctive as any species, so of course there is a human nature—whatever makes us a species is our nature. If there is any human nature, it is found in our *capacities*—our excellent memories, our higher intelligence, our ability to create symbols and cultures, our statistical learning prowess. But our higher cognitive powers, deriving to a large extent from the cortex, are more general and show no neurophysical or evolutionary evidence—nor even much behavioral evidence—for innate, hardwired specialization. There are many, many attempts to come to grips with the notion of human nature in the literature (Degler 1992; Stevenson 2000; Downes and Machery 2013; Enfield and Levinson 2006).

In Islam, human nature is referred to as *fitrah*. It is the belief that all humans are born in a natural state of submission to Allah, though they may fall away. "I created My servants in the right religion but the devils made them go astray" (*Sahih Muslim*). But Islamic writings also seem to assume that people will go astray, that it is not hard for people to go astray, and so on. Like some versions of Christianity, because Islam believes that Allah is all-knowing and all-seeing (i.e., omniscient), then predestination for damnation or heaven is the lot of all humans.

The Islamic view of human nature thus seems quite different from the Christian view. The Islamic view is that people are born as they should be, while the (evangelical, fundamentalist) Christian view is that they are born flawed and need to be fixed. Some people will find salvation for their souls and direction for their lives as they turn to Jesus, the theology goes. But at core, before Jesus, all humans are rebels against God and evil.

The Christian view of what is often known as the doctrine of the "total depravity of man" is partially constructed from verses like the following, Matthew 19:16–17: "And, behold, one came and said unto him, Good Master, what good thing shall I do, that I may have eternal life? And he said unto him, Why callest thou me good? There is none good but one, that is, God." This is a reflection of the idea that all humans are born as sinners, unable to win God's favor on their own, because of the sin of their proto-parents, Adam and Eve.

The Christian and Islamic views of human nature are compatible in a broad conceptual sense with Cartesian dualism, since they believe that the soul is eternal and capable of moral perfection, while the body is temporal and largely a source of evil. These religious views are not dualism so much in an opposition between mind and body but between spiritual vs. physical existence.

The Hindu view of human nature, on the other hand, is that there are two layers to the self (Kupperman 2012, 1), the *jiva* and the *atman*. The jiva is the "superficial self," the body, the personality, one's social self, and so on. The atman is the deeper self, a spiritual and impersonal self. The Upanishads outline this view of the self as they also prescribe a life that leads to liberation and ultimate satisfaction. "The knowing (Self) is not born, it dies not; it sprang from nothing" (*Second Valli*, 18). Or "The wise knows the self as bodiless within the bodies, as unchanging among changing things, as great and omnipresent, does never grieve" (22). Or "But he who has not first turned away from his wickedness, who is not tranquil, and subdued, or whose mind is not at rest, he can never obtain the Self (even) by knowledge" (24).

In the Buddhist *Dhammapada*, chapter 1, we encounter a religious view that is radically different from that of the three religions just mentioned, as it is different from most other religions. The Buddhist view strikingly reminiscent of the idea of humans that has emerged from our discussion until this point: "All that we are is the result of what we have thought: it is founded on our thoughts, it is made up of our thoughts." Buddhism does away with the eternal self, *atman*, of Hinduism, replacing it with "no-self" or *anatman*. Our identities are based on the *skandhas*: physical, feelings, perceptions, mental constructions/interpretations, and consciousness. Material forms come in six basic forms (the heard, the seen, the smelled, the tasted, the felt, and the thought). Sensations are seeing, hearing, tasting, smelling, feeling, and thinking. Perceptions are versions of these. From these we construct mental formations, including attention, will, determination, confidence, concentration, wisdom, energy, desire, hate, ignorance, conceit, self-illusion, and so on (see Mitchell 2002). We will return to the Buddhist view directly. The religious views, except for Buddhism, are primarily built around the perceived moral nature of humans. Humans are born moral or immoral. Buddhism doesn't participate in this true-or-false quiz, instead proposing the radical idea that humans make themselves. By the end of life, we are all self-evolved people, that is, the result of our cumulative experiences and memories. This is very similar to the Pirahã worldview.

However, religions certainly do not provide the only ideas about human nature. Many scientists offer their own views of human nature. We briefly examined the views of Freud, Jung, and others in chapter 1. More recent and sophisticated attempts to defend the idea of human nature exist, however. E. O. Wilson won the Pulitzer Prize with a book entitled *On Human Nature* (Wilson 1978). According to Wilson (3), any view from religion is suspect because "religions, like other human institutions, evolve so as to enhance the persistence and influence of their practitioners."[1] Wilson recognizes that humans are the output of millions of years of evolution. We are purposeless, aside from the purposes we give ourselves. But we are tightly constrained machines controlled by evolved psychological and emotional mechanisms that give us similar motives—violence, jealousy, self-preservation, desire for monogamy among females, desire to have sex with as many women as possible for men, and so on. Certainly the evolutionary concept of humans as biological machines has to be accepted at some level. For we are indeed the output of some one or another evolutionary process. On the other hand, although Wilson discusses what he believes to be hardwired characteristics of humans in each of his nine chapters, he offers no convincing account of the role of culture nor psychology in forming human nature. Of course, he is not alone in this. Steven Pinker and his fellow evolutionary psychologists also take a strong evolutionary approach, they believe, to human nature.

Before going into more detail, let me emphasize that the question we are trying to answer is whether there is such a thing as "human nature" and, if so, what kind of thing this is. Human nature has exercised scientists, philosophers, theologians, and others, likely as long as humans have reflected on the world. Yet ideas about it have only occasionally been based on anything we might today call science. E. O. Wilson must be credited with one of the first scientifically based claims that human nature is based on evolution emerged when he published *Sociobiology: The New Synthesis* (Wilson 1975). Like all of Wilson's work, the proposals in this book were based on painstaking research over many years. Wilson argued that just as genes are responsible for much of observed behaviors across nonhuman animals, so we might naturally expect them to be responsible for much of human behavior. According to sociobiology, human nature exists and is largely a function of the human genome. The idea that humans' and other animals' behavior are influenced by their evolution is such a natural and obvious idea that one struggles to understand the vitriol it engendered. Nevertheless, Wilson famously had water poured on his head at one lecture, received hateful messages, and became an object of scorn for many—both in and outside the academy—because of this, to me innocuous, hypothesis.

Human Nature and Evolution

I agree to some degree with Wilson's claim that because humans are animals, evolution is important to our biological makeup, much as it is for any other animal. But as we have seen throughout this study, biological evolution is just one force operating on human thought and values. Cultural evolution is crucial. And Wilson misses this entirely. To a large degree, Darwin ([1874] 1998) made Wilson's point more than a hundred years before Wilson, but also missed the contribution of society and culture to human nature. On the other hand, it is incontrovertible that a great deal of the human psyche is shared with other animals, especially our deeply influential emotional makeup (see Panksepp and Biven 2012, among many others).

Controversy always ensues when a bold proposal places some human behavior or predisposition closer to one side or the other of the nature-nurture continuum, as sociobiology and its offshoot, evolutionary psychology, do—especially when the proposal challenges popular notions of "free will," "dignity," and so on. A large number of people—for religious, social, or philosophical reasons—just do not want to believe that humans lack "total freedom" or that some of our most important choices in life are influenced by the environment in which our minds evolved more than one hundred thousand years ago. And yet simultaneously, another large group of people seem to find it appealing to think we can trace our behaviors back to our genes.

They believe that this would somehow explain us better than the complex, nuanced, and often imprecise notions of cultural shaping of human personalities, as advocated by behaviorists, some anthropologists, philosophers, and others. Moreover, there may even be some who believe that if "my genes made me do it," then we cannot really be blamed for our actions (just as there some who believe the same if "culture made me do it").

Yet in spite of its inherent scientific appeal, I have several objections to Wilson's sociobiology. First, as we have seen, the very idea of "instincts" is at best weakly supported as they are normally presented, as is the reasoning that is often used to establish these. In the hands of some major proponents of either sociobiology or evolutionary psychology, the reasoning that humans must have instincts takes the form of a crude syllogism: "All animals have instincts. Humans are animals. Therefore humans have instincts." One could just as easily argue in a very different direction: "All animals lack instincts for higher cognitive functions. Humans are animals. Therefore humans lack instincts for higher cognitive functions." Or, more weakly, "Humans seem to lack instincts. Humans are animals. Therefore perhaps all animals lack instincts." Contrary to the opinion of some, none of these premises or arguments have overwhelming support. There is room for a great deal of empirical work on instincts (or their absence) across all species.

However, the idea of instincts per se is not my objection to many of the models that propose them, such as sociobiology, evolutionary psychology, and their ilk. In fact, I believe that there is evidence that humans and other animals *may be* born with some instincts (candidates include grasping, breathing, making sounds—all non-epistemic). Rather, my objection is that the vast majority of research on human instincts *is looking in the wrong place.* It is looking for knowledge instead of more basic capacities, such as emotions. Research by Panksepp and Biven (2012), among many others, makes the case that our emotions are instinctual, comprising a small set of basic human drives or needs that have evolved over hundreds of thousands of years. Evidence for the evolutionary antiquity of emotions come from cross-species comparisons, locations of emotions in the brains of Homo sapiens and many other species, unconditioned responses directly observable via electrode probes, and so on. Our emotional responses to the world around us are shared across a large swath of the animal kingdom and heavily influence our cortical, intellectual, social, and cultural development. As one example of the physiology-emotion connection, let's consider the relationship between blood pressure and the emotion of rage.

For example, according to Panksepp and Biven:[2]

> Much of the intermingling of emotional feelings and physiological arousal could be because the primary-process emotional systems are situated in the same brain regions that regulate the activities of our viscera, our hormonal secretions, and our capacities for attention and action . . . Blood pressure . . . exerts influences on affect, as any chemical agent that raises blood pressure will make an angered person or animal feel more enraged. (2012, 31ff)

If physiology affects our emotions, then there is no a priori reason to expect that it could not affect our cognition. However, the interaction of physiology, emotions, and cognition is quite complex. It is the proximity and connections of the control centers, as well as their hormonal relationships (hormones generated by one affecting/effecting the other) that allows us to talk a bit more confidently about emotions as hardwired. On the other hand, the cognitive interpretation, expression, and reasoning regarding emotions varies from culture to culture, in minor and major ways. Thus, to my mind, the reason that the "cognitive revolution" is an illusionary revolution is that it became enamored with the idea that the mind is a digital computer operating on "representations" that are either formed by the brain synchronically or diachronically by evolution. If we ask how this analogy could lead to a "cognitive revolution" and then to human nature, there appear to be significant gaps in the reasoning.

INSTINCTS AND THE "COGNITIVE REVOLUTION"

To see this, let's review the standard theory of cognition birthed more than fifty years ago, the theory that has dominated discussion of human thinking for so many decades. We can better evaluate some of this theory by considering the summary of the cognitive theory offered by Steven Pinker (2003, 31ff). He lists five primary theses of the standard theory:

1. "The mental world can be grounded in the physical world by the concepts of information, computation, and feedback."
2. "The mind cannot be a blank slate because blank slates don't do anything."
3. "An infinite range of behavior can be generated by finite combinatorial programs in the mind."
4. "Universal mental mechanisms can underlie superficial variation across cultures."
5. "The mind is a complex system composed of many interacting parts."

To these, we might add the following:

6. A solution space is necessary for any learner.
7. A simple nature-vs.-nurture dichotomy is wrong.

8. The crucial focus of research is the learner's job in acquiring competence in all specific and general tasks.

Let's consider each of these points in turn. I want to first list my principal objections to each point, then examine each one in more depth. Point 1 is the foundational problem, the primary mistake of the cognitive sciences (Panksepp and Biven 2012; Paul Churchland 2013; Searle 1980b; and many others). The mind is grounded in the physical world by the physical world. To say that the mind is computationally grounded misses the point of evolution, of emotions, of the physiology and hormones that are active in, present in our every thought. The brain is an organ. It does some computation. It does lots of other things. We would be less than human if the mind were simply grounded in the world by computation. In fact, this first point is simply a rehash of Cartesian dualism, which Damasio (2005)—among many others (especially see Patricia Churchland 2013)—correctly labels a mistake.

Point 2 says nothing, really. Of course human brains are capable of things at birth. And the things that we are capable of distinguish us from other animals. The idea that humans learn to do things that other animals do not has long been recognized (Descartes goes on at length about this, for example, but so does the Bible). Thus humans must have a different biology than other animals or we would not be cognitive or physically different from them. Therefore this point adds nothing revolutionary that we did not know before. No one has ever denied—not even Locke or Aristotle, who were the principal sources of the metaphor of the human mind as a "blank slate"—that the human mind has innate characteristics in this sense. The question is not whether the mind is actually blank but how prespecified it is for higher cognitive functions; or to put it another way, where are the blank spots at and a few months prior to birth?

Point 3 likewise says little. Linguists—before Chomsky and before the cognitive revolution—have long known that human languages have no upper limit and that this astounding linguistic virtuosity is a product of a finite mind (though some linguists would have said "of a finite mind and a specific culture"). Leonard Bloomfield (1933), Edward Sapir (1921), Franz Boas ([1940] 1982), Kenneth Pike (1967), and multitudes of other linguists would have said the same thing decades before this revolution.

With regard to point 4, "universal mental mechanisms can underlie superficial variation across cultures," it is again not at all clear what kind of break with the past this statement represents, aside from the word "mechanisms." It is nearly inconceivable that any scientist would believe that different populations of humans have radically different brains, any more than they have different bodies (i.e., three or four arms, two heads, etc.) "Mechanisms" is potentially different because of the computational metaphor it derives from (again, the foundational mistake of the cognitive revolution). There is a certain physiological and evolutionary foundation of the brain. It is even conceivable that the ancient foundation has been altered in subtle or major ways in modern populations, due to recent cultural or biological pressures that alter relative fitness within that particular population. My principal objection, then, to this thesis is that it reinforces the fundamental error (Damasio 2005; D. Everett 2012a) of the metaphor of the mind as a computer, rather than understanding it as a biological entity. The cognitive revolution avoids reference to the brain by and large via a Whorfian sleight of hand whereby proponents refer to the understanding of cognition via the brain and its physiology as "reductionism," when in fact it is more appropriate to characterize talk of the mind as "additionalist."

The idea that a cognitive revolution took place is thus interesting because it is hard to see where it has claimed anything different from what we knew before—at least, according to the summary from Pinker (2003). But in one of the features I added to Pinker's list, "solution space," we can locate the cognitive revolution's significant departure from past theories of cognitive abilities and the mind—several (but not all) of the leading lights (especially Miller and Chomsky) of this revolution claimed that there are highly specialized solution spaces for human cognitive capacities—a priori, hardwired "modules" (as they came to be called after Fodor [1983]). But there are many alternatives, such as Goldsmith (2015). The questions, of course, are how large the solution space needs to be, how many such spaces the brain has and so on. And my answer throughout our discussion has been that the solution space can be boot-strapped. There is simply no interesting application of the concept of "human nature," other than physical attributes of humans, that make them stand out in a vertebrate lineup.

Points 6 through 8 are accepted by most researchers in cognitive science and by researchers on human abilities long before the so-called cognitive revolution. They are important for understanding and making progress in the study of human cognition, but further illustrate that what we need to know about human thinking and mental capacities are the same kinds of things that researchers have long been aware of. With regard to point 6, no one seriously believes that learners search randomly. They must work with some a priori solution space before they can learn languages and other milestones in their intellectual development. But this of course does not mean that the solution space is innate or that this idea is revolutionary. It means, though, that understanding how learners limit their solution spaces (where to look for answers in learning tasks) is a problem for all researchers, as it long has been. Point 7 says in effect that whatever human cognition in particular domains turns out to be, human biology and the environment interact in numerous and complicated ways to produce such cognition. Point 8 is just the idea that learners need to learn things well across all the domains crucial to healthy interactions with their environments, whether social, linguistic, or physical environments.

How does our behavior and talk reveal the dark matter in which our morality is formed and exercised? What is the source of our reactions and judgments to different situations, from gum chewing to disapproval reflexes and considered judgments? To propose a human nature, one must study the natures of a large variety of humans. For example, Margaret Mead's fame initially resulted from her research, which contradicted the genetic determinism of her day. This determinism was rejected by Mead's advisor (Franz Boas) and other colleagues as severely misguided, based on their field research, philosophies, and theories. Mead's ([1928] 2001) findings that sexual activity in Samoa was dramatically different, healthier, and resulting from and contributing to a less oppressive society flew in the face of that determinism—just as her subsequent work and many others over the years still do. We earlier discussed contrastive views of the morality of poverty and wealth between US culture and caboclo culture of Brazil. Whereas wealth is seen as a sign of God's blessing among many US Christians—or at least the result of dedication and shrewdness ("working smart")—it is seen as a sign of dishonesty, laziness, and greed among caboclos. Morality is culturally determined, based on emotions and ranking of values.

Further to point 8 above, consider Napoleon Chagnon's work among the Yanomami of South America. Chagnon's work is another attempt to support human nature. It purports to show a natural, gene-driven violence in human nature. And yet it fails to come to grips with the fact that similar societies of the Amazon, with similar material environments and shortages, lead very different lives, based on very different culture-apperception-dark-matter histories and rankings. Once we understand the basic biological and cognitive platforms of the species, the tasks that we need to perform, and the cultures in which we exist, it is easy enough to see that there is no need for instinct or human nature.

Related to this is what we know about the human genome—namely, that humans have fewer genes than corn. In my opinion, corn has more genes because it has fewer choices and is unable to learn. Humans have fewer not because we have more modules and instincts, but because we have fewer. We are designed to be flexible. We are the living creatures possessing the greatest degree of "free" will in the known universe. This is our evolutionary legacy and our greatest hope as a species.

What, then, is the "nature of human nature?" Does it exist? Is it knowledge? Is it a priori knowledge? The arguments presented above are that there is no human nature, if by that we mean inborn knowledge or concepts (which are just special forms of knowledge). There could be an anemic idea of human nature that is acceptable; namely, the biological differences between humans and nonhumans, whatever those differences are—though they will not, if this book is correct, involve biologically determined knowledge of any kind.

Conclusion

We began this book with some provocative statements on knowledge and learning: (i) minds do not learn; (ii) brains do not learn; (iii) societies do not learn; and (iv) cultures do not learn. Only individuals learn. And what individuals learn is largely in the form of a culturally articulated dark matter.

Our tripartite thesis, stated at the beginning of our discussion, was (i) that the unconscious of all humans falls into two categories, the unspoken and the ineffable; (ii) that all human unconscious is shaped by individual apperceptions in conjunction with a ranked-value, linguistic-based model of culture, and (iii) that the role of the unconscious in the shaping of cognition and our sense of self is not the result of instincts or human nature, but is articulated by our learning as cultural beings. In the above discussion, we offered and defended a novel conceptualization of the unconscious as "dark matter," which we defined as follows:

> Dark matter of the mind is any knowledge-how or any knowledge-that that is unspoken in normal circumstances, usually unarticulated even to ourselves. It may be, but is not necessarily, *ineffable*.[1] It emerges from acting, "languaging," and "culturing" as we learn conventions and knowledge organization, and adopt value properties and orderings. It is shared and it is personal. It comes via emicization, apperceptions, and memory, and thereby produces our sense of "self."

What we have found is that culture articulates the unconscious via emicization, learning how and that, value rankings, knowledge structures, and social relations and roles across various groups. What has been most emicized is most deeply unconscious by and large. The etic parts are lost in the concrete of emic gestalts. So we do have both conscious and unconscious, yet the latter is not innately structured in ways that Bastian, Chomsky, Freud, and others would have had us believe. Rather, it is structured by doing, thinking, talking, experiencing, and then interpreting those experiences to ourselves. It is learning social constraints and the meanings behind those constraints (ranked values). These are not claims about representations. Paul Churchland's recent work offers a direction for thinking about representations, should those be desired.

We explored the thesis about dark matter by looking at the historical, philosophical development of the concept and how it was appealed to account for a range of phenomena that are often no longer accepted. But what does exist is the contrast between distinct historical traditions regarding tacit knowledge, the Platonic and the Aristotelian. We next investigated a new, ranked-value, linguistically based theory of culture and argued that this model enables us to account for a range of phenomena that other theories struggle with (such as some of Harris's discussions about Indian agricultural families). From this the discussion moved to how the child learns about the world around it, attaching to its culture, via its mother, family, and wider connections, as a series of concentric circles. We showed how attachment precedes language and yet provides a range of profound apperceptions and exposure to culture that play a significant role in the construction of culture and a sense of self. Next, we examined how our thesis is supported by the visual, textual, and general interpretations of the world around us, looking in detail at Pirahã perception of two-dimensional figures. From this we moved to the purportedly most successful exemplar of the Platonic view of dark matter, Chomsky's universal grammar, showing how this model seems to be wrong, and almost certainly superfluous in accounting for the emergence of grammars from individual-culture connections that are motivated by the emotional need to form social bonds, the "interactional instinct," and cultural factors. We then examined an often neglected area of paragrammar, the use of gestures in language. We considered in detail the most comprehensive research program on gesture, the work of David McNeill and his colleagues, arguing that all of McNeill's work can be embedded into a culturally articulated conception of dark matter, offering little support for nativism. In particular, we considered and rejected the idea that the homesigns studied by Goldin-Meadow and her colleagues offered support for the Platonic view of dark matter (that it is not learned) as opposed to the more plausible Aristotelian view (that it is learned). From this we came to the last two chapters in the book and made the case that the concepts of instincts and human nature hinder our understanding of human behavior, society, and cognition, urging us to move beyond these notions that our "designer bias" leads us very often to favor for no rational reason.

Thus, we have seen that we learn things by the experiences we have from the womb to the coffin. Some of our learning begins as conscious engagement with etic experiences, eventually disappearing into the dark matter of our minds via emicization—building an insider perspective of a whole from outsider perspectives of its parts. By interpreting our experiences, by building a set of etic memories into an emically constructed whole, a gestalt being, we create ourselves. As Buddhism understands us to be the sum total of our thoughts, so we have arrived via our long, circuitous path through philosophy, visual perception, culture, linguistics, translation, and hermeneutics, at the understanding that our mental and social experiences lead to our emergent selves and that this vision is ill served by the proposal of mysterious genetically hardwired knowledge that only obscures and hinders our study of the individual in culture. Thus we must move beyond instincts and beyond human nature if we are to understand how our social lives and individual bodies work together to build our roles, understandings, values, knowledge structures, conventions, norms, and our very place in the world.

I have argued that Sapirian and Pikean linguistics provides an insightful model for investigating our symbiotic relationship to culture, through Pike's notions of etic, emic, slot, and filler.

Our discussion has deliberately examined very different epistemological domains in order to demonstrate the pervasive influence of the articulated unconscious I refer to as dark matter of the mind.

Notes

Preface

1. I should perhaps add to this introduction the prediction that anthropologists will almost certainly resist my postulation of culture as the centerpiece of cognition, just as linguists will oppose placing culture at the center of language, and psychologists will oppose the denial that individual cognition is all there is to the study of psychology. I can only hope to persuade some that the issues are worth exploring further and still others that the theory of culture and the mind developed in chapter 2 is worth pursuing further.

Introduction

1. On the surface, my definition of dark matter is reminiscent of Bourdieu (1977) and Bourdieu and Wacquant (1992), among others. While there are clearly similarities and affinities with the work of Bourdieu and my theory of dark matter, his theory of habitus is not, in my opinion, articulated in the same way as dark matter and thus lacks the ability to capture some of the ranked values constraints, knowledge structures, and social roles as precisely as the current model.

2. E.g., Majid and Levinson 2011; Polanyi (1966) 2009, 1974; Collins 2010; Gascoigne and Thorton 2013; Turner 2013.

3. See Lakoff and Johnson 1980 and Lakoff and Nuñez 2001, among others.

4. In the cognitive sciences, more generally, the idea of tacit knowledge has been around for a while. The online *Dictionary of Philosophy of Mind* offers a summary of much of the literature on this in philosophy: https://sites.google.com/site/minddict/knowledge-tacit.

5. Often confused, deep structure and universal grammar are not the same. However, the later theory of generative semantics did blur this distinction considerably for the lay reader by proposing that deep structure was in fact the same for all languages, what came to be known as the universal base hypothesis.

6. These are both idealizations. Our understanding of "phonetics" seen, for example, in something so erstwhile objective or etic as the International Phonetic Alphabet, is shaped by our emic perspectives and most "etic" categories are themselves idealized (affected by researchers' experience with emic categories) in cultural ways. So there is no truly objective vantage point, just ones less contaminated in ways we know of.

7. The standard convention in linguistics is that slashes // are used to enclose phonemic sounds and brackets [] to enclose phonetic sounds.

8. McQuown (1957) offers a critical review of Pike's *emic/etic* distinction.

9. See Panksepp and Biven 2012, LeDoux 2015, Bruusgaard et al. 2010, Sommer 1992, and Costandi 2012, respectively.

10. Though others—e.g., Lieberman (2013)—refer to this work in less complimentary terms as "warmed over" phrenology. I must admit to some irritation at the co-opting of the term *evolutionary* in the label of this research program, suggesting as it does that other forms of psychology are not based on evolutionary theory or that it is, for that matter.

11. Of course, authors as varied as Garfinkel (1991, 2002), Pike (1967), Boas ([1940] 1982), Rogers ([1961] 1995), Read (2011), Mead ([1928] 2001), Silverstein (2003, 2004), Tomasello (1999), Parsons (1970), Patricia Churchland (2013), White (1949), Coulter (1979, 1983), Bateson (2000), and Jesse Prinz (2002, 2011, 2014), among many others, have written on related topics. But their conclusions, evidence, and lessons learned also vary—usually profoundly—from the current work.

Chapter One

1. Other treatments of culture as grammar failed because they lacked the notion of ranked, violable values and more articulated concept of culture that I attempt to develop here.

2. See the bicycle-riding robot on designboom: http://www.designboom.com/technology/bicycle-riding-robot/.

3. Searle's (1980b) "Chinese room" analogy is famous in linguistics, artificial intelligence, and other scientific communities for underscoring the semantics problem for artificial intelligence, that is, that computers can produce forms but do not know what they mean. Some have claimed that this problem has been solved. I don't agree in the slightest with this assertion. There can be no semantics without culture, and until the culture problem is solved, the semantics problem will be intractable.

4. Pike did not recognize distinctive features and worked with the notion of phonemes. What I say in what follows works as well for the latter as for the former, but is perhaps clearer with features.

5. A *mora* is a relative unit of length proposed by many phonologists and phoneticians. A vowel would be one mora, a consonant either zero moras or some fraction thereof, in the simpler cases.

6. Anticipating our discussion of Aristotle, it would be anachronistic and incorrect to attribute to him the view that there is in fact no mind at all, only a body with a thinking organ that carries out its tasks and takes its form as part of its body and history, though Aristotle is my own inspiration for this line of thought. Just as one's heart may be damaged by diet or the liver by drink, so is the brain shaped, strengthened, or damaged by our diet, exercise, recreational activities, entertainments, etc. Our brain is just an organ of our body, merely a part of the whole.

7. Kant ([1903] 2001, 120) talks about "imagination" as "imagination has to bring the manifold of intuitions into the form of an image." That is, in this sense, his notion of imagination is to create an image—an image based on experiences and apperceptions that have been internalized. This is in fact not far off from the notion of dark matter that emerges from the present discussion. However, this Kantian notion of imagination requires, or coexists with, the kind of innate conceptual knowledge that is incompatible with our view, so Kant still lies outside the tradition in which I believe I am constructing my own ideas. Nevertheless, Kant's work on the imagination

does show that it is nearly impossible to pigeonhole him in any way (thanks to Yaron Senderowicz for bringing my attention to Kant's work on the imagination).

8. There are many reasons to entertain a robust skepticism of Chomsky's nativist theory of tacit knowledge of language, however, as per D. Everett (2012a) and V. Evans (2014), among others.

9. There is no pure version of either rationalism or empiricism; such distinctions are rarely embraced as we come to see more and more of the complexity of the natural world. I refer to them here as a simplification of the debate that aids the exposition and does no harm to the overall argument.

10. By referring to specific names or samples of any individual's work, I do not intend to create a false reification or synecdochical misrepresentation of an author that might diminish the subtlety of their thought in any way. Every author cited in this chapter held views that were wide ranging and impressively rich and nuanced, even to the degree of penning apparently self-contradictory passages. My appeal to their authorship and insights is intended neither to box them in nor to speak for them. It is clearly selective. Yet, I believe that my selections are consistent with a broader interpretation of their work.

11. All learners, of course, must have learning biases. But these biases need not take the form of innate knowledge, other than general perceptual constraints.

Chapter Two

1. Anthropology is a vast field, including social anthropology of the British variety, scientific anthropology, and various other approaches in a range of countries. Since my efforts here are directed toward developing my own theory, I have interacted primarily with anthropologists whose concerns with the intersection of language and culture have been a bigger influence on my development, namely, American linguistic anthropologists for the most part. This is not to slight other traditions intentionally, just to say that I made a decision to focus on a particular literature at the expense of others (and my own development, I am sure). Many will strongly disagree with my conception of culture. But disagreement is not new in anthropology.

2. A few definitions and descriptions of culture—in addition to those in the text—worth highlighting include the following (collected by Hervé Varenne, professor of education at Columbia University, all taken from http://varenne.tc.columbia.edu/hv/clt/and/culture_def.html):

Franz Boas: "Culture may be defined as the totality of the mental and physical reactions and activities that characterize the behavior of individuals composing a social group collectively and individually in relations to their natural environment, to other groups, to members of the group itself and of each individual to himself. It also includes the products of these activities and their role in the life of the groups. The mere enumerations of these various aspects of life, however, does not constitute culture. It is more, for its elements are not independent, they have a structure."

Ruth Benedict: "What really binds men together is their culture,—the ideas and the standards they have in common."

Margaret Mead: "Culture means the whole complex of traditional behavior which has been developed by the human race and is successively learned by each generation. A culture is less precise. It can mean the forms of traditional behavior which are characteristics of a given society, or of a group of societies, or of a certain race, or of a certain area, or of a certain period of time."

James Baldwin: "Culture was not a community basket weaving project, nor yet an act of God; being nothing more or less than the recorded and visible effects on a body of people of the vicissutes with which they had been forced to deal."

Antonio Gramsci: "One might say 'ideology' here, but on condition that the word is used in its highest sense of a conception of the world that is implicitly manifest in art, in law, in economic activity and in all manifestations of individual and collective life."

Lionel Trilling: "When we look at a people in the degree of abstraction which the idea of culture implies, we cannot but be touched and impressed by what we see, we cannot help being awed by something mysterious at work, some creative power which seems to transcend any particular act or habit or quality that may be observed. To make a coherent life, to confront the terrors of the outer and the inner world, to establish the ritual and art, the pieties and duties which make possible the life of the group and the individual—these are culture, and to contemplate these various enterprises which constitute a culture is inevitably moving. [. . .] [Freud] does indeed see the self as formed by its culture. But he also sees the self as set against the culture, struggling against it, having been from the first reluctant to enter it. Freud would have understood what Hegel meant by speaking of the 'terrible principle of culture.'"

Ward Goodenough: "The term culture [refers to] what is learned, . . . the things one needs to know in order to meet the standards of others." "[. . .] Therefore, if culture is learned, its ultimate locus must in individuals rather than in groups. If we accept this, then cultural theory must explain in what sense we can speak of culture as being shared or as the property of groups at all, and it must explain what the processes are by which 'sharing' arises."

Clifford Geertz: "[the culture concept] denotes an historically transmitted pattern of meanings embodied in symbols, a system of inherited conceptions expressed in symbolic forms by means of which men communicate, perpetuate, and develop their knowledge about and attitudes toward life . . . The point is sometimes put in the form of an argument that cultural patterns are "models," that they are sets of symbols whose relations to one another "model" relations among entities, processes . . . The term "model" has, however, two senses—and "of" sense and a "for" sense . . . Unlike genes, and other nonsymbolic information sources, which are only models for, not models of, culture patterns have an intrinsic double aspect: they give meaning, that is, objective conceptual form, to social and psychological reality both by shaping themselves to it and by shaping it to themselves."

Richard Shweder: "Culture refers to the intentional world. Intentional persons and intentional worlds are interdependent things that get dialectically constituted and reconstituted through the intentional activities and practices that are their products, yet make them up . . . Culture is the constituted scheme of things for intending persons."

Bruno Latour: "the set of elements that appear to be tied together when, and only when, we try to deny a claim or to shake an association."

3. It is trivially true that if you can understand a movie, you must understand something of the culture that produced it, but this doesn't mean that you are a member of that culture. Not even momentarily. And it is not obvious that a serious film *can* be understood well by those outside the culture.

4. Obviously DNA studies would be interesting and necessary scientifically before saying anything confident on this score, but it is difficult politically to carry out such studies because

in Brazil, those delegated to protect indigenous peoples are wary of anything that could be perceived as racist studies, especially studies carried out by "gringo" scientists.

5. One of the most common objections critics raise against N. Evans and Levinson (2009)—and my own work—is that the superficial absence of a particular feature does not mean that the feature is not present in the language abstractly. This is correct. But the critics then make the mistake of moving from this banal observation to conclude that they/we are either deliberately or ignorantly failing to understand the difference between Greenbergian and Chomskyan universals. This is an old accusation—one that I, among many others, have rebutted in numerous publications, but apparently one that gives some critics comfort. However, it is worth considering in detail here to see how the debate has been manipulated/misunderstood.

6. An isogloss is a geographic boundary marked by a particular linguistic feature, such as pronunciation of specific segments, a different word form (e.g., *gumband*, which means "rubberband" in the Pittsburgh area and environs), or different constructions (e.g., the geographical distribution of "My car needs wash*ed*" vs. "My car needs wash*ing*").

7. One of the more interesting immediate constituents of Pirahã society are *kaoáíbógí*, literally, "fast mouths," which represent a largish group of human-like entities that the Pirahãs claim to interact with. At one time I confused these with the concept of "spirits" (see D. Everett 2008), but they are in fact seen as living creatures of a different order than animals or humans.

8. This is another reason why I do not believe that there was any "cognitive revolution" on September 11, 1956 (Gardner 1987), but—to put it humorously—simply a highly celebrated meeting of secular dualists.

9. For example, see DTN's "Core Values": http://www.custompage.reslisting.com/apartments/mi/lansing/dtn-corporate-site-o/ourvalues.aspx.

10. "Man in a can" was the way that my brother-in-law, Dr. Hugh Behling, described the cans of horse semen he used to artificially inseminate mares.

Chapter Three

1. A couple that stand out are Rogoff 2003, 2011; Keller 2007; Levine 2013; and Otto and Keller 2014. What follows borrows from Everett 2014a.

2. See Bruner 1979, 1987, 1993, and 1997, inter alia.

3. As exemplified in work by Ochs and Capps (2002), Silverstein and Urban (1996), Urban (2000), and Sherzer (1991), among others.

4. Such as in research at the Max Planck Institute in Leipzig: http://www.eva.mpg.de/psycho/dog-cognition.php.

5. I use "attachment" here in a potentially nonstandard meaning that includes "identification with."

6. A Pirahã garden will usually be located within a thirty-minute walk from the nearest village. This distance is partly to discourage casual theft from the garden.

7. Toddlers rarely come to me—they look at my beard and say I look like a dog. If I insist, they scream.

8. For example, children experience and see others experience toothaches on a regular basis. A toothache, where there is no dentist, is agonizing. One must simply tolerate the pain until the nerve dies and the tooth has rotted out.

9. I recall one night with a visiting American dentist when a child near where we were sleeping was screaming and crying all night. The dentist asked me, "Dan, what is wrong with that baby? It sounds horribly ill and in pain." I replied, "If it were ill, the parents would have already

come to us for help. It is just pissed off about something." After about an hour, around 3 a.m., the dentist said "Dan, I have had medical training and I know a sick baby when I hear one. We have to go over there now."

So I wearily grabbed at my flashlight, slipped on my flip-flops and said halfheartedly, "Let's go." We walked to the small hut near the path that was the source of the siren-like wailing. When we arrived, the parents were feigning sleep, with the toddler sitting up by his father, screaming at the top of its lungs. I asked the father, "What is wrong with the baby? Is he sick?" The father ignored me, pretending to sleep. I turned to the dentist, "They want us to leave and are pretending to be asleep to communicate this." He said, "I am not leaving until the father tells us what is wrong."

So I shook the father. He looked at me as though he wanted to tell me to go fug myself, but said, "What?"

"What is wrong with the baby?"

"Nothing. It wants tit."

I communicated this to the dentist and said, "Satisfied? Let's go back to bed." The dentist said, "It is hard to believe. I would have sworn that this baby is very ill."

As we started to leave, the mother pulled the toddler across the new infant at her side and began to nurse it—clearly to keep us from coming back and to communicate to me that the message had been received. Still, however ignorant of the Pirahãs, the dentist was well-meaning.

10. A favorite pastime of preadolescent boys is to sit and shoot small lizards running zigzags more than five meters away with arrows while holding the bows in their feet.

11. After completing this chapter, I sent a draft to Steve Sheldon, a missionary who lived with his family among the Pirahãs from 1967–1976 and learned to speak the language well. Sheldon's comments are given below (e-mail from Sheldon, December 14, 2012):

> We felt they preferred to give birth in the river if possible [not relevant to this paper]. When Linda was expecting Scott, they kept on us about her having the baby in the river like they did.
>
> We saw some instances where a mother would help her daughter give birth.
>
> One young girl was having a very hard time with a birth and Linda wanted to go "help." The people did not want her to do so, nor would any of them go help. This young girl's mother had died not long before. Normally they would let us do whatever strange things we wanted to do. Not in this case.
>
> When our boys were young and nursing, the people did not like them to cry and would say things like: "We cannot dance if they are crying." This in spite of some of their weaned children carrying on just as you described.
>
> Our problem was we were following an American cultural norm of putting the boys to bed and "letting them cry" till they learned.
>
> Once the boys were weaned, they could also scream and carry on with no frustration on anyone's part—except ours.

Chapter Four

1. See William Barclay: http://www.dannychesnut.com/Bible/Barclay/First%20and%20Second%20Corinthians.htm.

2. Original caption for figure 4.1 (figure 1 in original volume): "All stimuli used in the ex-

periment. Left column from top to bottom: houseboat, jaguar, alligator, woman in hut, sloths, older man. Right column from top to bottom: squirrel monkey ocelot, howler monkey, toucan, tapir, fisherman. Cue items are shown to the left of test items. Houseboat and jaguar are warm-up items with simpler transformations. For full size stimuli, see online supplemental materials in order to recreate viewing conditions under which recognition is trivial for western adults."

3. "Following the Gestalt school, we use the terms 'perceptual organization' and 'perceptual reorganization' to emphasize the process by which local image features are appropriately integrated ('grouped') or segregated in order to arrive at a meaningful interpretation of the image—a 'gestalt'" (Kohler 1929).

4. "Participants included adult members of the Pirahã tribe (n = 9, mean estimated age = 30y) and as controls tested with the same stimuli, Stanford University students, faculty, and staff (n = 8, mean age = 26y). An additional control task with additional stimuli was tested on Stanford students (n = 10, mean age = 19y). The visual acuity of the Pirahã population was tested by DE and others some years earlier as part of a basic screen for medical services; the population was on the whole normal, with no cataracts and a small incidence of nearsightedness."

5. "The amount of blur and the black/white threshold points were set independently for each photograph based on a repeated trial and error procedure until we were satisfied with the subjective impressions that the two-tone was (a) hard to recognize without first seeing the photograph from which it was derived ('uncued') and (b) easy to see after seeing the photograph ('cued'). This stimulus creation and selection were guided by the perceptual judgment of the experimenters. Images were printed onto 12x12cm cards. These two-tones are similar in appearance, but different in method of stimulus creation (as well as experimental purpose) from the stimuli known as 'Mooney faces.' Mooney himself used the stimuli to study 'closure'-based recognition of individual images, analogous to our 'stage 1 uncued' presentation. Mooney's faces were hand-drawn artist's renderings of human faces under extreme illumination conditions (Mooney, 1957), so there is no corresponding photo from which the images were derived."

6. Original caption for figure 4.2 (figure 2 in original volume): "Upper left: an example two tone-stimulus from the Pirahã study. Subjects were first presented the two-tone alone and asked to point to the location of an eye or person in the image. Red circles mark where Pirahã participants indicated an eye, and numbers indicate individual participants. Circles outside the image show responses of the form 'there are no eyes here.' Only two participants (2, 8) correctly pointed to an eye in this image during Stage 1. Upper right: Performance of Pirahã participants on the original photo, which was presented alone after the two-tone image was removed from view. All participants correctly pointed to one of the two eyes. Bottom row: performance of Pirahã participants on the two-tone image during Stage 3, when it was shown side-by-side with the photo. Two Pirahã participants succeeded uncued (2 and 8), two more succeeded with the photo present, indicating reorganization (3 and 4), and five did not show evidence of photo-triggered perceptual reorganization (1, 5, 6, 7, 9)."

7. Original caption for figure 4.3 (figure 3 in original volume): "Summary of results from the Pirahã and the two U.S. control groups. Bars show participants' accuracy on photographs, practice items, and candidate reorganization trials (those trials on which the two-tone image was not recognized uncued). Error bars show the standard error of the mean."

8. Yoon, Whitthoft, et al. (2014) continue: "Control participants in the misaligned condition—like the controls in the main experiment but unlike Pirahã participants—showed near perfect performance on candidate reorganization trials (94.2%). This result would be ex-

pected if control participants experienced reorganization, and their performance did not depend solely on a spatial alignment strategy to localize features.

"To summarize these observations statistically, we conducted a repeated measures ANOVA with a 3-level within-subject factor (trial type: practice items, photos, candidate reorganization trials) and a 3-level between subject factor (group: Pirahã, U.S. controls, U.S. misaligned condition). There was a main effect of group ($F(2,23) = 32.6$, $\eta^2 = 0.74$, $p<0.001$) and a trial type x group interaction ($F(4,48) = 8.35$, $\eta^2 = 0.41$, $p<0.001$). Pairwise comparisons reveal that the Pirahã differ from both U.S. groups ($ps<0.001$), who do not differ from each other (Bonferroni corrected). Similarly, accuracy on candidate reorganization trials differs from accuracy on practice trials and photo recognition ($ps<0.001$), which do not differ from each other. A follow-up t-test compared Pirahã candidate reorganization performance to U.S. control performance in the misaligned condition (when U.S. controls do not have access to a non-recognition-based location matching strategy), showing that Pirahã performance was significantly lower ($t(17) = 8.26$, $p<0.001$)."

9. Orignal caption for figure 4.4 (figure 4 in original volume): "Example of misaligned photo and two-tone image pair. This image shows the actual degree of misalignment, but participants were never informed about the degree or direction of misalignment or even if any misalignment or distortion occurred between the images. The intersection of the horizontal and vertical red lines shows the same geometric point relative to the frame of the images. Participants could identify these matching points even without reference to the underlying images (for example, if the cards were blank). Actual corresponding features are shown with the red dot, and were displaced 1.8cm in a direction that varied from image pair to image pair."

10. [coronal] means produced with the tip of the tongue; [voiced] refers to the vibrations of the vocal folds ([+voiced] = vocal cords vibrating; [−voiced] = vocal cords not vibrating during production of sound; [−continuant] means that the air is blocked before continuing out the mouth.

Chapter Five

1. Take from Ernst and Young's website, "Our people—a diverse 21st-century workforce": http://www.ey.com/US/en/About-us/Our-people-and-culture.

2. See http://www.holacracy.org.

3. I think that hierarchy is always a value in some sense, e.g., the owner's decision to implement or halt holacracy, or to sell the company.

4. In this example, PROFIT and SHAREHOLDER are identical.

5. See in particular McCawley (1982) as a pioneering study of parentheticals in syntactic theory.

6. Though this word has been adapted to Pirahã phonology, it is a loanword of uncertain provenance.

7. This is not to say that there could not be other popular songs with conflicting messages, e.g., blaming the husband instead of the mistress. But hearers will pick and choose or try to blend conflicting ideas into a coherent whole. The effects of songs on dark matter are likely profound.

Chapter Six

1. See, for example, Halliday and Hasan 1976, Grimes 1975, Longacre 1976, Pike and Pike 1976, and Givón 1983.

2. In 2003 I filmed the entire process, from gathering the raw materials in the jungle to mixing the poison, to making the darts and the blowgun. Returning to the University of Manchester I placed my video recordings in the care of a postdoctoral associate to copy. Before she was able to work on them, however, all of my field data and equipment for that year, the tapes, cameras, and computer, were stolen from her ground-floor office.

3. Alternations with /t/s or involving different values for [continuant] or [voicing] are unattested.

4. Merge is a function that takes two objects (α and β) and merges them into an unordered set with a label. The label identifies the properties of the phrase. In Minimalism, no phrase structure can be formed without undergoing Merge. Since Merge is by definition a recursive operation, no language can exist without recursion, QED.

For example: Merge (α, β) → {α, {α, β}}

If α is a verb, e.g., *eat* and β a noun, e.g., *eggs*, then this will produce a verb phrase (i.e., where alpha is the head of the phrase) "eat eggs." As I said in D. Everett (2012b), "The operation Merge incorporates two highly theory-internal assumptions that have been seriously challenged in recent literature. The first is that all grammatical structures are binary branching, since Merge can only produce such outputs. The second is that Merge requires that all syntactic structures be endocentric (i.e. headed by a unit of the same category as the containing structure, e.g. a noun heading a noun phrase a verb a verb phrase, etc.)."

5. Some linguists, in commenting on the Pirahã data, indicate that it is irrelevant to even the theory of grammar which says that recursion is the building block of all languages. This is because they are prepared to adopt a string of ancillary hypotheses to artificially render Pirahã unable to express its recursion. See D. Everett (2012b) for details.

6. It has long amused me that Nevins, Pesetsky, and Rodrigues's (2009) paper is an attempt to refute me using my own earlier analysis and data. Replying to them is like having a debate with myself—the Dan Everett of thirty years of field experience among the Pirahãs vs. the Dan Everett of fourteen months of field experience, writing a PhD dissertation.

7. I use role and reference grammar here because to my mind it most effectively blends structural and functional-semantic principles into a theory of grammar. Nothing crucial hangs on this, however, and other theories might be compatible with the analysis offered here.

Chapter Seven

1. This is one reason that computers cannot be said to have language, in my opinion. For others, see D. Everett (2012a).

2. Sign languages, as McNeill makes clear in all of his work, represent a different case of combining static and dynamic gestures and diachronic as well as synchronic formation.

3. On the one hand, prosodies (pitch, length, intensity) often do have gesture-like functions (as many, including McNeill [1995, 2005, 2012], have observed). Thus based on work by Bolinger (1985, 1989), Lieberman (1967), Watson and Gibson (2004) and others, much of what is said about manual gestures also would apply to prosody. Thus there are both a functional and a formal distinction between speech and gesture.

4. Efron's need for the services of Van Veen was due, of course, to the lack of adequate technology for recording gestures. With the advent of certain types of new equipment, e.g. the cassette recorder, the laptop, and digital sound analysis, and—especially—quality portable video equipment, and so on, the course of linguistic history, as that of any science (or any culture in fact), took a different course, directed by the objects the culture produces for specific purposes. We tend to do more of the things we have more or better tools for. The better and more accessible the tools, the more the research they come to facilitate is undertaken. Tools alter science.

5. This study was part of the *First Annual Report of the Bureau of Ethnology to the Secretary of the Smithsonian Institution, 1879–1880*, 1881, 263–552 (Washington: Government Printing Office).

6. I was perhaps incorrect in asserting that separate domains of study were called for to understand the dynamic vs. static. That remains to be seen.

7. The research program known as linguistic typology might be considered the field perspective on language, since it examines all languages as part of a vast linguistic matrix.

8. See D. Everett 1994 and 2015 for more discussion. At that time I argued for distinct theories for the analysis of each type of cognition, though I now believe that separate approaches may not be needed, especially in light of Pike's urging us to study behavior as particle, wave, and field.

9. In this sense catchments are like discontinuous constituents in the syntax, such as *ne . . . pas* in French or even, *Who* did you wonder whether *they* were upset?

10. See Reinbold 2004 for an excellent analysis of discourse embedding in conversations in Banawá.

11. To say that this also entails that we use gestures without learning them, as McNeill seems to suggest, is unwarranted.

12. As Sascha Griffiths pointed out to me (e-mail, May 4, 2014) the gesture/speech equiprimordiality hypothesis avoids the nagging question of how long after we had gesture languages the human larynx would have waited to descend (see Lieberman 2007).

13. It is also worth noting that if McNeill's story is on the right track, the contribution of Merge to the evolution of language was neither necessary nor sufficient.

14. Seen engaged in a discourse and using hand gestures of different types, in the YouTube video "Spoken Pirahã with Subtitles": https://www.youtube.com/watch?v=SHv3-U9VPAs.

15. For example, one plausible line of reasoning is that an alliance with dogs is what gave Homo sapiens the final advantage over neanderthalensis (Shipman 2015).

16. Taken from https://www.hcii.cmu.edu/people/justine-cassell, last accessed January 2016.

Chapter Eight

1. An example of homopraxis in conventions.

2. Sexual harassment cases often entail this kind of misunderstanding—that the offended party should have understood the offender's emic perspective and then would know that there was no harm intended. But of course, one cannot assume in a society of strangers that one's psyche is an open book.

3. See the YouTube video "'Monolingual Fieldwork' Demonstration" (https://www.youtube.com/watch?v=sYpWp7g7XWU), sponsored by the Linguistic Society of America, for one such demonstration.

4. As David King (pers. comm.) points out to me, my conclusion here regarding indeterminacy of translation could be open to criticism because it would be possible—and this seems to be what Quine had in mind—for the translator not to know what they had misunderstood, because

the native speaker and translator could respond behaviorally in the same way ostensibly, but with different mental maps from experience to meaning—one person responding to rabbit parts and the other to a whole rabbit. In the field, however, this is a difference without a difference. The field researcher does not nor would not notice this problem. But of course it is entirely possible that two people can talk in translatable ways, each one of them buttressed and understanding via distinct dark matters. It is not merely possible but, if I am correct, inescapable.

5. Religions are important to any discussion of translation because they perhaps rely on it and promote it more than any other entities. And their views on it are that it has eternal significance.

6. Wikipedia, s.v. "Quran translations," last modified December 19, 2015, https://en.wikipedia.org/wiki/Quran_translations.

Chapter Nine

1. Whether Wittgenstein's later work was also dogmatic but not recognized to be so by him is not a matter we address here.

2. In fact—as an aside for linguists—all grammatical theories can be summarized (with a bit of humor) as:

Summary of Relevance Theory: Don't say more than you need to, but say what you have to.
Summary of Conversational Analysis: You can turn. You can keep going straight. You can stop.
Summary of Formal Linguistics and March Madness: It's all in the brackets.
Summary of Nativism: It was there all along.
Summary of Construction Grammar: If some things must be memorized, why not most things?
Summary of Everett's Cultural Grammar: Form is driven by meaning and meaning by culture.
Summary of Functional Linguistics: Form is driven by meaning.
Summary of Discourse Analysis: Know what you are talking about and how to make it clear.
Summary of Sociolinguistics: How many people really do that, when, and where?
Summary of Historical Linguistics: What was that?

3. Endocentricity is the property of a phrase headed by a word or phrase of the same category—*VP* headed by *V*, *NP* headed by *N*, etc.

4. And field researchers often have little time for arcane theoretical issues (though some make time), struggling with overwhelming complexity, disease, isolation, threats to their safety, and self-doubt, in easily the most challenging intellectual task of linguistics.

5. "It is perfectly safe to attribute this development [of innate language structures] to "natural selection", so long as we realize that there is no substance to this assertion, that it amounts to nothing more than a belief that there is some naturalistic explanation for these phenomena" (Chomsky 1972, 97).

In his book *Science of Language* (Chomsky 2012, 105), Chomsky says much the same: "Tell them the truth about evolution, which is that selection plays some kind of role, but you don't know how much until you know. It could be small, it could be large; it could [in principle] even be nonexistent."

6. The Wikipedia definition of grammaticalization is "a process of language change by which words representing objects and actions (nouns and verbs) transform to become grammatical markers (affixes, prepositions, etc.)" (Wikipedia, s.v. "grammaticalization," last updated January 4, 2016, https://en.wikipedia.org/wiki/Grammaticalization). For many researchers, however, grammaticalization has a meaning broader than simply *word→affix* changes. It can also mean anything that has become part of the rules or constraints that are posited as part of universal grammar or, even, anything that has become part of the grammatical rules or constraints of a particular language.

7. A syllable is a unit into which phonemes are arranged. There have been attempts to define the syllable phonetically (i.e., physically) as a single "chest pulse." Many phoneticians believe that the syllable is poorly motivated either acoustically or physiologically. On the other hand, phonologists have developed elaborate models of syllables that analyze the unit into hierarchical structures of various types. The overall (but especially linear) arrangement of phonemes into syllables is referred to as phonotactics. Berent argues in effect that the organization of the syllable is based on a type of phonotactic instinct often referred to as the sonority sequencing generalization.

8. Sonority is a formal property of sounds in which it is easier to produce "spontaneous voicing" (vibration of the vocal folds while producing the sound), though the layperson can refer to sonority as relative loudness with little loss of accuracy.

9. For independent reasons—but reasons that once again show the inadequacy of the SSG—onsets are preferred to codas, thus favoring the syllabification of *o.pa* over *op.a*. The reason that a simple preference such as "prefer onsets" is a problem for the SSG is that the preference clearly shows that SSG is unable to provide an adequate theory of syllabification (at least, on its own).

10. Formants are caused by resonance in the vocal tract. They are concentrations of acoustic energy around a particular frequency in the speech stream. Different formant frequencies and amplitudes result from changing shapes of the tract. For any given segment there will be several formants, each spaced at 1000 Hertz intervals. By resonance in the vocal tract, I mean a place in the vocal apparatus where there is a space for vibration—the mouth, the lips, the throat, the nasal cavity, and so on.

11. *F2* and *F3* refer to the second and third formants of the spectrographic representation or acoustic effects of producing sounds.

12. Ironically, there is an excellent neuroscientific study of emotions, Patricia Churchland (2012), which is all too often neglected by the proponents of nativist approaches to morality.

13. This is not referring to attempts to apply an original analysis crosslinguistically on the assumption that it must be universal. There are a multitude of studies that assume that structures found in one language will be found in another for principled reasons. Those principles might derive from UG (though this does not mean that the analysis could not have otherwise been done without appeal to UG). Or they might derive from functional reasons (communication ease, computational efficiency, meaning driving syntax, etc.). Using UG (or any other theory or set of hypotheses) as a source of ideas is fine and useful, but unless some UG principle *leads to* a particular analysis—one that cannot be captured under non-UG assumptions and otherwise accepted by most linguists, regardless of theoretical assumptions—UG is not causally implicated in the analysis.

14. For example, Rebecca Saxe's lab (Bedney et al. 2011) shows that the "visual cortex" can be used for language in visually impaired subjects.

Chapter Ten

1. Frankly, I am not sure how this characterization of religions differs from scientific theories.

2. But cf. LeDoux (2015) for a very different view of the neuroscientific understanding of human emotions.

Conclusion

1. E.g., Majid and Levinson 2011; Polanyi (1966) 2009, 1974; Collins 2010; Gascoigne and Thorton 2013; Turner 2013.

References

Adelson, E. H. 1993. "Perceptual Organization and the Judgment of Brightness." *Science* 262:2042–44.

Aikhenvald, Alexandra Y. 2003. *Language Contact in Amazonia*. Oxford University Press.

Albahari, Miri. 2006. *Analytical Buddhism: The Two-Tiered Illusion of Self*. Palgrave-Macmillan, New York.

Anderson, Michael L. 2014. *After Phrenology: Neural Reuse and the Interactive Brain*. Cambridge, MA: MIT Press.

Andrén, Mats. 2010. "Children's Gestures from 18 to 30 Months." PhD diss., Centre for Languages and Literature, Cognitive Semiotics, Lunds Universitet, Sweden.

Arbib, Michael A. 2005. "From Monkey-Like Action Recognition to Human Language: An Evolutionary Framework for Neurolinguistics." *Behavioral and Brain Sciences* 28 (2): 105–24.

———. 2012. *How the Brain Got Language: The Mirror System Hypothesis*. Oxford University Press.

Aristotle. 2007a. *Metaphysics*. Great Books of the Western World, edited by Mortimer Adler, Clifton Fadiman, and Philip W. Goetz. Volume 7, *Aristotle I*, 499–630. Chicago: Encyclopedia Britannica.

———. 2007b. *On the Soul*. Great Books of the Western World, edited by Mortimer Adler, Clifton Fadiman, and Philip W. Goetz. Volume 7, *Aristotle II*, 631–72. Chicago: Encyclopedia Britannica.

———. 2007c. *Politics*. Great Books of the Western World, edited by Mortimer Adler, Clifton Fadiman, and Philip W. Goetz. Volume 8, *Aristotle II*, 445–552. Chicago: Encyclopedia Britannica.

———. 2007d. *Posterior Analytics*. Great Books of the Western World, edited by Mortimer Adler, Clifton Fadiman, and Philip W. Goetz. Volume 7, *Aristotle I*, 97–142. Chicago: Encyclopedia Britannica.

Austin, John L. 1975. *How to Do Things with Words*. Cambridge, MA: Harvard University Press.

Ayer, Alfred Jules. 1940. *The Foundations of Empirical Knowledge*. London: Macmillan.

Bakhtin, Mikhail. 1982. *The Dialogic Imagination: Four Essays*. Austin: University of Texas Press.

———. 1984. *Problems of Dostoevsky's Poetics*. Edited and translated by Caryl Emerson. Minneapolis: University of Minnesota Press.

Bateson, Gregory. 2000. *Steps to an Ecology of Mind: Collected Essays in Anthropology, Psychiatry, Evolution, and Epistemology*. Chicago: University of Chicago Press.

Becker, Alton L. 2000. *Beyond Translation: Essays towards a Modern Philology*. Ann Arbor: University of Michigan Press.

Bedney, Marina, Alvaro Pascual-Leone, David Dodell-Feder, Evelina Fedorenko, and Rebecca Saxe. 2011. "Language Processing in the Occipital Cortex of Congenitally Blind Adults." *Proceedings of the National Academy of Sciences of the USA (PNAS)* 108 (11): 4429-34. doi:10.1073/pnas.1014818108.

Benedict, Ruth. 1934. *Patterns of Culture*. New York: Houghton Mifflin.

Berent, Iris. 2013a. *The Phonological Mind*. Cambridge: Cambridge University Press.

———. 2013b. "The Phonological Mind." *Trends in Cognitive Sciences* 17 (7): 319–27.

Berkeley, George. (1709) 2011. *Essay towards a New Theory of Vision*. London: Aeterna.

———. (1710) 1990. *A Treatise Concerning the Principles of Human Knowledge*. Great Books of the Western World, edited by Mortimer Adler, Clifton Fadiman, and Philip W. Goetz. Chicago: Encyclopedia Britannica.

———. (1721) 1990. *De Motu*. Great Books of the Western World, edited by Mortimer Adler, Clifton Fadiman, and Philip W. Goetz. Chicago: Encyclopedia Britannica.

Berlin, Brent, and Paul Kay. 1969. *Basic Color Terms: Their Universality and Evolution*. Berkeley: University of California Press.

Bloom, Paul. 2013. *Just Babies: The Origins of Good and Evil*. New York: Crown.

Bloomfield, Leonard. 1933. *Language*. New York: Holt, Rinehart, and Winston.

Blumberg, Mark S. 2006. *Basic Instinct: The Genesis of Behavior*. New York: Basic Books.

Boas, Franz. (1911) 1991. *Introduction to Handbook of American Indian Languages*. Lincoln: University of Nebraska Press.

———. 1912a. *Changes in Bodily Form of Descendants of Immigrants*. New York: Columbia University Press.

———. 1912b. "Changes in the Bodily Form of Descendants of Immigrants." *American Anthropologist* n.s. 14 (3): 530–62.

———. (1940) 1982. *Race, Language, and Culture*. Chicago: University of Chicago Press.

Boden, Margaret. 2006. *Mind as Machine: A History of Cognitive Science*. New York: Oxford University Press.

Bolinger, Dwight. 1985. *Intonation and Its Parts: Melody in Spoken English*. Stanford, CA: Stanford University Press.

———. 1989. *Intonation and Its Uses: Melody in Grammar and Discourse*. Stanford, CA: Stanford University Press.

Bourdieu, Pierre. 1977. *Outline of a Theory of Practice*. Cambridge: Cambridge University Press.

Bourdieu, Pierre, and Loïc J. D. Wacquant. 1992. *An Invitation to Reflexive Sociology*. Chicago: University of Chicago Press.

Bowlby, J. 1969. *Attachment*. Vol. 1 of *Attachment and Loss*. New York: Basic Books.

Boyd, Robert, and Peter J. Richerson. 1988. *Culture and the Evolutionary Process*. Chicago: University of Chicago Press.

———. 2005. *The Origin and Evolution of Cultures*. Oxford: Oxford University Press.

Brandom, Robert B. 1998. *Making It Explicit: Reasoning, Representing, and Discursive Commitment*. Cambridge, MA: Harvard University Press.

Brennan, Geoffrey, Lina Ericksson, Robert E. Goodin, and Nicholas Southwood. 2013. *Explaining Norms*. Oxford: Oxford University Press.

Broadbent, D. 1958. *Perception and Communication.* London: Pergamon Press.
Bruner, Jerome. 1979. *On Knowing: Essays for the Left Hand.* Cambridge, MA: Belknap Press of Harvard University Press.
———. 1987. *Actual Minds, Possible Worlds.* Cambridge, MA: Harvard University Press.
———. 1993. *Acts of Meaning: Four Lectures on Mind and Culture.* Cambridge, MA: Harvard University Press.
———. 1997. *The Culture of Education.* Cambridge, MA: Harvard University Press.
Bruusgaard, J. C., I. B. Johansen, I. M. Egner, Z. A. Rana, and K. Gundersen. 2010. "Myonuclei Acquired by Overload Exercise Precede Hypertrophy and Are Not Lost on Detraining." *Proceedings of the National Academy of Sciences of the USA (PNAS)* 107 (34): 15111–16. doi:10.1073/pnas.0913935107.
Buller, David J. 2006. *Adapting Minds: Evolutionary Psychology and the Persistent Quest for Human Nature.* Cambridge, MA: MIT Press / A Bradford Book.
Buller, Barbara, Ernest Buller, and Daniel Everett. 1993. "Stress Placement, Syllable Structure, and Minimality in Banawá," *International Journal of American Linguistics,* 59 (3): 280–93.
Calvin, John. (1536) 2013. *The Institutes of the Christian Religion.* Grand Rapids, MI: William B. Eerdmans.
Campbell, Joseph. 2003. *The Hero's Journey: Joseph Campbell on His Life and Work.* Novato, CA: New World Library.
Carey, Susan. 2009. *The Origin of Concepts.* Oxford: Oxford University Press.
Carruthers, Peter, Stephen Laurence, and Stephen Stich, eds. 2005. *The Innate Mind: Structure and Contents.* Oxford: Oxford University Press.
———. 2007. *The Innate Mind: Culture and Cognition.* Oxford: Oxford University Press.
———. 2008. *The Innate Mind: Foundations and the Future.* Oxford: Oxford University Press.
Cashdan, Elizabeth. 2013. "What Is a Human Universal? Human Behavioral Ecology and Human Nature." In *Arguing about Human Nature: Contemporary Debates,* edited by S. M. Downes and E. Machery. New York: Routledge.
Chagnon, Napoleon A. 1984. *Yanomamo: The Fierce People.* New York: Holt McDougal.
Chemero, Anthony. 2011. *Radical Embodied Cognitive Science.* Cambridge, MA: MIT Press.
Chomsky, Noam. 1956. "Three Models for the Description of Language." *IRE Transactions on Information Theory* IT-2 (3): 113–124.
———. 1959. "A Review of Skinner's *Verbal Behavior.*" *Language* 35 (1): 26–58.
———. 1965. *Aspects of the Theory of Syntax.* Cambridge, MA: MIT Press.
———. 1967. "A Review of B. F. Skinner's *Verbal Behavior.*" In *Readings in the Psychology of Language,* edited by Leon A. Jakobovits and Murray S. Miron, 142–43. Englewood Cliffs, NJ: Prentice-Hall.
———. 1972. *Language and Mind.* Enlarged edition. New York: Harcourt Brace Jovanovich.
———. 1984. "On Language and Culture." In *Contrasts: Soviet and American Thinkers Discuss the Future,* edited by Wiktor Osiatyński, 95–101. New York: MacMillan.
———. 1986. *Knowledge of Language: Its Nature, Origins, and Use.* New York: Praeger.
———. 1995. *The Minimalist Program.* Cambridge, MA: MIT Press.
———. 2010. "Against the Tide." Interview about D. Everett's work in article dedicated to Everett's research on language and culture. *GEO Magazine,* Indian edition, July 2010, 52.
———. 2012. *The Science of Language: Interviews with James McGilvray.* Cambridge: Cambridge University Press.
———. 2014. "Minimal Recursion: Exploring the Prospects." In *Recursion: Complexity in Cogni-*

tion, edited by Margaret Speas and Tom Roeper, 1–15. Cham: Springer International Publishing.

Churchland, Paul. 2013. *Plato's Camera: How the Physical Brain Captures a Landscape of Abstract Universals*. Cambridge, MA: MIT Press.

Churchland, Patricia S. 2012. *Braintrust: What Neuroscience Tells Us about Morality*. Princeton, NJ: Princeton University Press.

———. 2013. *Touching a Nerve: The Self as Brain*. New York: W. W. Norton.

Clark, Herbert H., and B. C. Malt. 1984. "Psychological Constraints on Language: A Commentary on Bresnan and Kaplan and on Givón." In *Method and Tactics in Cognitive Science*, edited by W. Kintsch, J. R. Miller, and P. Polson, 191–214. Hillsdale, NJ: Erlbaum.

Clearfield, M., and K. S. Mix. 2001. "Amount versus Number: Infants' Use of Area and Contour Length to Discriminate Small Sets." *Journal of Cognition and Development* 2 (3): 243–60.

Collins, Harry. 2010. *Tacit and Explicit Knowledge*. Chicago: University of Chicago Press.

Corballis, Michael C. 2002. *From Hand to Mouth*. Princeton, NJ: Princeton University Press.

———. 2007. "Recursion, Language, and Starlings." *Cognitive Science* 31 (4): 697–704.

Costandi, Moheb. 2012. "Microbes Manipulate Your Mind: Bacteria in Your Gut May Be Influencing Your Thoughts and Moods." *Scientific American Mind* 23 (3): 32–37.

Coulter, Jeff. 1979. *The Social Construction of Mind: Studies in Ethnomethodology and Linguistic Philosophy*. London: Rowman and Littlefield.

Coulter, Jeff. 1983. *Rethinking Cognitive Theory*. London: Palgrave Macmillian.

Croft, William. 2001. *Radical Construction Grammar: Syntactic Theory in Typological Perspective*. Oxford: Oxford University Press.

Culicover, Peter W., and Ray Jackendoff. 2005. *Simpler Syntax*. Oxford: Oxford University Press.

Cutler, Anne. 2012. *Native Listening: Language Experience and the Recognition of Spoken Words*. Cambridge: MA: MIT Press.

Damasio, Anthony. 2005. *Descartes' Error: Emotion, Reason, and the Human Brain*. London: Penguin.

D'Andrade, Roy. 1995. *The Development of Cognitive Anthropology*. Cambridge: Cambridge University Press.

———. 2008. *A Study of Personal and Cultural Values: American, Japanese, and Vietnamese*. London: Palgrave Macmillan.

Darwin, Charles. (1874) 1998. *The Descent of Man*. Great Minds Series. Amherst, NY: Prometheus Books.

Dascal, M. 2002. "Language as a Cognitive Technology." *International Journal of Cognition and Technology* 1 (1): 35–89.

Dascal, Marcelo, ed. 2012. "Culture, Language, Cognition." *Pragmatics and Cognition* 20 (2). Twentieth anniversary special edition (entire issue dedicated to a discussion of *Language: The Cultural Tool*).

Davidson, Donald. 1967. "Truth and Meaning." *Synthese* 17 (1): 304–23.

———. 1973. "On the Very Idea of a Conceptual Scheme." *Proceedings and Addresses of the American Philosophical Association* 47 (1973–1974): 5–20.

———. 2004. *Problems of Rationality*. Vol. 4. Oxford: Oxford University Press.

Deacon, Terrence W. 1998. *The Symbolic Species: The Co-evolution of Language and the Brain*. New York: W. W. Norton.

Degler, Carl N. 1992. *In Search of Human Nature: The Decline and Revival of Darwinianism in American Social Thought*. Oxford: Oxford University Press.

Deleuze, Gilles, and Felix Guattari. 1996. *What Is Philosophy?* New York: Columbia University Press.

Deloache, J. S. 1997. “The Credible Shrinking Room: Very Young Children's Performance with Symbolic and Nonsymbolic Relations.” *Psychological Science* 8 (4): 308–13.

———. 2000. “Dual Representation and Young Children's Use of Scale Models. *Child Development* 71 (2): 329–38.

De Ruiter, J. P., and D. Wilkins. 1998. “The Synchronization of Gesture and Speech in Dutch and Arrernte (an Australian Aboriginal Language).” In *Oralité et Gestualité*, edited by S. Santi, I. Guaïtella, C. Cavé, and G. Konopczynski, 603–07. Paris: L'Hamattan.

Descartes, Rene. 2007. *Meditations*. Great Books of the Western World, edited by Mortimer Adler, Clifton Fadiman, and Philip W. Goetz. Volume 28, *Bacon, Descartes, Spinoza*, 295–330.

Descola, Philippe. 2013. *Beyond Nature and Culture*. Chicago: University of Chicago Press.

Dolan, R. J., G. R. Fink, E. Rolls, M. Booth, A. Holmes, R. S. Frackowiak, et al. 1997. “How the Brain Learns to See Objects and Faces in an Impoverished Context.” *Nature* 389: 596–99.

Donadio, Rachel. 2013. “When Italians Chat, Hands and Fingers Do the Talking.” *New York Times*, July 1. http://www.nytimes.com/2013/07/01/world/europe/when-italians-chat-hands-and-fingers-do-the-talking.html.

Downes, Stephen M., and Edouard Machery, eds. 2013. *Arguing about Human Nature: Contemporary Debates*. New York: Routledge.

Dreyfus, Hubert L. 1965. *Alchemy and Artificial Intelligence*. Santa Monica, CA: Rand.

———. 1994. *What Computers* Still *Can't Do: A Critique of Artificial Reason*. Cambridge, MA: MIT Press.

Duncan, Susan. 2002. “Gesture, Verb Aspect, and the Nature of Iconic Imagery in Natural Discourse.” *Gesture* 2 (2): 183–206.

———. 2006. “Co-expressivity of Speech and Gesture: Manner of Motion in Spanish, English, and Chinese.” *Proceedings of the Berkeley Linguistics Society* 27:353–70.

Dutton, Denis. 2010. *The Art Instinct: Beauty, Pleasure, and Human Evolution*. Bloomsbury Press.

Eberhardt, Jennifer L., Valerie J. Purdie, Phillip Atiba Goff, and Paul G. Davies. 2004. “Seeing Black: Race, Crime, and Visual Processing.” *Journal of Personality and Social Psychology* 87 (6): 876–93.

Eckert, Penelope. 2008. “Variation and the Indexical Field.” *Journal of Sociolinguistics* 12 (4): 453–76.

Efron, David. (1942) 1972. *Gesture, Race and Culture: A Tentative Study of the Spatio-Temporal and “Linguistic” Aspects of the Gestural Behavior of Eastern Jews and Southern Italians in New York City, Living under Similar as Well as Different Environmental Conditions*. The Hauge: Mouton.

Ehrlich, Paul. R. 2001. *Human Natures: Genes, Cultures, and the Human Prospect*. New York: Penguin.

Ellwood, Robert. 1999. *The Politics of Myth: A Study of C. G. Jung, Mircea Eliade, and Joseph Campbell*. New York: State University of New York Press.

Elster, Jon. 2007. *Explaining Social Behavior: More Nuts and Bolts for the Social Sciences*. Cambridge: Cambridge University Press.

Enfield, Nick J. 2013a. “Language, Culture, and Mind: Trends and Standards in the Latest Pendulum Swing.” Review of *Language: The Cultural Tool*, by Daniel Everett. *Journal of the Royal Anthropological Institute* (N.S.) 19 (1): 155–69.

———. 2013b. *Relationship Thinking*. Oxford: Oxford University Press.

Enfield, Nick J., ed. 2002. *Ethnosyntax: Explorations in Grammar and Culture*. Oxford: Oxford University Press.

Enfield, Nick J., and Stephen C. Levinson, eds. 2006. *Roots of Human Sociality: Culture, Cognition, and Interaction*. New York: Berg.

Epps, Patricia. 2011. "Phonological Diffusion in the Amazonian Vaupes." Talk at CUNY Phonological Forum Conference on Phonology of Endangered Languages.

Ericksen, Kristen. 2015. "Dark Energy, Dark Matter." National Aeronautics and Space Administration (NASA) website, last updated June 15, 2015. http://science.nasa.gov/astrophysics/focus-areas/what-is-dark-energy/.

Evans, Nicholas, and Stephen C. Levinson. 2009. "The Myth of Language Universals: Language Diversity and Its Importance for Cognitive Science." *Behavior and Brain Sciences* 32:429–92. doi:10.1017/S0140525X0999094X

Evans, Vyvyan. 2014. *The Language Myth: Why Language Is Not an Instinct*. Cambridge: Cambridge University Press.

Everett, Caleb, Damián E. Blasi, and Seán G. Roberts. 2015. "Climate, Vocal Folds, and Tonal Languages: Connecting the Physiological and Geographic Dots." *Proceedings of the National Academy of Sciences of the USA (PNAS)* 112 (5): 1322–27. doi:10.1073/pnas.1417413112.

Everett, Caleb D. 2013a. "Evidence for Direct Geographic Influences on Linguistic Sounds: The Case of Ejectives." *PLOS ONE* 8 (6): e65275. doi:10.1371/journal.pone.0065275.

———. 2013b. *Linguistic Relativity: Evidence across Languages and Cognitive Domains*. Berlin: De Gruyter Moutin.

Everett, Caleb, and Keren Madora. 2012. "Quality Recognition among Speakers of an Anumeric Language." *Cognitive Science* 36:130–41.

Everett, Daniel L. 1979. "Aspectos da Fonologia do Pirahã." Master's thesis, Universidade Estadual de Campinas. http://ling.auf.net/lingbuzz/001715.

———. 1982. "Phonetic Rarities in Pirahã." *Journal of the International Phonetics Association* 12:94–96.

———. 1983. "A Lingua Pirahã e a Teoria da Sintaxe." PhD diss., Universidade Estadual de Campinas. Published as *A Lingua Pirahã e a Teoria da Sintaxe*, Editora da UNICAMP, 1992.

———. 1985. "Syllable Weight, Sloppy Phonemes, and Channels in Pirahã Discourse." *Proceedings of the Berkeley Linguistics Society* 11:408–16.

———. 1986. "Pirahã." In *Handbook of Amazonian Languages I*, edited by Desmond Derbyshire and Geoffrey Pullum, 200–326. Berlin: Mouton de Gruyter.

———. 1988. "On Metrical Constituent Structure in Pirahã Phonology." *Natural Language and Linguistic Theory* 6:207–46.

———. 1994. "The Sentential Divide in Language and Cognition: Pragmatics of Word Order Flexibility and Related Issues." *Journal of Pragmatics and Cognition* 2 (1) : 131–66.

———. 1995. "Optimality Theory and Arawan Prosodic Systems." Unpublished paper. http://roa.rutgers.edu/files/121-0496/121-0496-EVERETT-0-0.PDF.

———. 2001. "Monolingual Field Research." In *Linguistic Fieldwork*, edited by Paul Newman and Martha Ratliff, 166–88. Cambridge: Cambridge University Press.

———. 2004. "Coherent Fieldwork." In *Linguistics Today*, edited by P. van Sterkenberg, 141–62. Amsterdam: John Benjamins.

———. 2005a. "Cultural Constraints on Grammar and Cognition in Pirahã: Another Look at the Design Features of Human Language." *Current Anthropology* 76:621–46.

———. 2005b. "Periphrastic Pronouns in Wari'." *International Journal of American Linguistics* 71 (3): 303–26.

———. 2008. *Don't Sleep, There Are Snakes: Life and Language in the Amazonian Jungle.* New York: Pantheon.

———. 2009a. "Pirahã Culture and Grammar: A Response to Some Criticisms." *Language* 85:405–42.

———. 2009b. "Wari' Intentional State Construction Predicates." In *Investigations of the Syntax-Semantics-Pragmatics Interface*, edited by Robert D. Van Valin Jr., 381–409. Amsterdam: John Benjamins.

———. 2010a. "The Shrinking Chomskyan Corner in Linguistics: A Final Reply to Nevins, Pesetsky, Rodrigues." Response to the criticisms Nevins, Pesetsky, and Rodrigues raise against various papers of Everett on Pirahã's unusual features, published in *Language* 85 (3). http://ling.auf.net/lingbuzz/000994.

———. 2010b. "You Drink. You Drive. You Go to Jail. Where's Recursion?" Paper originally presented at the 2009 University of Massachusetts Conference on Recursion. http://ling.auf.net/lingbuzz/001141.

———. 2012a. *Language: The Cultural Tool.* New York: Pantheon Books.

———. 2012b. "What Does Pirahã Have to Teach Us about Human Language and the Mind?" *WIREs Cognitive Science* 3:555–63. doi:10.1002/wcs.1195.

———. 2013a. "A Reconsideration of the Reification of Linguistics." Paper presented at The Cognitive Revolution, 60 Years at the British Academy, London.

———. 2013b. "The State of Whose Art?" Reply to Nick Enfield's review of *Language: The Cultural Tool* in *Journal of the Royal Anthropological Institute* 19 (1).

———. 2014a. "Concentric Circles of Attachment in Pirahã: A Brief Survey" In *Different Faces of Attachment: Cultural Variations of a Universal Human Need*, edited by Heidi Keller and Hiltrud Otto, 169–86. Cambridge: Cambridge University Press.

———. 2014b. "The Role of Culture in Language Emergence." In *The Handbook of Language Emergence*, edited by Brian MacWhinney and William O'Grady, 354–76. Hoboken, NJ: Wiley-Blackwell.

———. 2015. "A Cultural Context." *Edge*. What Do You Think about Machines That Think? Annual Question Series. http://edge.org/response-detail/26103.

———. Forthcoming. *How Language Began* (working title). New York: W. W. Norton / Liveright.

Everett, Daniel L., and Keren Everett. 1984. "On the Relevance of Syllable Onsets to Stress Placement." *Linguistic Inquiry* 15:705–11.

Everett, Keren M. 1998. "The Acoustic Correlates of Stress in Pirahã." *Journal of Amazonian Languages* 1 (2): 104–62.

Faller, Martina. 2007. "The Cusco Quechua Reportative Evidential and Rhetorical Relations." *Linguistische Berichte Sonderheft Special Issue on Endangered Languages* 14:223–51.

Feyerabend, Paul. 2010. *Against Method*. 4th ed. London: Verso.

Fillmore, Charles J. 1988. "The Mechanisms of 'Construction Grammar.'" *Proceedings of the Berkeley Linguistics Society* 14:35–55.

Fitch, W. Tecumseh. 2010. *The Evolution of Language*. Cambridge: Cambridge University Press.

Flanagan, Owen. 2013. *The Bodhisattva's Brain: Buddhism*. Cambridge, MA: MIT Press / A Bradford Book.

Fleming, Amy. 2014. "What Does Meat Taste Of?" *Guardian*, June 3. http://www.theguardian.com/lifeandstyle/wordofmouth/2014/jun/03/what-does-meat-taste-of-flavours.

de Fockert, J., J. Davidoff, J. Fagot, C. Parron, and J. Goldstein. 2007. "More Accurate Size Contrast Judgments in the Ebbinghaus Illusion by a Remote Culture." *Journal of Experimental Psychology: Human Perception and Performance* 33 (3): 738–42.

Fodor, Jerry A. 1983. *The Modularity of the Mind: An Essay on Faculty Psychology*. Cambridge, MA: MIT Press.

———. 1998. *Concepts: Where Cognitive Science Went Wrong*. Oxford: Oxford University Press.

———. 2003. *Hume Variations*. Oxford: Oxford University Press.

Foley, William A. 1997. *Anthropological Linguistics: An Introduction*. London: Wiley-Blackwell.

———. 2007. "Reason, Understanding and the Limits of Translation." In *Language Documentation and Description*, edited by Peter K. Austin, vol. 4, 100–19. London: SOAS.

Frank, Michael, Daniel L. Everett, Evelina Fedorenko, and Edward Gibson. 2008. "Number as a Cognitive Technology: Evidence from Pirahã Language and Cognition." *Cognition* 108, 819–24.

Frank, Stefan L., Rens Bod, and Morten H. Christiansen. 2012. "How Hierarchical Is Language Use?" *Proceedings of the Royal Society B*. doi:10.1098/rspb.2012.1741.

Frankl, Viktor E. (1946) 2006. *Man's Search for Meaning*. Beacon Press. Boston.

Freedman, David A. 2010. *Statistical Models and Causal Inference: A Dialogue with the Social Sciences*. Edited by David Collier, Jasjeet S. Sekhon, and Philip B. Stark. Cambridge: Cambridge University Press.

Frege, Gottlob. 1980. *Philosophical Writings: Translations*. Edited by P. T. Geach and Max Black. London: Blackwell.

Freud, Sigmund. (1916) 2009. *A General Introduction to Psychoanalysis*. Eastford, CT: Martino Fine Books.

Freyd, J. J. 1983. "Shareability: The Social Psychology of Epistemology." *Cognitive Science* 7:191–210.

Futrell, Richard, Steve T. Piantadosi, Laura Stearns, Daniel L. Everett, and Edward Gibson. Forthcoming. "A Corpus Analysis of Pirahã Grammar: An Investigation of Recursion." *PLOS ONE*.

Gallagher, Tom. 2001. "Understanding Other Cultures: The Value Orientations Method." Paper presented at the Association of Leadership Educators Conference, Minneapolis, MN.

Gardner, Howard E. 1987. *The Mind's New Science: A History of the Cognitive Revolution*. New York: Basic Books.

Garfinkel, Harold. 1991. *Studies in Ethnomethodology*. Cambridge, UK: Polity.

———. 2002. *Ethnomethodology's Program: Working out Durkheim's Aphorism*. Edited by Anne Rawls. Lanham, MD: Rowman and Littlefield.

Gascoigne, Neil, and Tim Thornton. 2013. *Tacit Knowledge*. London: Routledge.

Geertz, Clifford. 1973. *The Interpretation of Cultures*. New York: Basic Books.

Gellatly, Angus. 1986. *The Skillful Mind*. London: Open University Press.

Gentner, T. Q., K. M. Fenn, D. Margoliash, and H. C. Nubaum. 2006. "Recursive Syntactic Pattern Learning by Songbirds." *Nature* 440:1204–07.

Gibbs, Raymond G. 2005. *Embodiment and Cognitive Science*. Cambridge: Cambridge University Press.

Gibson, E., S. Piantadosi, K. Brink, L. Bergen, E. Lim, and R. Saxe. 2013. "A Noisy-Channel Account of Cross-Linguistic Word Order Variation." *Psychological Science* 4 (7):1079–88. doi:10.1177/0956797612463705.

Gibson, J. J. 1966. *The Senses Considered as Perceptual Systems*. Boston: Houghton Mifflin.

———. 1979. *The Ecological Approach to Visual Perception*. Boston: Houghton Mifflin.

Gil, David. 1994. "The Structure of Riau Indonesian." *Nordic Journal of Linguistics* 17:179–200.

Giorgolo, Gianluca. 2010. "A Formal Semantics for Iconic Spatial Gestures." In *Logic, Language and Meaning*, edited by M. Aloni, B. Harald, T. de Jager, and K. Schulz, 305–14. Berlin: Springer.

Givón, Talmy. 1979. *On Understanding Grammar*. New York: Academic Press.

Givón, Talmy, ed. 1983. *Topic Continuity in Discourse: A Quantitative Cross-Language Study*. Amsterdam: John Benjamins.

Golani, Ilan. 2012. "Recursive Embedding in mouse free exploration?" Paper presented at conference on *Language: The Cultural Tool*, by Daniel Everett, Tel Aviv University.

Gold, Joel. 2012. "The Dark Matter of the Mind." *Edge*. What Is Your Favorite Deep, Elegant, or Beautiful Explanation? Annual Question Series. http://edge.org/response-detail/11095.

Gold, Joel, and Ian Gold. 2014. *Suspicious Minds: How Culture Shapes Madness*. New York: Free Press.

Goldberg, Adele. 1995. *Constructions: A Construction Approach to Argument Structure*. Chicago: University of Chicago Press.

———. 2006. *Constructions at Work: The Nature of Generalization in Language*. Oxford: Oxford University Press.

Goldin-Meadow, S. 2015. "Studying the Mechanisms of Language Learning by Varying the Learning Environment and the Learner." *Language, Cognition, and Neuroscience* 30 (8): 899–911. doi:10.1080/23273798.2015.1016978.

———. Forthcoming. "Homesign." In *Encyclopedia of Language Development*, edited by P. Brooks, V. Kempe, and J. G. Golson, Sage Publications.

Goldsmith, John. 2015. "Towards a New Empiricism for Linguists." In *Empiricism and Language Learnability* by Nick Chater, Alexander Clark, John Goldsmith, and Amy Perfors, 58–105. Oxford: Oxford University Press.

Gonçalves, M. A. 2005. Comment on "Cultural Constraints on Grammar and Cognition in Pirahã," by D. L. Everett. *Current Anthropology* 46, 636.

Goodenough, Ward H. 1981. *Culture, Language, and Society*. Menlo Park, CA: Benjamin-Cummings.

Gopnik, Alison. 2010. *The Philosophical Baby: What Children's Minds Tell Us about Truth, Love, and the Meaning of Life*. Picador: New York.

———. 2014. "Innateness." *Edge*. What Scientific Idea Is Ready for Retirement? Annual Question Series. http://edge.org/response-detail/25360.

Gopnik, Alison, Andrew N. Meltzoff, and Patricia K. Kuhl. 2001. *The Scientist in the Crib: Minds, Brains, and How Children Learn*. Reprint edition. New York: William Morrow Paper backs.

Gordon, Peter. 2004. "Numerical Cognition without Words: Evidence from Amazonia." *Science* 306, 496–99.

Graeber, David. 2001. *Toward an Anthropological Theory of Value: The False Coin of our Own Dreams*. New York: Palgrave.

Green, Jennifer. 2014. *Drawn from the Ground: Sound, Sign and Inscription in Central Australian Sand Stories*. Cambridge: Cambridge University Press.

Green, Leslie, ed. 2013. *Contested Ecologies: Dialogue in the South on Nature and Knowledge*. Cape Town: Human Sciences Research Council Press.

Greenberg, Joseph H. 1966. *Universals of Language*. Cambridge, MA: MIT Press.

Greenberg, Michael. 2012. "The Problem of the New York Police." *New York Review of Books*, October 25. http://www.nybooks.com/articles/2012/10/25/problem-new-york-police/.

Gregory, R. L. 1970. *The Intelligent Eye*. London: Weidenfeld and Nicolson.

———. 2005. "The Medawar Lecture 2001: Knowledge for Vision: Vision for Knowledge." *Philosophical Transactions of the Royal Society B: Biological Sciences* 360 (1458): 1231–51.

Grice, Paul. 1991. *Studies in the Way of Words*. Cambridge, MA: Harvard University Press.

Grimes, Joseph Evans. 1975. *The Thread of Discourse*. Berlin: Walter de Gruyter.

Grunbaum, Adolf. 1985. *The Foundations of Psychoanalysis: A Philosophical Critique*. Berkeley: University of California Press.

Hacking, Ian. 2000. *The Social Construction of What?* Cambridge, MA: Harvard University Press.

Haidt, Jonathan. 2013. *The Righteous Mind: Why Good People Are Divided by Politics and Religion*. New York: Vintage.

Hall, Edward T. (1959) 1973. *The Silent Language*. Reissue ed. New York: Anchor Books.

———. 1976. *Beyond Culture*. Garden City, NY: Anchor Press.

———. 1990. *The Hidden Dimension*. New York: Anchor Books.

Halliday, M. A. K., and Ruqaiya Hasan. 1976. *Cohesion in English*. English Language Series. London: Routledge.

Harman, Gilbert, and Ernie Lepore, eds. 2014. *A Companion to W. V. O. Quine*. London: Wiley-Blackwell.

Harris, Marvin. 2001. *Cultural Materialism: The Struggle for a Science of Culture*. Walnut Creek, CA: AltaMira.

———. 2006. *Cultural Anthropology*. Boston: Allyn and Bacon.

Harris, Sam. 2013. "The Roots of Good and Evil: An Interview with Paul Bloom." *Sam Harris: The Blog*, November 12. http://www.samharris.org/blog/item/the-roots-of-good-and-evil.

Harris, Zellig. 1951. *Methods in Structural Linguistics*. Chicago: University of Chicago Press.

Haugeland, John. 1998. *Having Thought: Essays in the Metaphysics of Mind*. Cambridge, MA: Harvard University Press.

Hauser, Marc D. 2006. *Moral Minds: How Nature Designed Our Universal Sense of Right and Wrong*. New York: Ecco.

Hauser, Marc, Noam Chomsky, and Tecumseh Fitch. 2002. "The Faculty of Language: What Is It, Who Has It, How Did It Evolve?" *Science* 298: 569–1579.

Hefner, Robert W. 1993. *Conversion to Christianity: Historical and Anthropological Perspectives on a Great Transformation*. Berkeley: University of California Press.

Heine, Steven J. 2011. *Cultural Psychology*. New York: W. W. Norton.

Helmholtz, H. 1878. *Selected Writings of Hermann Helmholtz*. Translated by R. Kahl. Middletown, CT: Wesleyan University Press.

Henrich, Natalie, and Joseph Henrich. 2007. *Why Humans Cooperate: A Cultural and Evolutionary Explanation*. Oxford: Oxford University Press.

Hewes, Gordon W. 1973. "Primate Communication and the Gestural Origins of Language." *Current Anthropology* 14:5–24.

Ho, Karen. 2009. *Liquidated: An Ethnography of Wall Street*. Raleigh, NC: Duke University Press.

Hopfield, J. J. 1982. "Neural Networks and Physical Systems with Emergent Collective Computational Abilities." *Proceedings of the National Academy of Sciences of the USA (PNAS)* 79 (8): 2554–58.

Hopper, Paul. 1988. "Emergent Grammar and the A Priori Grammar Postulate." In *Linguistics in Context: Connecting Observation and Understanding*, edited by Deborah Tannen, 117–34. New York: Ablex.

Hsieh, P. J., E. Vul, and N. Kanwisher. 2010. "Recognition Alters the Spatial Pattern of FMRI Activation in Early Retinotopic Cortex." *Journal of Neurophysiology* 103 (3): 1501–07.

Hume, David. (1739–40) 1978. *Treatise of Human Nature*. 2nd ed. Oxford: Oxford University Press.

Hymes, Dell. 1974. *Foundations in Sociolinguistics: An Ethnographic Approach*. Philadelphia: University of Pennsylvania Press.

Ingmire, Jann. 2014. "Gesturing with Hands Is a Powerful Tool for Children's Math Learning." *UChicagoNews*, March 10. http://news.uchicago.edu/article/2014/03/10/gesturing-hands-powerful-tool-children-s-math-learning.

Iverson, Jana M., and Susan Goldin-Meadow. 1997. "What's Communication Got to Do with It? Gesture in Congenitally Blind Children." *Developmental Psychology* 33:453–67.

Jackendoff, Ray. 2003. *Foundations of Language: Brain, Meaning, Grammar, Evolution*. Oxford: Oxford University Press.

Jackendoff, Ray, and Eva Wittenberg. 2012. "Even Simpler Syntax: A Hierarchy of Grammatical Complexity." Unpublished paper. https://depts.washington.edu/lingconf/abstracts/JackendoffandWittenberg.pdf.

———. Forthcoming. *A Hierarchy of Grammatical Complexity*.

Jakobson, Roman. 1990. *On Language*. Edited by Linda R. Waugh and Monique Monville-Burston. Cambridge, MA: Harvard University Press.

James, William. (1900) 2001. *Talks to Teachers on Psychology and to Students on Some of Life's Ideals*. Clear Spring, MD: Dorley House Books.

———. (1906) 1996. *Essays in Radical Empiricism*. Omaha: University of Nebraska Press.

———. 1907. *Pragmatism: A New Name for Some Old Ways of Thinking*. London: Longmans, Green.

Joaquim, Anna Dina L., and John H. Schumann. 2013. *Exploring the Interactional Instinct*. Oxford: Oxford University Press.

Jones, R. K., and M. A. Hagen. 1980. "A Perspective on Cross-cultural Picture Perception." In *The Perception of Pictures*, edited by M. A. Hagen, vol. 2, 193–226. New York: Academic Press.

Jung, Carl G. (1916) 2003. *Psychology of the Unconscious*. Mineola, NY: Dover Publications.

Kant, Immanuel. (1903) 2007. *Critique of Pure Reason*. A-ed. Great Books of the Western World, edited by Mortimer Adler, Clifton Fadiman, and Philip W. Goetz. Volume 39, *Kant*, 1–252 . Chicago: Encyclopedia Britannica.

———. (1904) 2007. *Critique of Pure Reason*. B-ed. Great Books of the Western World, edited by Mortimer Adler, Clifton Fadiman, and Philip W. Goetz. Volume 39, *Kant*, 291–364. Chicago: Encyclopedia Britannica.

Katz, Jerrold J. 1972. *Linguistic Philosophy: The Underlying Reality of Language and Its Philosophical Import*. London: Allen and Unwin.

Keller, Heidi. 2007. *Cultures of Infancy*. Sussex: Psychology Press.

Kendon, Adam. 2004. *Gesture: Visible Action as Utterance*. Cambridge: Cambridge University Press.

Khazan, Olga. 2014. "How We Get Tall." *Atlantic*, May 9. http://www.theatlantic.com/health/archive/2014/05/how-we-get-tall/361881/.

Kinsella, Anna R. 2009. *Language Evolution and Syntactic Theory*. Cambridge: Cambridge University Press.

Kita, Sotaro. 2000. "How Representational Gestures Help Speaking." In *Language and Gesture*, edited by David McNeill, 162–85. Cambridge: Cambridge University Press.

Kohler, W. 1929. *Gestalt Psychology*. Oxford: Liveright.

Kohn, Eduardo. 2013. *How Forests Think: Toward an Anthropology beyond the Human*. Berkeley: University of California Press.

Koster, Jan. 1992. "Against Tacit Knowledge." In *Language and Cognition 2: Yearbook 1992 of the Research Group for Linguistic Theory and Knowledge Representation of the University of Groningen*, edited by D. Gilbers and S. Looyenga, 193–204. Groningen: The Group.

Kovacs, I., and M. Eisenberg. 2004. "Human Development of Binocular Rivalry." In *Binocular Rivalry*, edited by D. Alais and R. Blake, 101–16. Cambridge, MA: MIT Press.

Kroeber, Alfred L., and Clyde Kluckhohn. 1952. *Culture: A Critical Review of Concepts and Definitions*. Harvard University Peabody Museum of American Archeology and Ethnology Papers 47.

Kuhn, Thomas. 1996. *The Structure of Scientific Revolutions*. 3rd ed. Chicago: University of Chicago Press.

Kuper, Adam. 2000. *Culture: The Anthropologists' Account*. Cambridge, MA: Harvard University Press.

Kupperman, Joel J. 2012. *Human Nature: A Reader*. Indianapolis: Hackett.

Kurzweil, Ray. 2012. *How to Create a Mind: The Secret of Human Thought Revealed*. New York: Viking.

Ladefoged, Peter, and Daniel Everett. 1996. "The Status of Phonetic Rareties." *Language* 72 (4): 794–800.

Ladefoged, Peter, Jenny Ladefoged, and Daniel L. Everett. 1997. "Phonetic Structures of Banawá, an Endangered Language." *Phonetica* 54:94–111.

Laland, Kevin N., Bennett G. Galef, Kim Hill, and Kristin E. Bonnie, eds. 2009. *The Question of Animal Culture*. Cambridge, MA: Harvard University Press.

Lakoff, George. 1977. "Linguistic Gestalts." In *Papers from the Thirteenth Regional Meeting of the Chicago Linguistic Society*. Chicago: University of Chicago.

Lakoff, George, and Mark Johnson. 1980. *Metaphors We Live By*. Chicago: University of Chicago Press.

Lakoff, George, and Rafael Nuñez. 2001. *Where Mathematics Comes From: How the Embodied Mind Brings Mathematics into Being*. New York: Basic Books.

Langley, Pat, Herbert A. Simon, Gary L. Bradshaw, and Jan M. Zytkow. 1987. *Scientific Discovery: Computational Explorations of the Creative Processes*. Cambridge, MA: MIT Press.

Lende, Daniel H., and Greg Downey. 2012. *The Encultured Brain: An Introduction to Neuroanthropology*. Cambridge, MA: MIT Press.

Lenneberg, Eric. 1967. *Biological Foundations of Language*. New York: John Wiley and Sons.

Latour, Bruno. 1986. *Laboratory Life: The Construction of Scientific Facts*. Princeton, NJ: Princeton University Press.

———. 2007. *Reassembling the Social: An Introduction to Actor-Network-Theory*. Oxford: Oxford University Press.

LeDoux, Joseph. 2015. *Anxious: Using the Brain to Understand and Treat Fear and Anxiety*. New York: Viking.

Lee, Namhee, Lisa Mikesell, Anna Dina L. Joaquin, Andrea W. Mates, and John H. Schumann. 2009. *The Interactional Instinct: The Evolution and Acquisition of Language*. Oxford: Oxford University Press.

Leibowitz, H., R. Brislin, L. Perlmutter, and R. Hennessy. 1969. "Ponzo Perspective Illusion as a Manifestation of Space Perception." *Science* 166:1174–76.

Levi-Strauss, Claude. (1949) 1969. *The Elementary Structures of Kinship*. Boston: Beacon Press.

———. (1978) 1995. *Myth and Meaning: Cracking the Code of Culture*. New York: Schocken.

Levine, Robert A. 2013. "Attachment Theory as Cultural Ideology." In *Different Faces of Attachment*, edited by Heidi Keller and Hiltrud Otto, 50–65. Cambridge: Cambridge University Press.

Levinson, Stephen C. 2006. "On the Human 'Interaction Engine.'" In *Roots of Human Sociality: Culture, Cognition, and Interaction*, edited by Nick J. Enfield and Stephen C. Levinson, 399–460. New York: Berg.

Levinson, Stephen C., and Pierre Jaisson. 2005. *Evolution and Culture: A Fyssen Foundation Symposium*. Cambridge, MA: MIT Press.

Levinson, Stephen C., and Asifa Majid. 2014. "Differential Ineffability and the Senses." *Mind and Language* 29 (4): 407–27.

Lewis, David. 2002. *Convention: A Philosophical Study*. London: Blackwell.

Lieberman, Philip. 1967. *Intonation, Perception, and Language*. Cambridge, MA: MIT Press.

———. 2007. "The Evolution of Human Speech: Its Anatomical and Neural Bases." *Current Anthropology* 48 (1): 39–66.

———. 2013. *The Unpredictable Species: What Makes Humans Unique*. Princeton, NJ: Princeton University Press.

LiPuma, Edward, and Benjamin Lee. 2004. *Financial Derivatives and the Globalization of Risk*. Raleigh, NC: Duke University Press.

Lobina, David J., and José E. Garcia-Albea. 2009. "Recursion and Cognitive Science: Data Structures and Mechanisms." In *Proceedings of the 31st Annual Conference of the Cognitive Science Society*, edited by Niels A. Taatgen and Henk van Rijn, 1347–52. Austin, TX: Cognitive Science Society.

Longacre, Robert E. 1964. *Grammar Discovery Procedures*. The Hague: Mouton.

———. 1976. *An Anatomy of Speech Notions*. Peter de Ridder Publi and Co cations in Tagmemics. Lisse: Peter de Ridder Press.

Ludmer, R., Y. Dudai, and N. Rubin. 2011. "Uncovering Camouflage: Amygdala Activation Predicts Long-Term Memory of Induced Perceptual Insight." *Neuron* 69 (5): 1002–14.

MacWhinney, Brian. 2004. "A Multiple Process Solution to the Logical Problem of Language Acquisition." *Journal of Child Language* 31 (4): 883–914.

———. 2005. "A Unified Model of Language Acquisition." In *Handbook of Bilingualism: Psycholinguistic Approaches*, edited by J. Kroll and A. De Groot, 49–67. New York: Oxford University Press.

———. 2006. "Emergentism—Use Often and With Care." *Applied Linguistics* 27 (4): 729–40. doi:10.1093/applin/aml035.

Majid, Asifa, and Stephen C. Levinson. 2011. "The Senses in Language and Culture." *Senses in Society* 6 (1): 5–18.

Mameli, Matteo. 2008. "On Innateness: The Clutter Hypothesis and the Cluster Hypothesis." *Journal of Philosophy* 105:719–36.

Mameli, Matteo, and Patrick Bateson. 2006. "Innateness and the Sciences." *Biology and Philosophy* 21:155–88.

Marcus, Gary. 2015. "Face It, Your Brain Is a Computer." *New York Times*, June 27. http://www.nytimes.com/2015/06/28/opinion/sunday/face-it-your-brain-is-a-computer.html.

Matthews, G. H. 1965. *Hidatsa Syntax*. Berlin: Mouton.

McCarthy, John. 1979. "Ascribing Mental Qualities to Machines." Stanford University Computer Science Department.

McCarthy, John, and Alan Prince. 1994. "The Emergence of the Unmarked: Optimality in Prosodic Morphology." In *Proceedings of the North Eastern Linguistic Society*. Amherst, MA: GLSA, University of Massachusetts.

McCawley, James D. 1982. "Parentheticals and Discontinuous Constituent Structure." *Linguistic Inquiry* 13 (1): 91–106.

McDowell, John. 2013. *The Engaged Intellect: Philosophical Essays*. Cambridge, MA: Harvard University Press.

McLaughlin, Brian, and Karen Bennett. 2011. "Supervenience," *Stanford Encyclopedia of Philosophy*, first published July 25, 2005, revised November 2, 2011. http://plato.stanford.edu/archives/win2011/entries/supervenience.

McNeill, David. 1992. *Hand and Mind: What Gestures Reveal about Thought*. Chicago: University of Chicago Press.

———. 2005. *Gesture and Thought*. Chicago: University of Chicago Press.

———. 2012. *How Language Began: Gesture and Speech in Human Evolution*. Cambridge: Cambridge University Press.

McNeill, David, ed. 2000. *Language and Gesture*. Cambridge: Cambridge University Press.

McQuown, Norman A. 1957. "Review of *Language in Relation to a Unified Theory of the Structure of Human Behavior* by Kenneth L. Pike." *American Anthropologist* 59 (1): 189–92. doi:10.1525/aa.1957.59.1.02a00640.

Mead, George Herbert. 1974. *Mind, Self, and Society from the Standpoint of a Social Behaviorist*. Edited by C. W. Morris. Chicago: University of Chicago Press.

Mead, Margaret. (1928) 2001. *Coming of Age in Samoa*. New York: Perennial Classics.

Messing, Joachim. 2001. "Do Plants Have More Genes Than Humans?" *TRENDS in Plant Science* 6 (5): 195.

Millikan, Ruth Garrett. 1998. "Language Conventions Made Simple." *Journal of Philosophy* 95 (4): 161–80.

Mitchell, Donald W. 2002. *Buddhism*. Oxford: Oxford University Press.

Mooney, C. M. 1957. "Closure as Affected by Viewing Time and Multiple Visual Fixations." *Canadian Journal of Psychology* 11 (1): 21–28.

Nagel, Thomas. 1974. "What Is It Like to Be a Bat?" *The Philosophical Review* LXXXIII (4): 435–50.

Nevins, Andrew, David Pesetsky, and Cilene Rodrigues. 2009. "Pirahã Exceptionality: A Reassessment." *Language* 85 (2): 355–404.

Newell, Allen, J. C. Shaw, and Herbert A. Simon. 1958. "Elements of a Theory of Human Problem Solving." *Psychological Review* 65 (3): 151–66.

Newson, Lesley, Peter J. Richerson, and Robert Boyd. 2007. "Cultural Evolution and the Shaping of Cultural Diversity." In *Handbook of Cultural Psychology*, edited by Shinobu Kitayama and Dov Cohen, 454–76. New York: Guilford Press.

Nida, Eugene A. 1964. *Toward a Science of Translating: With Special Reference to Principles and Procedures Involved in Bible Translating*. Leiden: Brill Academic Publishing.

Ochs, Elinor, and Lisa Capps. 2002. *Living Narrative: Creating Lives in Everyday Storytelling*. Cambridge, MA: Harvard University Press.

Ogden, C. K., and I. A. Richards. (1923) 1989. *The Meaning of Meaning*. San Diego: Harcourt Brace Jovanovich.

Ohala, John. 1992. "Alternatives to the Sonority Hierarchy for Explaining the Shape of Morphemes." *Papers from the Parasession on the Syllable*, 319–38. Chicago: Chicago Linguistic Society.

Olson, Randy. 2014. "Why the Dutch Are So Tall." *Randal S. Olson*, June 23. http://www.randalolson.com/2014/06/23/why-the-dutch-are-so-tall/.

Otto, Hiltrud, and Heidi Keller. 2014. *Different Faces of Attachment: Cultural Variations on a Universal Human Need.* Cambridge: Cambridge University Press.

Panksepp, Jaak, and Lucy Biven. 2012. *The Archaeology of Mind: Neuroevolutionary Origins of Human Emotions.* New York: W. W. Norton.

Parsons, Talcott. 1970. *Social Structure and Personality.* New York: Free Press.

Paul, Annie Murphy. 2011. *Origins: How the Nine Months before Birth Shape the Rest of Our Lives.* New York: Free Press.

Pearl, Lisa. 2013. "Evaluating Learning Strategy Components: Being Fair." Forthcoming in *Language.* http://ling.auf.net/lingbuzz/001940.

Peirce, C. S. 1977. *Semiotics and Significs.* Edited by Charles Hardwick. Bloomington: Indiana University Press.

———. 1992. *The Essential Peirce: Selected Philosophical Writings (1867–1893).* Indiana University Press. Bloomington, IN.

———. 1998. *The Essential Peirce, Volume 2: Selected Philosophical Writings, 1893–1913.* Indiana University Press. Bloomington, IN.

Pelli, D. 1999. "Close Encounters: An Artist Shows That Size Affects Shape." *Science* 285:844–46.

Pepperberg, Irene M. 1992. "Proficient Performance of a Conjunctive, Recursive Task by an African Gray Parrot (Psittacus erithacus)." *Journal of Comparative Psychology* 106:295–305.

Perfors, Amy, Josh B. Tenenbaum, Edward Gibson, and T. Regier. 2010. "How Recursive Is Language? A Bayesian Exploration." In *Recursion and Human Language*, edited by Harry van der Hulst, 159–75. Berlin: DeGruyter Mouton.

Perfors, Amy Francesca, and Daniel Joseph Navarro. 2012. "What Bayesian Modelling Can Tell Us about Statistical Learning: What It Requires and Why It Works." In *Statistical Learning and Language Acquisition*, edited by Patrick Rebuschat and John N. Williams, 383–408. Boston: De Gruyter Mouton.

Piaget, J. 1926. *The Language and Thought of the Child.* New York: Harcourt, Brace, Jovanovich.

Piantadosi, Steve T., Laura Stearns, Daniel L. Everett, and Edward Gibson. 2012. "A Corpus Analysis of Pirahã Grammar: An Investigation of Recursion." Accessed May 10, 2013; last accessed January 2016. https://colala.bcs.rochester.edu/papers/piantadosi2012corpus.pdf.

Piatelli-Palmarini, Massimo. 1980. *Language and Learning: The Debate between Jean Piaget and Noam Chomsky.* Cambridge, MA: Harvard University Press.

Pica, P., S. Jackson, R. Blake, and N. F. Troje. 2011. "Comparing Biological Motion Perception in Two Distinct Human Societies." *PLOS ONE* 6 (12): e28391.

Pike, Kenneth L. (1943) 1945. *Tone Languages: The Nature of Tonal Systems, with a Technique for the Analysis of Their Significant Pitch Contrasts.* Rev. ed. Glendale, CA: Summer Institute of Linguistics.

———. 1945. *The Intonation of American English.* University of Michigan Publications in Linguistics 1. Ann Arbor: University of Michigan Press.

———. 1952. *Grammatical Prerequisites to Phonemic Analysis. Word* 3:155–72.

———. 1962. *With Heart and Mind: A Personal Synthesis of Scholarship and Devotion.* Grand Rapids, MI: Eerdmans.

———. 1967. *Language in Relation to a Unified Theory of the Structure of Human Behavior*. 2nd rev. ed. Janua Linguarum, series maior, 24. The Hague: Mouton.

———. 1978. "Particularization versus Generalization, and Explanation versus Prediction." In *The Teaching of English in Japan*, edited by Ikuo Koike et al., 783–85. Tokyo: Eichosha.

———. 1998. "Semantics in a Holistic Context: With Preliminary Convictions and Approaches." In *Papers from the Fourth Annual Meeting of the Southeast Asian Linguistics Society 1994*, edited by Udom Warotamasikkhadit and Thanyarat Panakul, 177–97. Tempe: Program for Southeast Asian Studies, Arizona State University.

Pike, Kenneth L., and Evelyn G. Pike. 1976. *Grammatical Analysis*. Summer Institute of Linguistics Publications in Linguistics 53. Dallas: Summer Institute of Linguistics and the University of Texas at Arlington.

Pinker, Steven. 1995. *The Language Instinct: How the Mind Creates Language*. New York: W. W. Norton.

———. 2003. *The Blank Slate: The Modern Denial of Human Nature*. New York: Penguin Books.

Polanyi, Michael. (1966) 2009. *The Tacit Dimension*. Chicago: University of Chicago Press.

———. 1974. *Personal Knowledge: Towards a Post-Critical Philosophy*. Chicago: University of Chicago Press.

Postal, Paul M. 2009. "The Incoherence of Chomsky's 'Biolinguistic' Ontology." *Biolinguistics* 3 (1): 104–23.

Pratt, Scott. 2002. *Native Pragmatism: Rethinking the Roots of American Philosophy*. Bloomington: Indiana University Press.

Prince, Alan, and Paul Smolensky. (1993) 2004. *Optimality Theory: Constraint Interaction in Generative Grammar*. Malden, MA: Blackwell.

Prinz, Jesse J. 2002. *Furnishing the Mind: Concepts and Their Perceptual Basis*. Cambridge, MA: MIT Press / A Bradford Book.

———. 2011. *The Emotional Construction of Morals*. Oxford: Oxford University Press.

———. 2014. *Beyond Human Nature: How Culture and Experience Shape the Human Mind*. New York: W. W. Norton.

Prinz, Wolfgang. 2012. *Open Minds: The Making of Agency and Intentionality*. Cambridge, MA: MIT Press.

———. 2013. *Action Science: Foundations of an Emerging Discipline*. Cambridge, MA: MIT Press.

Quine, Willard Van Orman. 1951. "Two Dogmas of Empiricism." *Philosophical Review* 60:20–43.

———. 1960. *Word and Object*. Cambridge, MA: MIT Press.

———. 1985. *The Time of My Life: An Autobiography*. Cambridge, MA: MIT Press / A Bradford Book.

Quinn, Naomi. 2005. *Finding Culture in Talk: A Collection of Methods*. Culture, Mind, and Society. London: Palgrave Macmillan.

Read, Dwight W. 2011. *How Culture Makes Us Human: Primate Social Evolution and the Formation of Human Societies*. Walnut Creek, CA: Left Coast Press.

Reinbold, Julia Ulrike. 2004. "Intonation and Information Structure in Banawá." Master's thesis, Manchester University, Linguistics.

Rey, A., Perruchet, P., and Fagot, J. 2011. "Centre-Embedded Structures Are a By-Product of Associative Learning and Working Memory Constraints: Evidence from Baboons (Papio Papio)." *Cognition* 123 (1): 180–4.

Richardson, Don. 2005. *Peace Child*. 4th ed. Ventura, CA: Regal Books.

Richardson, Robert C. 2007. *Evolutionary Psychology as Maladapted Psychology*. Cambridge, MA: MIT Press / A Bradford Book.

Richerson, Peter J., and Robert Boyd. 2005. *Not by Genes Alone: How Culture Transformed Human Evolution*. Chicago: University of Chicago Press.

Rizzolatti, Giacomo, and Michael A. Arbib. 1998. "Language within Our Grasp." *Trends Neuroscience* 21:188–94.

Rock, I., S. Hall, and J. Davis. 1994. "Why Do Ambiguous Figures Reverse?" *Acta Psychologica* 87 (1): 33–59.

Rogers, Carl. (1961) 1995. *On Becoming a Person: A Therapist's View of Psychotherapy*. New York: Houghton Mifflin.

Rogoff, Barbara. 2003. *The Cultural Nature of Human Development*. Oxford: Oxford University Press.

———. 2011. *Developing Destinies: A Mayan Midwife and Town*. Oxford: Oxford University Press.

Rokeach, Milton. 1973. *The Nature of Human Values*. New York: Free Press.

Rorty, Richard. 1981. *Philosophy and the Mirror of Nature*. Princeton, NJ: Princeton University Press.

Rosenbaum, David A. 2014. *It's a Jungle in There: How Competition and Cooperation in the Brain Shape the Mind*. Oxford: Oxford University Press.

Ryle, Gilbert. (1949) 2002. *The Concept of Mind*. Chicago: University of Chicago Press.

———. 1968. *The Thinking of Thoughts*. Saskatoon: University of Saskatchewan.

Sakel, Jeanette. 2012a. "Acquiring Complexity: The Portuguese Spoken by Pirahã Men." *Linguistic Discovery* 10:75–88.

———. 2012b. "Transfer and Language Contact: The Case of Pirahã." *International Journal of Bilingualism* 16:307–40.

Sakel, Jeanette, and Daniel L. Everett. 2012. *Linguistic Field Work: A Student Guide*. Cambridge: Cambridge University Press.

Sakel, Jeanette, and Eugenie Stapert. 2010. "Pirahã—In Need of Recursive Syntax?" In *Recursion and Human Language*, edited by Harry van der Hulst, 3–16. Berlin: Mouton de Gruyter.

Sapir, E. 2001. "Why Cultural Anthropology Needs the Psychiatrist." *Psychiatry: Interpersonal and Biological Processes* 64 (1): 2–10.

Sapir, Edward. 1921. *Language: An Introduction to the Study of Speech*. New York: Harvest Books.

———. 1928. "The Unconscious Patterning of Behavior in Society." In *The Unconscious: A Symposium* by C. M. Child et al., with an introduction by Ethel S. Dummer, 114–42. New York: A. A. Knopf.

———. 1929. "The Status of Linguistics as a Science." *Language* 5 (4): 207–14.

———. 1934. "The Emergence of the Concept of Personality in the Study of Cultures." *Journal of Psychology* 5:408–15.

———. 1985. *Selected Writings in Language, Culture, and Personality*. Edited by David G. Mandelbaum. Berkeley: University of California Press.

———. 1993. *The Psychology of Culture: A Course of Lectures*. Edited by Judith Irvine. Berlin: Mouton de Gruyter.

de Saussure, Ferdinand. (1916) 2012. *A Course in General Linguistics*. Chicago: Open Court.

Saville-Troike, Muriel. 1982. *The Ethnography of Communication: An Introduction*. Malden, MA: Wiley-Blackwell.

Schama, Simon. 1997. *The Embarrassment of Riches: An Interpretation of Dutch Culture in the Golden Age*. New York: Vintage.

Schegloff, Emanuel A. 1984. "On Some Gestures' Relation to Talk." In *Structures of Social Action*, edited by J. M. Atkinson and J. Heritage, 266–98. Cambridge: Cambridge University Press.

Scheper-Hughes, Nancy. 2013. "Family Life as *Bricolage*: Reflections on Intimacy and Attachment in *Death without Weeping*." In *Different Faces of Attachment*, edited by Hiltrud Otto and Heidi Keller, 230–62. Cambridge: Cambridge University Press.

Schönpflug, Ute, ed. 2008. *Cultural Transmission: Psychological, Developmental, Social, and Methodological Aspects*. Cambridge: Cambridge University Press.

Searle, John. 1970. *Speech Acts: An Essay in the Philosophy of Language*. Cambridge: Cambridge University Press.

———. 1972. "Chomsky's Revolution in Linguistics." *New York Review of Books*, June 29, 16–24.

———. 1978. "Literal Meaning." *Erkenntnis* 1:207–24.

———. 1980a. "'Las Meninas' and the Paradoxes of Pictorial Representation." *Critical Inquiry* 6 (3): 477–88.

———. 1980b. "Minds, Brains and Programs." *Behavioral and Brain Sciences* 3 (3): 417–57. doi:10.1017/S0140525X00005756.

———. 1983. *Intentionality: An Essay in the Philosophy of Mind*. Cambridge: Cambridge University Press.

———. 1997. *The Mystery of Consciousness*. New York: New York Review of Books.

———. 2006. "What Is Language?" Unpublished paper. http://socrates.berkeley.edu/~jsearle/whatislanguage.pdf.

———. 2010. *Making the Social World: The Structure of Human Civilization*. Oxford: Oxford University Press.

Segall, M. H., D. T. Campbell, and M. J. Herskovits. 1966. *The Influence of Culture on Visual Perception*. Indianapolis: Bobbs-Merrill.

Selkirk E. 1984. "On the Major Class Features and Syllable Theory." In *Language and Sound Structures: Studies in Phonology*, edited by Mark Aronoff and Richard T. Oehrle, 107–36. Cambridge, MA: MIT Press.

Shannon, Claude E. 1949. "A Mathematical Theory of Communication." *The Mathematical Theory of Communication*. Urbana, IL: University of Illinois Press.

Shapiro, Lawrence. 2010. *Embodied Cognition*. London: Routledge.

Sherzer, Joel. 1991. *Verbal Art in San Blas: Kuna Culture through Its Discourse*. Cambridge: Cambridge University Press.

Shipman, Pat. 2015. *The Invaders: How Humans and Their Dogs Drove Neanderthals to Extinction*. Cambridge, MA: Harvard University Press.

Silverstein, Michael. 1979. "Language Structure and Linguistic Ideology." In *The Elements: A Parasession on Linguistic Units and Levels*, edited by R. Cline, W. Hanks, and C. Hofbauer, 193–247. Chicago: Chicago Linguistic Society.

———. 2003. "Indexical Order and the Dialectics of Sociolinguistic Life." *Language and Communication* 23:193–229.

———. 2004. "Cultural Concepts and the Language-Culture Nexus." *Current Anthropology* 45 (5): 621–52.

Silverstein, Michael, and Greg Urban, eds. 1996. *Natural Histories of Discourse*. Chicago: University of Chicago Press.

Simon, Herbert A. 1962. "The Architecture of Complexity." *Proceedings of the American Philosophical Society* 106(6): 467–82.

———. 1990. *Reason in Human Affairs*. Stanford, CA: Stanford University Press.

———. 1991. *Models of My Life: The Remarkable Autobiography of the Nobel Prize Winning Social Scientist and Father of Artificial Intelligence*. New York: Basic Books.

———. 1996. *The Sciences of the Artificial*. Cambridge, MA: MIT Press.

Skinner, B. F. 1957. *Verbal Behavior*. New York: Appelton-Century-Crofts.

Skipper, Jeremy I., Susan Goldin-Meadow, Howard C. Nusbaum, and Steven L. Small. 2009. "Gestures Orchestrate Brain Networks for Language Understanding." *Current Biology* 19 (8): 661–67.

Slatkin, Montgomery, and Laurent Excoffier. 2012. "Serial Founder Effects during Range Expansion: A Spatial Analog of Genetic Drift." *Genetics*. 191 (1): 171–81.

Soames, Scott. 2010. *What Is Meaning?* Princeton, NJ: Princeton University Press.

Sommer, Barbara. 1992. "Cognitive Performance and the Menstrual Cycle." In *Cognition and the Menstrual Cycle*, edited by John T. Richardson, 39–66. New York: Springer-Verlag.

Sontag, Susan. 2013. *Susan Sontag: Essays of the 1960s and 70s*. New York: Library of America.

Spelke, E. S. 1990. "Principles of Object Perception." *Cognitive Science* 14 (1): 29–56.

Spelke, Elizabeth S. 2013. "Developmental Sources of Social Divisions." In *Neurosciences and the Human Person: New Perspectives on Human Activities*. Pontifical Academy of Sciences, *Scripta Varia* 121, edited by A. M. Battro, S. Dehaene, and W. J. Singer. Vatican City. www.casinapioiv.va/content/dam/accademia/pdf/sv121/sv121-spelke.pdf.

Sperber, Dan, and Deidre Wilson. 1995. *Relevance: Communication and Cognition*. 2nd ed. Oxford: Blackwell Publishing.

Stanovich, Keith E. 2005. *The Robot's Rebellion: Finding Meaning in the Age of Darwin*. Chicago: University of Chicago Press.

Steels, L. 2005. "The Emergence and Evolution of Linguistic Structure: From Lexical to Grammatical Communication Systems." *Connection Science* 17:213–30.

Stenzel, Kristine. 2005. "Multilingualism in the Northwest Amazon, Revisted." *Memorias del Congresso de Idiomas Indigenas de Latinoamerica, II*. AILLA, University of Texas, Austin.

Sterelny, Kim. 2014. *The Evolved Apprentice: How Evolution Made Humans Unique*. Cambridge, MA: MIT Press / A Bradford Book.

Stevenson, Leslie. 2000. *The Study of Human Nature: A Reader*. Oxford: Oxford University Press.

Stivers, Tanya, N. J. Enfield, Penelope Brown, Christina Englert, Makoto Hayashi, Trine Heinemann, Gertie Hoymann, Frederico Rossano, Jan Peter de Ruiter, Kyung-Eun Yoon, and Stephen C. Levinson. 2009. "Universals and Cultural Variation in Turn-Taking in Conversation." *Proceedings of the National Academy of Sciences of the USA (PNAS)* 106 (26): 10587–92.

Tedlock, Dennis, and Bruce Mannheim. 1995. *The Dialogic Emergence of Culture*. Urbana: University of Illinois Press.

Tinbergen, Niko. 1963. "On Aims and Methods of Ethology." *Zeitschrift fur Tierpsychologie* 20:410–33.

Tomasello, Michael. 1999. *The Cultural Origins of Human Cognition*. Cambridge, MA: Harvard University Press.

———. 2008. *Origins of Human Communication*. Cambridge, MA: Harvard University Press.

———. 2014. *A Natural History of Human Thinking*. Cambridge, MA: Harvard University Press.

Thomason, Sarah Grey. 2008. "Does Language Contact Simplify Grammars?" Plenary address at the annual meeting of the Deutsche Gesellschaft für Sprachwissenschaft, Bamberg, Germany, February 29.

Tooby, Joel. 2014. "Learning and Culture." *Edge*. What Scientific Question Is Ready for Retirement? Annual Question series. http://edge.org/response-detail/25343.

Tooby, John, and Leda Cosmides. 1992. "The Psychological Foundations of Culture." In *The Adapted Mind: Evolutionary Psychology and the Generation of Culture*, edited by J. Barkow, L. Cosmides, and J. Tooby, 19–136. New York: Oxford University Press.

Treisman, A. 1991. "Search, Similarity and the Integration of Features between and within Dimensions." *Journal of Experimental Psychology: Human Perception and Performance* 17:652–76.

Trudgill, Peter. 2011. *Sociolinguistic Typology: Social Determinants of Linguistic Complexity*. Oxford: Oxford University Press.

Turing, Alan M. 1952. "The Chemical Basis of Morphogenesis." *Philosophical Transactions B* 237:37–72.

Turner, Stephen P. 2013. *Understanding the Tacit*. London: Routledge.

Twain, Mark. (1916) 1995. *The Mysterious Stranger*. Literary Classics. Amherst, NY: Prometheus Books.

Tylor, Edward. (1871) 1920. *Primitive Culture*. Vol. 1. New York: J. P. Putnam's Sons.

Unger, Peter. 2014. *Empty Ideas: A Critique of Analytical Philosophy*. Oxford: Oxford University Press.

Urban, Greg. 2000. *A Discourse-Centered Approach to Culture*. Austin: University of Texas Press.

Van Valin, Robert D., Jr. 2005. *Exploring the Syntax-Semantics Interface*. Cambridge: Cambridge University Press.

———. 2006. "Semantic Macroroles and Language Processing." In *Semantic Role Universals and Argument Linking: Theoretical, Typological and Psycholinguistic Perspectives*, edited by Ina Bornkessel, Matthias Schlesewsky, Bernard Comrie, and Angela D. Friederici, 263–302. Berlin: Mouton de Gruyter.

Van Valin, Robert D., Jr., and Randy LaPolla. 1997. *Syntax: Structure, Meaning, and Function*. Cambridge: Cambridge University Press.

Vogt, Evon Z., and Ethel M. Albert. 1966. *People of Rimrock: A Study of Values in Five Cultures*. Cambridge, MA: Harvard University Press.

Vygotsky, Lev S. 1978. *Mind in Society: The Development of Higher Psychological Processes*. Edited by Michael Cole. Harvard University Press.

Wallace, Anthony F. C. 1970. *Culture and Personality*. New York: Random House.

Walton, Kerry D., L. Benavides, N. Singh, and N. Hatoum. 2005. "Long-Term Effects of Microgravity on the Swimming Behaviour of Young Rats." *Journal of Physiology* 565 (2): 609–26. doi:10.1113/jphysiol.2004.074393.

Walton, Kerry, C. Hefferman, D. Sulica, and L. Benavides. 1997. "Changes in Gravity Influence Rat Postnatal Motor System Development: From Simulation to Space Flight." *Gravitational and Space Biology Bulletin* 10 (2):111–18.

Watson, Duane, and Edward Gibson. 2004. "The Relationship between Intonational Phrasing and Syntactic Structure in Language Production." *Language and Cognitive Processes* 19 (6): 713–55.

Webster, Richard. 1996. *Why Freud Was Wrong: Sin, Science, and Psychoanalysis*. New York: Basic Books.

Weinreich, Uriel, William Labov, and Marvin I. Herzog. 1968. "Empirical Foundations for a Theory of Language Change." In *Directions for Historical Linguistics*, edited by W. Lehmann and Y. Malkiel, 95–189. Austin: University of Texas Press.

Wells, Rulon S. 1947. "Immediate Constituents." *Language* 23:81–117.

Whitaker, D., and P. V. McGraw. 2000. "Long-Term Visual Experience Recalibrates Human Orientation Perception." *Nature Neuroscience* 3 (1): 13.

White, Douglas R. and Ulla C. Johansen. 2006. *Network Analysis and Ethnographic Problems: Process Models of a Turkish Nomad Clan.* Lanham, MD: Lexington Books.

White, Leslie A. 1949. *The Science of Culture.* New York: Grove Press.

Whitehead, Hal, and Luke Rendell. 2014. *The Cultural Lives of Whales and Dolphins.* Chicago: University of Chicago Press.

Wierzbicka, Anna. 1996. *Semantics: Primes and Universals.* Oxford: Oxford University Press.

Wilke, Andreas, John M. C. Hutchinson, Peter M. Todd, and Daniel J. Kruger. 2006. "Is Risk Taking Used as a Cue in Mate Choice?" *Evolutionary Psychology* 4:367–93.

Wilkins, David P. 1999. "Spatial Deixis in Arrernte Speech and Gesture: On the Analysis of a Species of Composite Signal as Used by a Central Australian Aboriginal Group." Paper 6 in *Proceedings of the Workshop on Deixis, Demonstration and Deictic Belief in Multimedia Contexts, held on occasion of ESSLI XI,* edited by Elisabeth André, Massimo Poesio, and Hannes Rieser, 31–45. Workshop held in the section "Language and Computation" as part of the Eleventh European Summer School in Logic, Language and Information, August 9–20, 1999, Utrecht, The Netherlands.

Wilson, E. O. 1975. *Sociobiology: The New Synthesis.* Cambridge, MA: Harvard University Press / Belknap Press.

———. 1978. *On Human Nature.* Cambridge, MA: Harvard University Press.

Winawer, J., N. Witthoft, M. C. Frank, L. Wu, A. R. Wade, and L. Boroditsky. 2007. "Russian Blues Reveal Effects of Language on Color Discrimination." *Proceedings of the National Academy of Sciences of the USA (PNAS)* 104 (19): 7780–85.

Wittgenstein, Ludwig. (1922) 1998. *Tractatus Logico-Philosophicus.* Mineola, NY: Dover Publications.

———. (1953) 2009. *Philosophical Investigations.* Edited by P. M. S. Hacker and Joachim Schulte. Chichester: Wiley-Blackwell.

———. (1958) 1965. *The Blue and Brown Books: Preliminary Studies for the Philosophical Investigations.* New York: Harper Torchbooks.

Yang, Charles. 2015. "Negative Knowledge from Positive Evidence." *Language* 91 (4): 938–53.

Yoon, J. M. D. 2012. "Vision and Revision: Cue-Triggered Perceptual Reorganization of Two-Tone Images in U.S. Preschool Children and Adults." PhD diss., Stanford University.

Yoon, J. M. D., N. Whitthoft, J. Winawer, M. C. Frank, D. L. Everett, and E. Gibson. 2014. "Cultural Differences in Perceptual Reorganization in US and Pirahã Adults." *PLOS ONE* 9 (11): e110225. doi:10.1371/journal.pone.0110225.

Yoon, J. M. D., J. Winawer, N. Witthoft, and E. M. Markman. 2007. "Striking Deficiency in Top-Down Perceptual Reorganization of Two-Tone Images in Preschool Children." *Proceedings of the 6th IEEE International Conference on Development and Learning,* 181–86.

Zlatev, Jordan. Forthcoming. "The Emergence of Gestures." In *The Handbook of Language Emergence,* edited by Brian MacWhinney and William O'Grady. Hoboken, NJ: Wiley.

Index

Page numbers in bold indicate definitions, figures, and tables.

CPSIA information can be obtained
at www.ICGtesting.com
Printed in the USA
LVOW13s0507220118
563338LV00003B/6/P

9 780226 526782